FINITE MATHEMATICS

WITH APPLICATIONS

THIRD EDITION

FINITE MATHEMATICS

WITH APPLICATIONS

THIRD EDITION

DAVID E. ZITARELLI
RAYMOND F. COUGHLIN
Temple University
JONATHAN C. KNAPPENBERGER
Holy Family College

Raymond-Reese Book Company
Elkins Park, PA

To Paul, Nichole, Virginia, Christina, Sara and Tyler

Printed in the United States of America

Finite Mathematics with Applications, Third Edition
0-9647077--5-6

Photography by

Robert Leahy

Bob is an award-winning underwater photo-journalist who has recently turned his attention to "land-based" artistry. The photographs in this text are from his country collection that is featured in his latest book, *Wisdom of the Plain Folk: Songs and prayers from the Amish and Mennonites*, for which Bob's wife, Donna, compiled the script. It is published by Penguin Studio.

Contents

Preface

Students often ask, "Why do I have to study this subject?" And as programs and careers require more and more mathematics, especially now in this hi-tech age, the question is asked in increasing numbers. The goal of this book is to answer this question. Although the table of contents is fairly standard for a finite mathematics textbook, the means of achieving the goal differ from other texts in two ways.

Applications

Every section contains several referenced applications of mathematics to disciplines outside of mathematics, and usually outside of science. These applications appear in examples and in exercises. Each one of these applications is accompanied by a footnote so that you can pursue the article in greater depth. The instructor can assign the article as a special project. Because such applications deal with real-life situations, calculators are often necessary to carry out the computations.

Every chapter also contains at least one Case Study that treats a particular application in greater depth than do the referenced exercises and examples. The topics have been chosen not only to illustrate instances in which mathematics has played a crucial role in solving an important problem, but to also highlight areas of particular human interest.

Even if an instructor does not have time to cover the referenced exercises and examples or the Case Studies, it is hoped that their relevance and wealth, both in diversity and number, will make a lasting impression.

Organization

The text can be divided into three parts: linear mathematics (Chapters 1 and 2), probability (Chapters 3, 4 and 5), and statistics (Chapter 6). Chapter 1 introduces the Cartesian coordinate system and linear equations. Systems of two equations in two variables are used extensively in Chapter 2 so they are treated in depth in Chapter 1. One important use of linear mathematics in business and industry is to approximate growth by using a straight line and then predicting future growth. Finding the appropriate straight line is a process called least squares. It is given a thorough treatment in Chapter 1. The Case Study shows how it is possible to predict when Olympic records are likely to be broken. Section 1-5 introduces word problems, which are the essence of Chapter 2.

Chapter 2 deals with another mathematical process that is used extensively in business and industry. It is called linear programming. The method takes complicated real-life problems, translates them into mathematics, and then solves the mathematical problem. The ability to translate a lengthy paragraph into mathematics is an extremely useful tool in many careers. It teaches you how to think abstractly by visualizing an abstract construct - the solution to the general linear programming problem - and then manipulating the specific numbers and concepts of a particular problem within the framework of the abstract form. The ability to transfer your thought processes between specific, concrete ideas and general, abstract ideas is not only necessary in mastering mathematics but it is crucial to most disciplines and career paths.

Chapter 3 lays the foundation for the rest of the text. It reviews the basic aspects of set theory and counting techniques, including Venn diagrams, permutations and combinations. In Chapter 4 the basics of probability theory are motivated by a study of automobile accidents on a major highway in Boston. After the foundation of probability experiments are described using set theory an counting techniques, laws governing addition, subtraction, multiplication and division of probabilities are covered. Chapter 5 covers some additional topics from probability theory.

Chapter 6 presents the rudiments of statistics, including graphical methods of studying data, measures of central tendency, measures of spread of data, the normal curve and the binomial approximation to the normal curve.

David E. Zitarelli
Raymond F. Coughlin
Jonathan C. Knappenberger

1 LINEAR EQUATIONS

Chapter Overview

A linear equation in two variables is an equation whose graph is a straight line. The association of a curve (like a line) with an equation (like a linear equation) was a revolutionary idea that occurred during the early part of the 17th century. The first two sections discuss the correspondence between curves in two dimensions and equations in two variables. Section 1-1 introduces the Cartesian coordinate system and sketches the graphs of various curves. Section 1-2 discusses three standard forms of linear equations.

Section 1-3 deals with systems of linear equations. It presents several methods for solving systems of two linear equations in two variables, emphasizing the method of elimination. It also discusses the advantages of the elimination method. Section 1-4 discusses least squares, featuring an examination of voting trends in presidential elections. This section lays the foundation for the Case Study that is based on an Olympic event. Section 1-5 demonstrates how to translate word problems into mathematical language, the essence of Chapter 2.

Case Study Preview

The women's 100-meter dash has been contested in the Olympics since 1928. Florence "Flo-Jo" Griffith-Joyner won the event in flamboyant style at the 1988 Olympics in Seoul, Korea. She became a leader not only of sprints but of clothing fashions in track-and-field as well.

The world has experienced many drastic changes since the event was added to the modern Olympics in 1928, yet the winning times have been lowered with remarkable regularity in almost every quadrennial affair. The Case Study examines this surprising phenomenon. It derives an equation that projects the year in which the event will be run under ten seconds, a performance that seems out of reach at present. The equation also predicts what the winning time will be in any future Olympic games. You can test its validity for the most recent Olympics leading up to the year you are reading this book.

1-1 THE CARTESIAN COORDINATE SYSTEM

About 1630 two Frenchmen, René Descartes and Pierre Fermat, independently revolutionized the way of expressing scientific laws. The idea they hit upon has come to be called "analytic geometry" or "algebraic geometry" because it is a blend of algebra and geometry. Descartes is usually given credit for being the founder of this field, and the term "Cartesian coordinate system" reflects our debt to him, but Fermat deserves at least as much credit for his pioneering work.[1] Sections 1-1 and 1-2 are devoted to their discovery.

The Graph of an Equation

Consider the formula $y = 3x + 1$. Although it is an algebraic expression that describes the relationship between the variables x and y, it can also be viewed as a geometric object, namely a line in the plane. To achieve this viewpoint both Descartes and Fermat introduced the idea of a *coordinate system*, which we describe briefly.

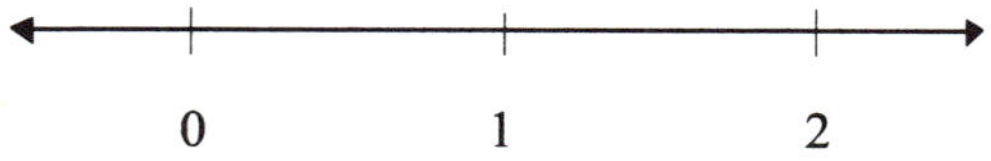

Figure 1–1

We start with the **real line**. It is a straight line with one point marked as the **origin**, corresponding to the number 0 (see Figure 1–1). A point to the right of 0 is marked and it corresponds to the number 1. The distance from 0 to 1 is called a **unit distance**. Points that are located at one-unit intervals to the right of 1 are marked 2, 3, etc. Negative numbers are located to the left of 0. In this way, each real number a corresponds to the point that is $|a|$ units to the right of 0 if a is positive or to the left if a is negative. Additional points, corresponding to some rational numbers, are marked off in Figure 1–2.

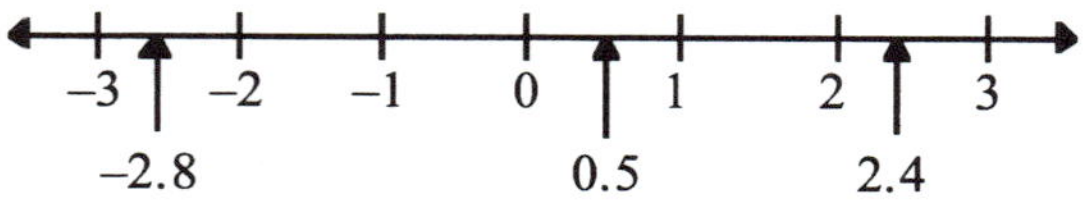

Figure 1–2

Next construct a vertical line perpendicular to the horizontal real line at the origin, and mark off a unit length on it. The positive direction is upward. Usually the horizontal line is called the ***x*-axis** and the vertical line the ***y*-axis**. Every point in the plane is represented by an ordered pair of numbers. The ordered pair (1, 2) corresponds to the

[1]For the historical details, see Section 17.11 (pp. 386-388) of Carl B. Boyer and Uta C. Merzbach, *A History of Mathematics* (second edition), New York, John Wiley & Sons, 1989.

point that is one unit to the right of the y-axis and two units above the x-axis. (See Figure 1–3.) In the ordered pair (1, 2), 1 is called the ***x*-coordinate** and 2 is called the ***y*-coordinate**. The numbers in the ordered pairs are also called the first and second **coordinates**

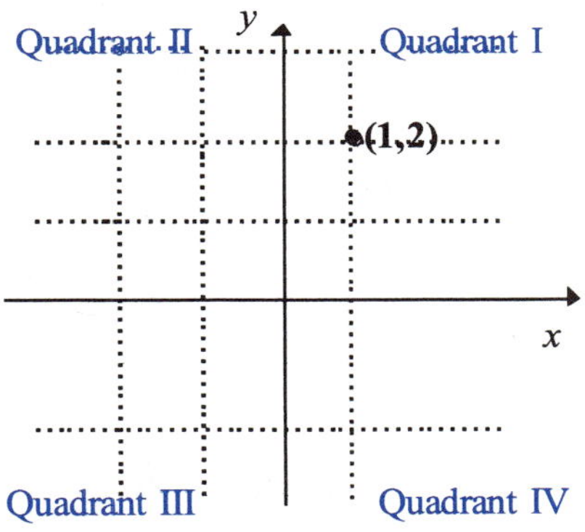

Figure 1–3

This coordinate system is called the **Cartesian coordinate system** after Descartes. The choice of the letters x and y is arbitrary; we will change as dictates. The Cartesian coordinate system divides the plane into four distinct **quadrants** denoted by the Roman numerals shown in Figure 1–3. In Figure 1–4 additional points are plotted in the coordinate system.

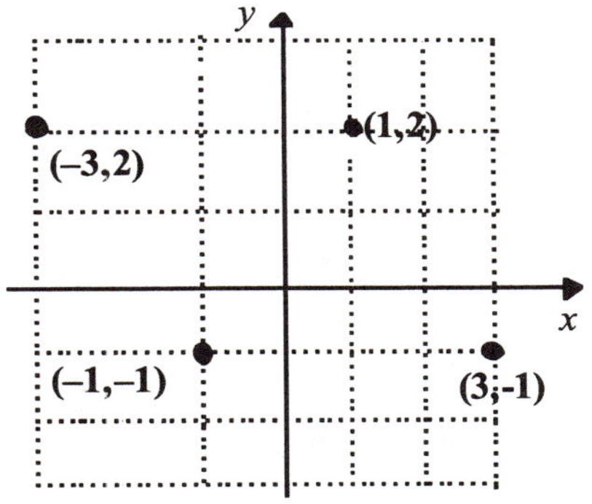

Figure 1–4

Consider the equation $y = 3x + 1$. If $x = 1$ then $y = 4$. These values are treated as the point (1, 4) in the Cartesian coordinate system. The point (1, 4) is said to "satisfy" the equation $y = 3x + 1$. The set of all points which satisfy an equation is called the **graph** of the equation. The graph is obtained from a table of values of x and the corresponding values of y. Example 1 illustrates the procedure for drawing the graph of an equation.

EXAMPLE 1

Sketch the graph of the equation $y = 3x + 1$.

Solution. Consider the equation $y = 3x + 1$. To find the graph of $y = 3x + 1$, make a table of some values of x and the corresponding values of y. If $x = -2$, then $y = 3x + 1 =$

$3(-2) + 1 = -5$, meaning that the point $(-2, -5)$ lies on the graph. The table lists four additional points corresponding to $x = -1$, 0, 1, and 2.

x	y	Point
−2	−5	(−2, −5)
−1	−2	(−1, −2)
0	1	(0, 1)
1	4	(1, 4)
2	7	(2, 7)

Table 1–1

Next plot the five points (−2, −5), (−1, −2), (0, 1), (1, 4), and (2, 7) on a Cartesian coordinate system. When these points are joined the resulting graph appears to be the line drawn in Figure 1–5.

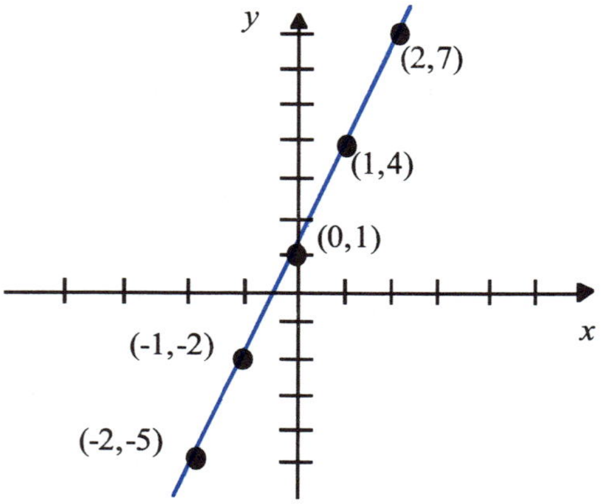

Figure 1–5

How can we be sure that the graph of $y = 3x + 1$ is a line? One way is to plot additional points. By letting x be *any* real number in the equation $y = 3x + 1$ there will be an infinite number of points of the form $(x, 3x + 1)$ on the graph. These points constitute the graph of $y = 3x + 1$ shown in Figure 1–5.

The five values that we chose for x in the table were selected arbitrarily. The graph of $y = 3x + 1$ will be the same line even if other values of x are selected.

Example 1 illustrates how algebra and geometry were molded into one mathematical discipline called analytic geometry. The essence is that equations in two variables correspond to curves in two dimensions. The procedure for sketching the graph of an equation is called the **plotting method**. It consists of three steps:

1. Make a table of values of x and y.
2. Treat the values as ordered pairs on a coordinate system and plot them.
3. Connect the points with a smooth curve.

The graph of the equation is the curve that results from step 3. In Section 1-2 we will see how to proceed in the opposite direction, from the graph to the equation.

This simple idea allows algebraic equations to be viewed geometrically, reflecting the old adage that "a picture is worth a thousand words." Example 2 illustrates it.

EXAMPLE 2

Sketch the graph of the equation $4x + 2y = 1$ by the plotting method.

Solution. Construct a table for the equation $4x + 2y = 1$ by choosing a value for x and substituting it into the equation to get the corresponding value of y. If $x = 1$ then $4(1) + 2y = 1$, so $y = -3/2$. Therefore $(1, -3/2)$ is a point on the graph of $4x + 2y = 1$. The table provides four additional points.

x	y	Point
−2	9/2	(−2, 9/2)
−1	5/2	(−1, 5/2)
0	1/2	(0, 1/2)
1	−3/2	(1,−3/2)
2	−7/2	(2, −7/2)

Plot these points and draw the curve through them. The graph is shown in Figure 1–6.

Notice that the graph of the equation $4x + 2y = 1$ shown in Figure 1–6 is a line. Plotting additional points will confirm this fact. It is possible to prove the following statement.

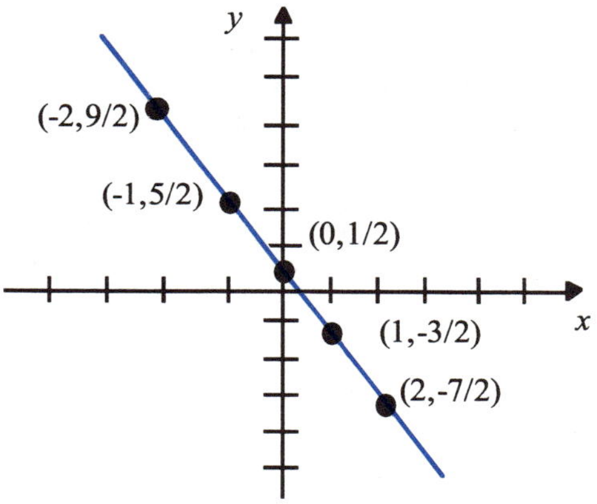

Figure 1–6

> **The Equation of a Line**
>
> The graph of an equation of the form $Ax + By = C$ is a line when A, B, and C are real numbers such that A and B are not both zero.

An equation in two variables whose graph is a line is called a **linear equation**. Examples 1 and 2 show that $y = 3x + 1$ and $4x + 2y = 1$ are linear equations.

Applications

For many businesses, the total cost C (in dollars) of producing a commodity and the number x of units produced can be written as a linear equation. For instance, let

$$C = 500 + 4x$$

If no units are produced, then $x = 0$ so $C = \$500$. This amount is called the overhead cost. If $x = 25$ then 25 units are produced so the total cost is $C = \$600$.

Often the total revenue R (in dollars) derived from selling the commodity can be written as a linear equation. For instance, let

$$R = 8x$$

If 25 units are produced then $R = \$200$, in which case the revenue is \$400 less than the cost. The business will have to produce more units in order to turn a profit. In Section 1-3 we will explain how to compute the number of units needed to derive a profit from the sales, but for now we lay the foundation for the solution.

EXAMPLE 3

Sketch the graphs of the cost equation, $C = 500 + 4x$, and the revenue equation, $R = 8x$, on the same coordinate axis.

Solution. Use the plotting method to sketch the graph of the cost $C = 500 + 4x$. A calculator can be quite helpful in making the table. For instance, if $x = 150$ then the value of C can be obtained from the following sequence of key strokes:

4 [×] 150 [+] 500 [=]

The calculator should then display 1100, which is the value of C. The points will then be comprised of the x-coordinate and the C-coordinate.

x	C	Point
0	500	(0, 500)
50	700	(50, 700)
100	900	(100, 900)
150	1100	(150, 1100)

Sketch the graph of $R = 8x$ for the same values of x by first completing a similar table for the total revenue.

x	R	Point
0	0	(0, 0)
50	400	(50, 400)
100	800	(100, 800)
150	1200	(150, 1200)

The graphs of the two equations are shown in Figure 1–7.

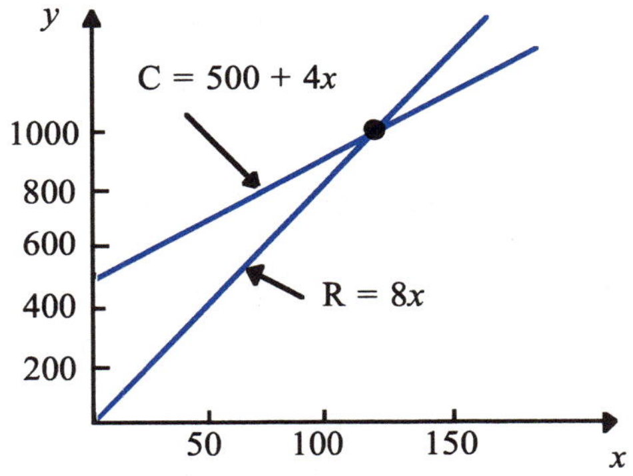

Figure 1–7

Most large firms have a planning group which constructs sales forecasts. These forecasts are critical for production scheduling, plant design, and financial planning. Graphs are often used for viewing the relationship between sales growth and external financial requirements. Example 4 examines the Arinson Products Company, whose owner, John Arinson, sought funds to maintain exclusive control of the company.[2]

EXAMPLE 4

Let x represent the sales growth rate of the Arinson Products Company and let y represent the external funds required (in thousands of dollars), where

$$y = 494x - 16$$

(a) Draw the graph of the equation.
(b) If the sales growth rate is 10%, what external funding is required?
(c) If \$107,500 is available for external funding, what is the sales growth rate?

Solution.
(a) If $x = 0.1$ (corresponding to a 10% sales growth rate), then $y = 494(0.1) - 16 = 33.4$. The following table lists several other values of x and the corresponding values of y.

x	y	Point
–0.2	–114.8	(–0.2, –114.8)
–0.1	–65.4	(–0.1, –65.4)
0	–16	(0, –16)
0.1	33.4	(0.1, 33.4)

The graph is sketched in Figure 1–8.

(b) A sales growth rate of 5% means that $x = 0.05$. Then $y = 494(0.05) - 16 = 8.7$. Since the units for y are thousands of dollars, \$8700 in external funds are required.

---

[2] B. Campsey and E. Brigham, *Introduction to Financial Management*, The Dryden Press, 1985, pp. 190-204.

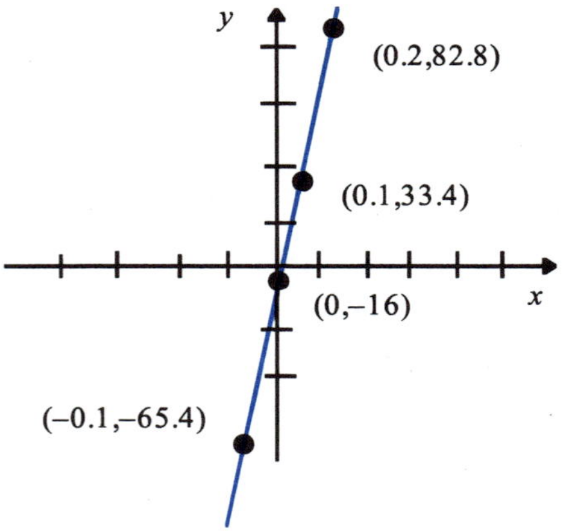

Figure 1–8

(c) If \$107,500 is available for external funding, set $y = 107.5$. The sales growth rate is obtained by substituting $y = 107.5$ into the equation and solving for x.

$$494x - 16 = y$$

$$494x - 16 = 107.5$$

$$494x = 107.5 + 16 = 123.5$$

$$x = \frac{123.5}{494} = 0.25$$

Therefore the sales growth rate is $x = 25\%$.

The following example uses a linear equation to estimate the population of California for any year during the 1980's.

EXAMPLE 5

Based on figures supplied by the U.S. Census Bureau, the population of California in the decade 1980-1990 can be described by the equation $C = 617{,}134.8x + 23{,}667{,}902$, where x is the number of years after 1980.

(a) What was the population of California in 1980?
(b) What was the population of California in 1990?
(c) What was the increase in California's population in the 1980s?

Solution.

(a) Since x is the number of years after 1980, the population in 1980 can be found by setting $x = 0$. Then $C = 23{,}667{,}902$, which was California's population in 1980.
(b) For the year 1990 set $x = 10$. Then $C = 6{,}171{,}348 + 23{,}667{,}902 = 29{,}839{,}250$. This was California's population in 1990.
(c) The increase in population in the 1980s is found by subtracting the population in 1980 from the population in 1990. Hence the increase was

$$29{,}839{,}250 - 23{,}667{,}902 = 6{,}171{,}348.$$

Notice that the answer to Example 5(c) is ten times the coefficient of x in the equation $C = 617{,}134.8x + 23{,}667{,}902$. This coincidence is not an accident. The exercises in this section and the next two sections will explore the matter further.

What's in a Name?

All of the graphs in this section have been lines. Of course these are not the only graphs of algebraic equations according to the method of Descartes and Fermat; there are also circles, parabolas, and many others. It is useful to be able to picture a graph when its equation is given.

So far, graphs have been drawn by the plotting method, but this method has its shortcomings. Given the table of five points for the equation $y = 3x + 1$ in Example 1, it is conceivable that its graph could be the curve in Figure 1–9. In Section 1-2 we will see why it must be a straight line.

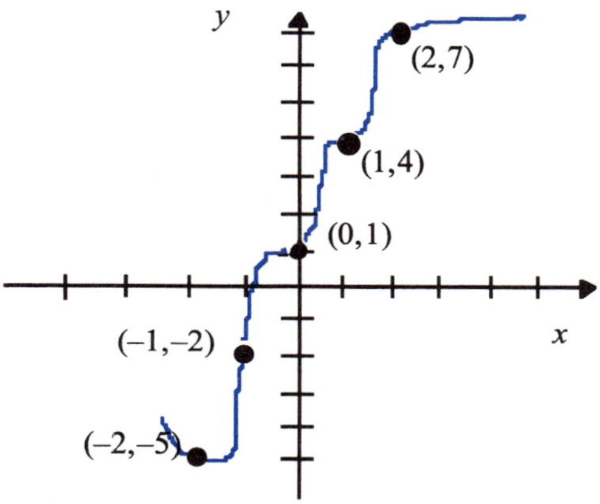

Figure 1–9

EXERCISE SET 1-1

In problems 1 to 4 fill in the table for the given values

1. $y = 2x + 3$

x	y	Point
−2		
−1		
0		
1		
2		

2. $y = x - 2$

x	y	Point
−2		
−1		
0		
1		
2		

3. $y = 4 - x$

x	y	Point
–2		
–1		
0		
1		
2		

4. $3x - y = 2$

x	y	Point
–2		
–1		
0		
1		
2		

In problems 5 to 10 use the plotting method to sketch the graph of the equation.

5. $y = 2x - 3$
6. $y = x - 2$
7. $y = 2x + 3$
8. $3x - y = 2$
9. $2x + y = 6$
10. $4x + y = 7$

In problems 11 to 14 the graphs of the given equations are straight lines. Plot two points and then sketch the graph.

11. $x + 4y = 7$
12. $2x + y = 3$
13. $3x + 4y = -12$
14. $3x + y = 1$

In problems 15 to 18 sketch the graphs of the total cost C and the total revenue R on the same set of coordinate axes.

15. $C = 100 + 3x$
    $R = 13x$
16. $C = 40 + x$
    $R = 5x$
17. $C = 25 + 3x$
    $R = 8x$
18. $C = 300 + x$
    $R = 31x$

Problems 19 and 20 refer to the equation $y = 494x - 16$ given in Example 4.

19. (a) If the sales growth rate is 20%, what external funding is required?
    (b) If \$132,200 is available for external funding, what is the sales growth rate?

20. (a) If the sales growth rate is –5%, what external funding is required?
    (b) If \$82,800 is available for external funding, what is the sales growth rate?

In problems 21 to 24 use the plotting method to sketch the graph of the equation.

21. $x - y = 3$
22. $x + 3y = 3$
23. $3x - y = -1$
24. $3x - 4y = 6$

In problems 25 to 34 the equations are nonlinear. Sketch the graph using the plotting method.

25. $y = x^2 - 3$
26. $y = x^2 + 1$
27. $y = -x^2$
28. $y = -2x^2$
29. $y = 1 - 2x^2$
30. $y = 1 - x^2$
31. $y = x^3$
32. $y = x^3 - 8$
33. $y = \sqrt{x}$
34. $y = \sqrt{x} - 1$

In problems 35 and 36 sketch the graphs of the two equations on the same set of coordinate axes.

35. $x + y = 120$

 $100x + 40y = 6000$

36. $45x + 9y = 45$

 $10x + 6y = 20$

Problems 37 to 40 refer to the equation 5F – 9C = 160, where F is degrees Fahrenheit and C is degrees Celsius.

37. Sketch the graph of the equation.
38. Find the degrees Celsius when F = 68 degrees.
39. Find the degrees Celsius when F = 98.6 degrees.
40. Find the degrees Fahrenheit when C = 20 degrees.

Problems 41 and 42 refer to a company whose total cost is $C = 0.5x + 20{,}000$, where x is the number of sales and the fixed cost is $20,000.

41. If the company sells $60,000 worth of goods, what is the total cost?
42. If the company sells $30,000 worth of goods, what is the total cost?

Referenced Exercise Set 1-1

1. The percentage share of the national radio audience for FM radio stations (contrasted with AM radio stations) in any year since 1972 can be approximated by the formula

 $$S = 3.514286Y - 226.2953$$

 where S is a percent and Y is the last two digits in the year.[3] (For instance, for the year 1989, Y = 89.)
 (a) What was the approximate percentage share in 1985?
 (b) In what year was the percentage share about 50%?

2. The amplifier in a stereo system must be strong enough to produce loud peak volumes level (106 decibels) in the speakers. The strength needed depends on the character of the room. For instance, a "live" room is one in which a hand clap will cause a fluttering echo. For Acoustic Research loudspeakers[4], the linear relationship between the power y (in watts) and the volume of a "live" room (in cubic feet) is

 $$y = \frac{x}{180} + \frac{80}{9}$$

 (a) What power is needed for a 2000-cubic-feet room?
 (b) What power is needed for a 6500-cubic-feet room?
 (c) If a system produces 40 watts, what size room will allow it to produce loud peak levels?

---

[3]This equation was derived from data supplied by Statistical Research, Incorporated.

[4]Source: Teledyne Acoustic Research of Canton, MA. This company manufactures the Acoustic Research loudspeakers.

VCRs are now in most American homes. Most VHS systems have a speed switch that allows the viewer to record in three different speeds: 2-hour standard play (SP), 4-hour long play (LP), and 6-hour super long standard play (SLP). Problems 3 and 4 involve equations that allow a viewer to get the highest quality from a recording.[5]

3. Suppose a viewer wishes to record a program that lasts between 2 and 4 hours. Let x be the length of the program in hours. Then the highest quality recording using an entire T-120 videocassette can be obtained by recording in LP mode for $y = 2x - 4$ hours and in SP mode for 4 B y hours. How long should the viewer record in LP mode if the program lasts:
   (a) 3 hours?
   (b) 3 1/2 hours?
   (c) 2 hours and 40 minutes?

4. Many people see little difference in quality between LP and SLP modes. Suppose a viewer wishes to record a program that lasts between 2 and 6 hours using LP and SLP modes. Let x be the length of the program in hours. Then the highest quality recording using the entire T-120 videocassette can be obtained by recording in SLP mode for $y = 1.5x$ B 3 hours and in SP mode 4 B y for hours. How long should the viewer record in SLP mode if the program lasts:
   (a) 4 hours?
   (b) 3 1/2 hours?
   (c) 4 hours and 40 minutes?

5. The first federal income tax went into effect in 1862, one year after President Abraham Lincoln signed it into legislation. Phil Lapsansky, the curator of the Afro-American collection at the Library Company of Philadelphia, discovered an original tax form in the stacks of that library, which Benjamin Franklin founded in 1731.[6] Computing taxes was a lot easier then; the tax rate was 3% on income from \$600 to \$10,000 and 5% beyond that.
   (a) Write an equation for the tax T of a person whose income in 1862 was x dollars, where x is less than 10,000.
   (b) Write an equation for the tax T of a person whose income in 1862 was x dollars, where x is more than 10,000.

6. Consider the statement: "There are six times as many students as professors at this university." Let S stand for the number of students and P stand for the number of professors.
   (a) If P=1200 what is S?
   (b) If S=1200 what is P?
   (c) Write an equation using the variables S and P to represent the given statement.[7]
   (d) Substitute the values you obtained in parts (a) and (b) into the equation in part (c). If they do not satisfy the equation, revise the equation and repeat this part.

---

[5]G. N. Fiore, "An application of linear equations to the VCR," *Mathematics Teacher*, October 1988, pp. 570-572.

[6]Leonard W. Boasberg, "The tax that was due for 1862," *The Philadelphia Inquirer*, April 12, 1990, pp. 1-D, 4-D.

[7]It has been reported that only 63% of engineering students and 43% of social science students get this equation correct. Sources for the study addressing this issue can be found in the article by Annie A. Selden and John Selden, Jr., entitled "Do you know the students-and-professors problem?" that appeared in a former newsletter of the Mathematical Association of America called *UME Trends*, May 1989, p. 6.

Multiple-Choice Exercise Set 1-1

Exercises 1 to 5 refer to the real line below.

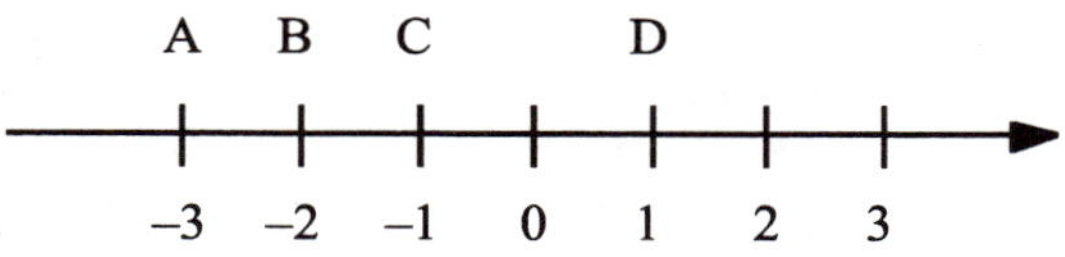

1. Which real number is labeled B on the real line?
(a) –3/2 (b) –1 (c) 1 (d) 3 (e) none of these

2. Which real number is labeled D on the real line?
(a) –3/2 (b) –1 (c) 1 (d) 2 (e) none of these

3. Which letter represents the real number 0.5 on the real line?
(a) A (b) B (c) C (d) D (e) none of these

4. Which letter represents the real number 2 on the real line?
(a) A (b) B (c) C (d) D (e) none of these

5. Which letter represents the real number –5/2 on the real line?
(a) A (b) B (c) C (d) D (e) none of these

6. If $y = 2x + 6$ and $x = -4$ then $y =$
(a) 14 (b) 2 (c) –14 (d) –2 (e) none of these

7. If $y = 2x + 6$ and $x = 3/2$ then $y =$
(a) 9 (b) 12 (c) 22/3 (d) 0 (e) none of these

8. If $y = 2x + 6$ and $x = 0.1$ then $y =$
(a) 6.2 (b) 5.8 (c) 6.1 (d) 5.9 (e) none of these

9. If $2x + y = 1$ and $x = -4$ then $y =$
(a) –7 (b) 7 (c) 9 (d) –3/2 (e) none of these

10. A point that is not on the line $y = -x$ is
(a) (–1, 1) (b) (1, –1) (c) (0, 0) (d) (2, 2)

11. A point that is not on the line $2x + y = 3$ is
(a) (–1, 1) (b) (1, 1) (c) (0, 3) (d) (2, –1)

12. A point that is not on the line $2x - y = 3$ is
(a) (–1, –5) (b) (–1, 5) (c) (0, –3) (d) (2, 1)

13. The graph of the equation $-2x + 9y = 0$ is a
(a) line (b) parabola (c) circle (d) point (e) none of these

14. The graph of the equation $x = 2y + 1$ is a
(a) line (b) parabola (c) circle (d) point (e) none of these

15. Which equation has the property that its graph passes through quadrants I and III only?
(a) $x + y = 0$ (b) $y = x + 1$ (c) $y = x - 1$ (d) $x - y = 0$ (e) none of these

1-2 LINES

There are two basic aspects to the Descartes-Fermat invention of analytic geometry. One is a fixed frame of reference, called the Cartesian coordinate system. The other is the relationship between equations and curves. Section 1-1 showed how to begin with a linear equation in two variables and draw its graph. This section illustrates the reverse direction: given information about the graph of a line, construct its equation.

General Form

In Section 1-1 we stated that a linear equation in two variables is an equation whose graph is a line. In this section we examine three forms of linear equations. Recall that an equation in two variables x and y is linear if and only if it can be written in the form $\mathrm{A}x + \mathrm{B}y = \mathrm{C}$. This form of a linear equation is called the *general form*. Therefore the equation $4x + 3y = 12$ is linear, but the equations $x^2 - y = 0$ and $\sqrt{x} - y = 0$ are nonlinear.

DEFINITION

The **general form** of a linear equation is

$$\mathrm{A}x + \mathrm{B}y = \mathrm{C}$$

where A, B, and C are real numbers, with A and B not both equal to 0.

The ***x*-intercept** of a line is the point where the line crosses the x-axis, and the ***y*-intercept** is the point where it crosses the y-axis. Thus an x-intercept must have a y-coordinate equal to 0 and a y-intercept must have an x-coordinate equal to 0.

Example 2 in Section 1-1 showed how to sketch the graph of $4x + 2y = 1$ by the plotting method. Example 1 below illustrates a method for graphing linear equations in general form that uses the two intercepts and is more efficient than the plotting method.

EXAMPLE 1

Sketch the graph of the linear equation $3x + 2y = 4$.

Solution. Find the y-intercept by setting $x = 0$. Then $y = 2$, so the y-intercept is $(0, 2)$. Find the x-intercept by setting $y = 0$. Then $x = 4/3$, so the x-intercept is $(4/3, 0)$. Draw the line through the two intercepts as shown in Figure 1–10.

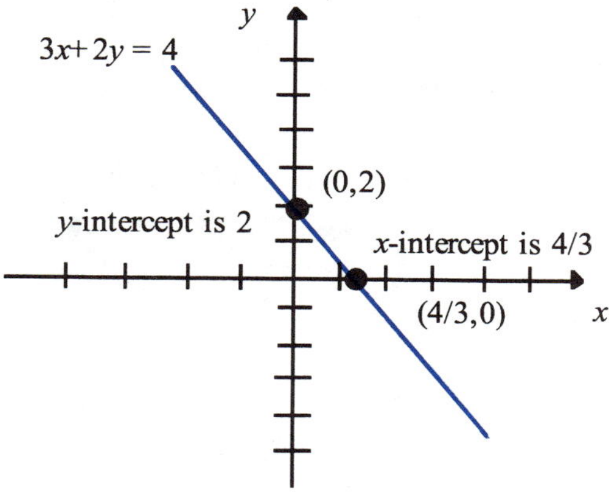

Figure 1–10

Slope-Intercept Form

Every equation of the form $y = mx + b$, where m and b are numbers, is linear since it can be written in the general form $mx - y = -b$. Therefore the graph of the equation $y = 2x - 3$ is a line with $m = 2$ and $b = -3$. In addition, the graph of the equation $y = 3x + 1$ is a straight line. The expression "the line $y = mx + b$" refers to the graph of the linear equation $y = mx + b$.

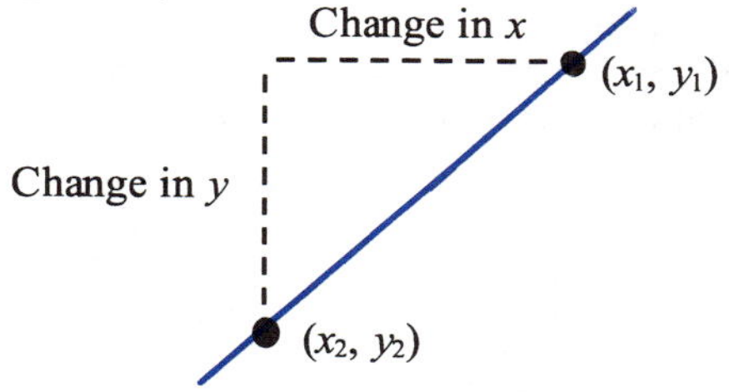

Figure 1–11

The numbers m and b in the linear equation $y = mx + b$ play an important role. The property measured by m is intuitively described as "steepness." It is called the *slope* of the line. To measure the slope, start at some point (x_1, y_1) on the line, proceed in the vertical direction some distance, then proceed in the horizontal direction until the line is intersected at a point (x_2, y_2). (See Figure 1–11.) The vertical distance is called the "change in y" (or the "rise") and the horizontal distance is called the "change in x" (or the "run"). The slope is defined as the ratio of the change in y to the change in x (or the ratio of the rise to the run).

DEFINITION

Let (x_1, y_1) and (x_2, y_2) be distinct points on a line with $x_1 \neq x_2$. The slope of the line is the number

$$m = \frac{\text{change in} y}{\text{change in} x} = \frac{y_2 - y_1}{x_2 - x_1}$$

This definition indicates that the method for computing the slope of a line is to select two points on the line and then to use the formula in the definition. For example, consider the line $y = 2x - 3$. (See Figure 1–12.) Two points on it are $(1, -1)$ and $(4, 5)$.

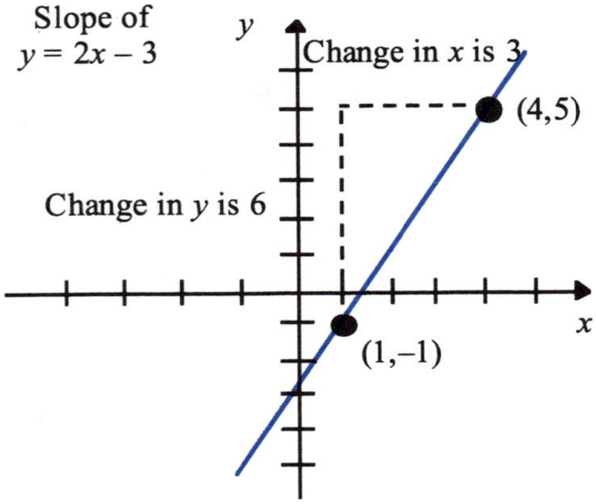

Figure 1–12

Therefore the slope of the line is

$$m = \frac{\text{change in}\,y}{\text{change in}\,x} = \frac{5-(-1)}{4-1} = \frac{5+1}{3} = \frac{6}{3} = 2$$

It was stated above that the equation $y = 2x - 3$ is linear with $m = 2$. It is no accident that this value of m is equal to the value of m computed by formula. We return to this property later in this section.

Figure 1–13 displays the graphs of several lines. As the values of x increase from left to right, the lines with positive slope tend upward (or increase), while the lines with negative slope tend downward (or decrease.)

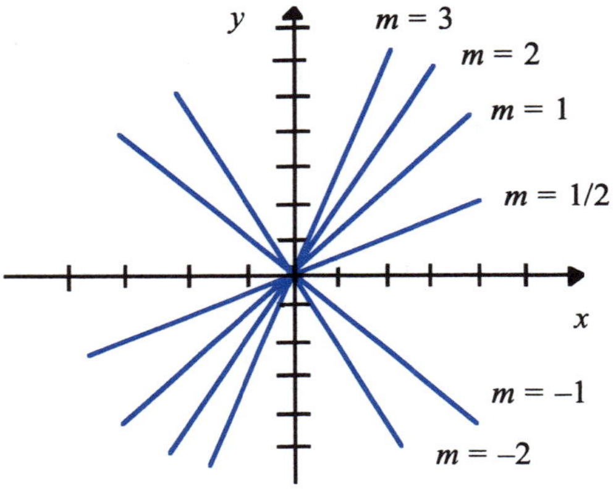

Figure 1–13

Example 2 investigates the slope determined by two distinct pairs of points on the same line.

EXAMPLE 2

Find the slope of the line $y = 3x + 1$ through each pair of points.
(a) (0, 1) and (3, 10)
(b) (–1, –2) and (4, 13)

Solution.
(a) The slope of the line through the points (0, 1) and (3, 10) is

$$m = \frac{10-1}{3-0} = \frac{9}{3} = 3$$

(b) The slope of the line through the points (–1, –2) and (4, 13) is

$$m = \frac{13-(-2)}{4-(-1)} = \frac{13+2}{4+1} = \frac{15}{5} = 3$$

The result of Example 2 is not surprising because a line has only one slope, so no matter which pair of points is chosen on the line, the formula for the slope will always produce the same number. Therefore, as illustrated in Example 2, the slope of a line is independent of the two points chosen on it. Another way of viewing the slope can be seen from its table of points:

x	y
–2	–5
–1	–2
0	1
1	4
2	7

Each time the value of x increases by 1, the value of y increases by the slope 3.

The number b in the linear equation $y = mx + b$ can be obtained by setting $x = 0$. Then $y = \text{m}\cdot 0 + b = b$, so the line passes through the point $(0, b)$. The number b is called the y-intercept of the line $y = mx + b$ even though formally the y-intercept is the point $(0, b)$.

The linear equation $y = mx + b$ is called the *slope-intercept form* of a line because m is the slope, and b is the y-intercept.

DEFINITION

The **slope-intercept form** of a line is

$$y = mx + b$$

where m is the slope and b is the y-intercept.

EXAMPLE 3

Sketch the graph of the equation $y = -2x + 3$.

The y-intercept is $b = 3$ so the line passes through the point (0, 3). The slope can be obtained directly from the form of the equation: $m = -2$. It is not necessary to pick two points on the line and to compute the slope m from them. Since m is negative, the line slopes downward (from left to right). The graph is sketched in Figure 1–14.

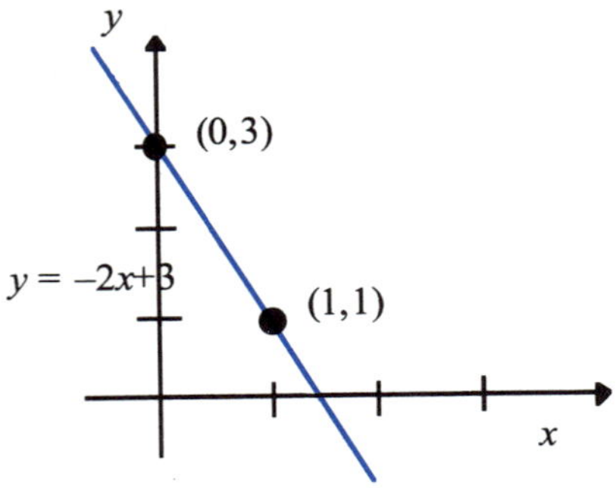

Figure 1–14

Parallel Lines

Two lines are **parallel** if they do not meet in either direction when extended indefinitely. Since parallel lines have the same steepness, it follows that two lines are parallel if and only if they have the same slope.

EXAMPLE 4

Write the equation of the line that is parallel to the line $y = -2x + 2$ and passes through the origin.

Solution. The graph of the linear equation $y = -2x + 2$ is a line with slope $m = -2$. Therefore the new line will have the same slope $m = -2$. Since the new line passes through (0, 0), its y-intercept is $b = 0$. Substituting the values $m = -2$ and $b = 0$ into the slope-intercept form $y = mx + b$ yields $y = -2x$. This is the equation of the line that is parallel to $y = -2x + 2$ and passes through the origin. The lines are sketched in Figure 1–15.

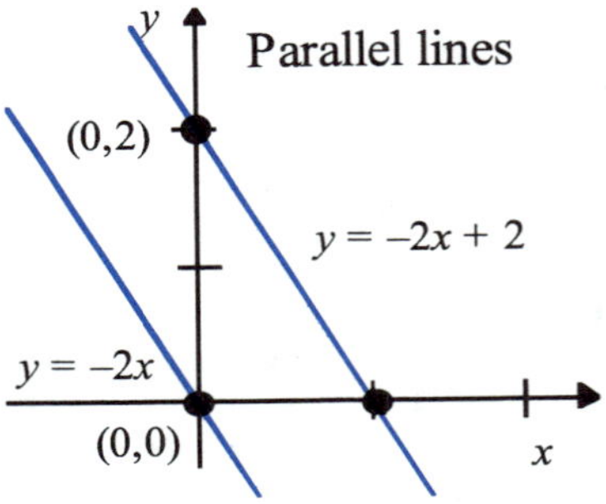

Figure 1–15

There are two special kinds of lines. **Horizontal lines** are parallel to the x-axis, and **vertical lines** are parallel to the y-axis. The slope of a horizontal line can be determined by choosing any two points $(x_1, 0)$ and $(x_2, 0)$ on the x-axis. By definition the slope of the line through these points is

$$m = \frac{\text{change in} y}{\text{change in} x} = \frac{0 - 0}{x_2 - x_1} = 0$$

Thus the slope of the x-axis is 0. Since parallel lines have the same slope, it follows that *the slope of any horizontal line is zero.*

Next consider vertical lines. If $(0, y_1)$ and $(0, y_2)$ are two points on the y-axis, the slope is

$$m = \frac{\text{change in} y}{\text{change in} x} = \frac{y_2 - y_1}{0 - 0}$$

Since division by 0 is not permitted, the slope of the y-axis is not defined. It follows that *the slope of any vertical line is not defined.*

EXAMPLE 5

Write the equations of the horizontal and vertical lines that pass through the point (2, 3).

Solution. The slope of the horizontal line is $m = 0$. Since the horizontal line passing through (2, 3) is parallel to the x-axis, its y-intercept is $b = 3$. Substituting $m = 0$ and $b = 3$ into the equation $y = mx + b$ yields $y = 3$.

The equations of vertical lines cannot be obtained from the equation $y = mx + b$ because m is not defined. The vertical line through (2, 3) is located 2 units to the right of the y-axis. Some additional points on it are (2, 2), (2, 1), (2, 0), and (2, –1). *Note that the value of x is always 2.* Hence the equation of the line is $x = 2$.

Both lines are sketched in Figure 1–16.

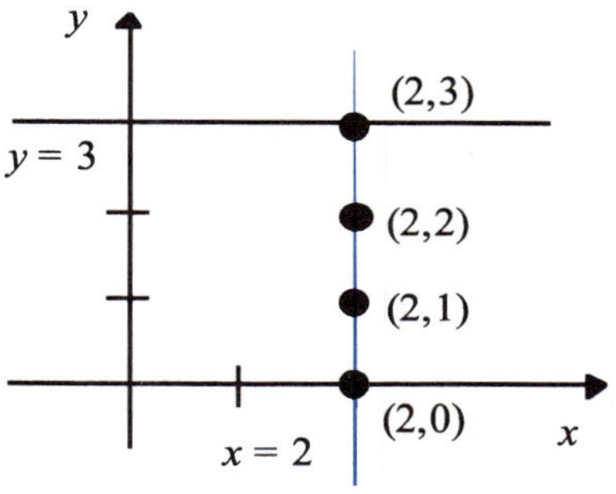

Figure 1–16

Example 5 suggests the following standard forms for all horizontal and vertical lines:

Horizontal & Vertical Lines

Horizontal lines are of the form $y = c$, where c is a number
Vertical lines are of the form $x = k$, where k is a number

Point-Slope Form

The equation of a line that is neither vertical nor horizontal can be expressed in several forms. We have already indicated that the slope-intercept form gives the equation of a line when its slope and y-intercept are known. But what if the slope m and another point (x_1, y_1) on the line are known? Then an arbitrary point (x, y) on the line can be used with the point (x_1, y_1) to compute the slope m:

$$m = \frac{y - y_1}{x - x_1}$$

Multiplying both sides of this equation by $x - x_1$ yields a new equation

$$y - y_1 = m(x - x_1)$$

which is aptly named the *point-slope form*.

DEFINITION

The point-slope form of a line is

$$y - y_1 = m(x - x_1)$$

where m is the slope and (x_1, y_1) is a given point on the line.

Example 6 shows how to convert from this form to slope-intercept form.

EXAMPLE 6

Find the point-slope form and the slope-intercept form of the line that passes through the point (3, 2) and has slope 4.

Solution. We have $x_1 = 3$, $y_1 = 2$, and $m = 4$. The equation of the line in point-slope form is $y - 2 = 4(x - 3)$. Use the properties of algebra indicated on the right of each form to write the equation of the line in slope-intercept form:

$y - 2 = 4(x - 3)$	point-slope form
$y = 4(x - 3) + 2$	add 2 to each side
$y = 4x - 12 + 2$	distributive law
$y = 4x - 10$	add −12 and 2

The equation $y = 4x - 10$ is the slope-intercept form with slope $m = 4$ (this slope was given) and y-intercept $b = -10$. The graph is sketched in Figure 1–17.

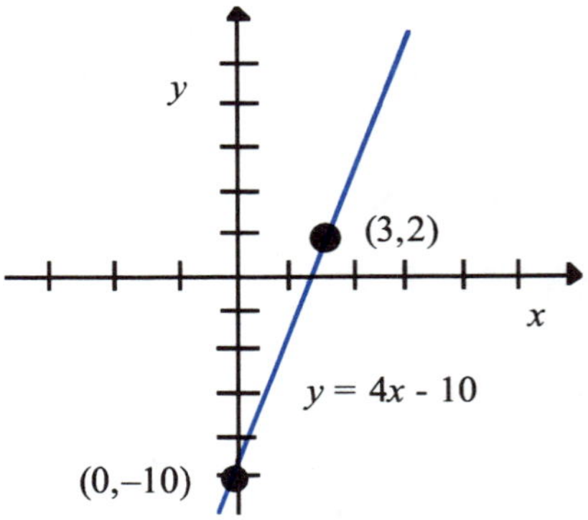

Figure 1–17

There is only one line that can pass through two given points. Example 7 shows how to determine the equation of that line when two points on it are known.

EXAMPLE 7

Find an equation of the line that passes through the points (1, 2) and (–1, 5).

Solution. The slope of the line is

$$m = \frac{5-2}{-1-1} = \frac{3}{-2} = -\frac{3}{2}$$

Now use the point-slope form with either point. We choose to work with (1, 2) but we could have chosen (–1, 5) instead. Using $x_1 = 1$, $y_1 = 2$, and $m = -3/2$, the point-slope form yields

$$y - 2 = -\frac{3}{2}(x-1)$$

The equation in slope-intercept form is then found by solving for y:

$$y = -\frac{3}{2}x + \frac{7}{2}$$

Linear Depreciation

Tax laws require businesses to determine the current value of every asset. Most assets lose value as they get older. It is said that they *depreciate* in value. There are several ways to compute the depreciation of an asset. One method, called **linear depreciation**, assumes that the current value y of the asset is related to the age x of the asset by a linear equation. This equation usually takes the form $y = b - mx$, where b is the purchase price of the asset, or the value when $x = 0$, and m is the rate of depreciation.

EXAMPLE 8

A firm purchases a machine for $50,000 and the machine has a useful life (determined by tax laws) of ten years. If the machine depreciates $4,000 per year, the value of the machine after x years is

$$y = 50{,}000 - 4{,}000x$$

The salvage value of an asset is its value when the asset reaches its useful life. Determine the salvage value of the machine.

Solution. The salvage value is equal to the value of y when $x = 10$. Substituting yields

$$y = 50{,}000 - (4{,}000)(10) = 50{,}000 - 40{,}000 = 10{,}000$$

Therefore the salvage value of the machine is $10,000.

The following is a summary of the main items in this section.

1. The slope of a line is given by $m = \dfrac{\text{change in } y}{\text{change in } x} = \dfrac{y_2 - y_1}{x_2 - x_1}$, unless $x_1 = x_2$.
2. The slope measures steepness. As x increases from left to right, y increases if the slope is positive and y decreases if the slope is negative.
3. Horizontal lines have form $y = c$, and their slope is 0. Vertical lines have form $x = k$, and their slope is undefined.
4. The three principal forms for the equation of a line are given in Table 1–2.

Name of form	Equation	Variables
general	$Ax + By = C$	A, B and C are real numbers; A and B cannot both be 0
slope-intercept	$y = mx + b$	m = slope, b = y-intercept
point-slope	$y - y_1 = m(x - x_1)$	m = slope, (x_1, y_1) = point

Table 1–2

EXERCISE SET 1-2

In problems 1 to 8 sketch the graph of the line.

1. $3x + 4y = 12$
2. $3x - y = 6$
3. $5x - 2y = 10$
4. $5x + 3y = 15$
5. $y = 4x - 5$
6. $y = 2x + 1$
7. $y = -2x + 3$
8. $y = -3x - 2$

In problems 9 to 16 find the slope of the line passing through the two points.

9. (–1, 2), (3, –4)
10. (–3, 8), (3, –8)
11. (1, 2), (2, 4)
12. (3, 1), (4, 3)
13. (1, 2), (2, –4)
14. (3, –1), (4, 3)
15. (1.6, 4.1), (2.5, 5.9)
16. (3/2, 5), (–1, –3/2)

In problems 17 and 18 determine if the lines are parallel.

17. Line 1 passes through (0, 1) and (1, 2); line 2 passes through (–1, 6) and (6, –1).
18. Line 1 passes through (1, 3) and (5, –1); line 2 passes through (1, 1) and (2, 0).

In problems 19 and 20 determine the slope of the line.

19. (a) $x = 2$ (b) $y = -7$
20. (a) $x = 15$ (b) $y = 2$

In problems 21 to 26 find the slope-intercept form of the line passing through the given points.

21. (1, 2), (2, 4)
22. (3, 1), (4, 3)
23. (1, 2), (2, –4)
24. (3, –1), (4, 3)
25. (2, 1), (3, –6)
26. (2, –1), (–3, 6)

In problems 27 to 42 find an equation of the line from the information given. Sketch the graph of the line.

27. $m = 2$, $b = 7$.
28. $m = -2$, $b = 7$.
29. $m = -2$, $b = -7$.
30. $m = -2$, $b = 0$.
31. $m = -2$; the line passes through (6, –1).
32. $m = 1$; the line passes through (6, –1).
33. $b = -1$ and the line is parallel to $y = 2x + 1$.
34. The line passes through (1, 1) and is parallel to the line $x - 3y = 4$.
35. The line passes through (–2, 0) and (0, 2).
36. The line passes through (2, 0) and (0, –2).
37. The line passes through (–2, 0) and (–2, –2).
38. The line passes through (0, 4) and (2, 4).
39. The horizontal line which passes through (–5, –3).
40. The vertical line which passes through (–5, –3).
41. The slope is undefined and the x-intercept is –2.
42. The slope is 0 and $b = -4$.

43. Consider the equation $3x + 2y = 12$.
(a) Put the equation in slope-intercept form.
(b) What is the slope?
(c) What is the y-intercept?

44. Consider the equation $5x + 3y = 15$.
(a) Put the equation in slope-intercept form.
(b) What is the slope?
(c) What is the y-intercept?

In problems 45 and 46 write the equation (in point-slope form) of the line containing the indicated points.

45. (–1, 5), (0, 1), (1, –3)
46. (–2, 5), (0, 1), (2, –3)

47. For tax purposes some firms assume that the current value y of an asset is related to the asset's age x by a linear equation. Suppose a firm buys a work station with 10 microcomputers, and assumes that after x months the work station's worth is given by

$$y = 60{,}000 - 15{,}000x$$

(a) Sketch the graph of the equation.
(b) What is the work station's worth after 3 years?
(c) What economic interpretation can be given to the y-intercept?

Problems 48 to 51 are based on figures supplied by the U.S. Census Bureau.

48. The population of Texas was 14,229,191 in 1980. The average increase for each year during the 1980's was 283,061.4.
(a) Write an equation to describe the population of Texas during that decade.
(b) Use this equation to determine the population of Texas in 1990.

49. The population of New York was 17,558,072 in 1980. The average increase for each year during the 1980's was 48,643.3.
(a) Write an equation to describe the population of New York during that decade.
(b) Use this equation to determine the population of New York in 1990.

50. In Section 1-1, the equation for the population of California in the 1980's was given as

$$C = 617{,}134.8x + 23{,}667{,}902$$

Explain the meaning of the slope of this equation.

51. The equation for the population of Iowa in the 1980's is given by

$$I = -12{,}638.4x + 2{,}913{,}808.$$

Explain the meaning of the negative slope.

52. Prove that the slope of the line $y = mx + b$ is m. *Hint*: Let (x_1, y_1) and (x_2, y_2) be points on the line with $x_1 \neq x_2$. Show that $m = \dfrac{y_2 - y_1}{x_2 - x_1}$

53. Is it true that every equation with variables x and y in which the highest power of x is one and the highest power of y is one must be linear?

Referenced Exercise Set 1-2

1. The petroleum industry uses "decline curve analysis" to estimate reserves of oil and gas wells, even though more sophisticated techniques are available. The reason for the continued emphasis on this method is that the required data is almost always available. Two petroleum engineers developed a simple graphing technique for evaluating production data from a linear equation.[8] Let y represent the loss ratio (a technical term measured here in months) and x represent the time (also in months).

[8]D. A. Rowland and C. Lin, "New Linear Method Gives Constants of Hyperbolic Decline," *Oil and Gas Journal*, January 14, 1985, pp. 86-90.

The line which the engineers discovered for a particular set of data has slope 0.5 and y-intercept 7.5.

(a) Write the equation of the line.

(b) Sketch the line on the interval from 0 months to 72 months.

(c) What is the loss ratio after 6 years?

2. This problem deals with four "rules of thumb" relating the height and weight of adults to those of children.[9] Write each relationship as an equation. (Writing this equation requires the introduction of appropriate variables.)

(a) The height of a male adult is double the height at age 2.

(b) The height of a female adult is double the height at age 1 1/2.

(c) The weight of a male adult is five times the weight at age 2.

(d) The weight of a female adult is five times the weight at age 1 1/2.

3. Figure 1–18 shows the power necessary to achieve loud peak volume levels (106 decibels) for a pair of Acoustic Research loudspeakers.[10] The three different types of rooms ("dead", "average", and "live") require different amplifier sizes to produce loud peak volume levels. Let y be the power (in watts) and x be the volume of the room (in cubic feet). For each type of room, express y in terms of x in slope-intercept form.

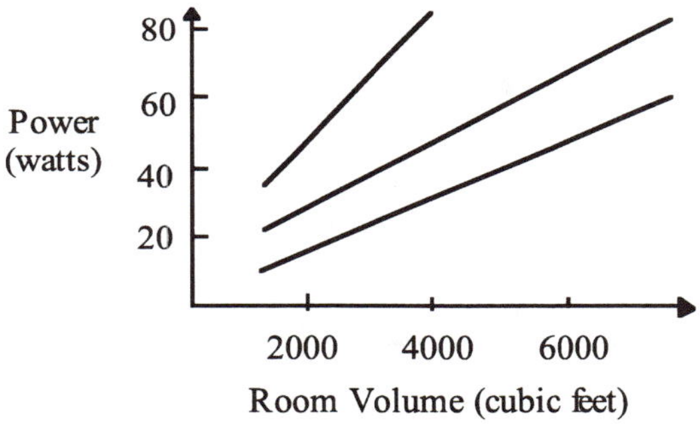

Figure 1–18

4. A carpenter asked University of Delaware professor Richard J. Crouse the following question.[11] "I am building roofs that are 30 feet long at the base. The pitch of each roof rises 4 inches for every horizontal foot. I have to install vertical supports every 16 inches. (See Figure 11.) This forces me to climb the ladder, measure 16 inches horizontally, measure the vertical height at that point, climb down the ladder, saw the support, and climb back up to put it in place. Is there a formula I can use to determine the lengths of the supports in advance?"

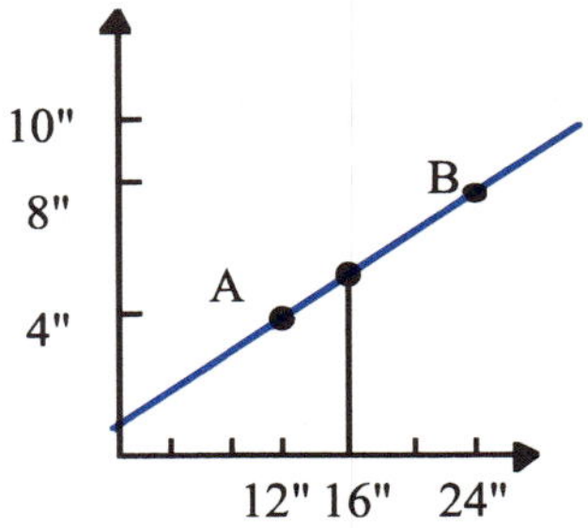

Figure 1–19

[9] Wilton M. Krogman, *Child Growth*, University of Michigan Press, 1972, p. 42.

[10] Source: Teledyne Acoustic Research of Canton, Massachusetts.

[11] R. J. Crouse, "Linear Equation Saves Carpenter's Time," *Mathematics Teacher*, May 1990, pp. 400-401.

(a) What are the coordinates of points A and B in Figure 1–19?
(b) Write the equation of the line through points A and B, where x is the horizontal distance and y is the height.
(c) Use the equation in (b) to find the length of the support at 16 inches.
(d) Use the equation in (b) to find the length of the support at 8 feet.

Problems 5 and 6 deal with the percentage of energy that is supplied to the body by fats and carbohydrates during treadmill running.[12] Figure 1–20 shows that each energy source is a linear function of time.

5. (a) Interpret the meaning of the fact that during the first few minutes of running the slope of the carbohydrate function is positive.
   (b) Interpret the meaning of the fact that after a few minutes of running the slope of the carbohydrate function is negative.

6. (a) Interpret the meaning of the fact that during the first few minutes of running the slope of the fat function is negative.
   (b) Interpret the meaning of the fact that after a few minutes of running the slope of the fat function is positive.

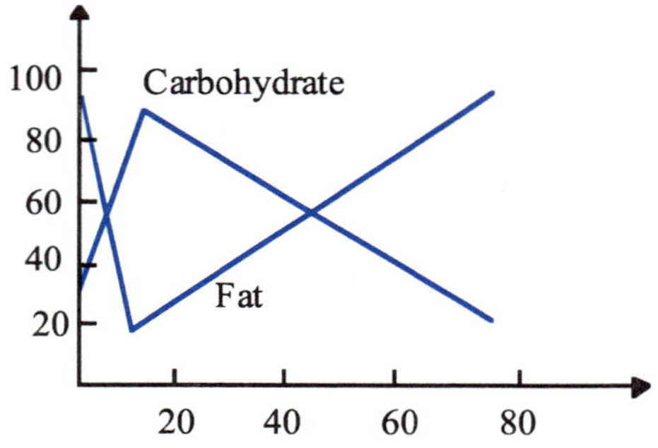

Figure 1–20

Cumulative Exercise Set 1-2

In problems 1 to 3 use the plotting method to sketch the graph of the equation.

1. $y = 2x - 3?$

2. $x + 4y = 4$

3. $y = -x^2 + 1$

4. Sketch the graphs of the total cost function $C = 50 + 2x$ and the total revenue function $R = 12x$ on the same set of coordinate axes.

5. Find the slope-intercept form of the line passing through the points $(-1, 4)$ and $(1, -4)$.

---

[12]Trevor Smith, "Chemistry, Exercise, and Weight Control," *Today's Chemist*, February, 1990, pp. 10-11, 21.

6. Find the equations of the vertical line and the horizontal line passing through the point (2, 5).

In problems 7 and 8 find an equation of the line from the given information.

7. The line has y-intercept 5 and is parallel to the line $2x - 4y = 1$.
8. The line has x-intercept 7 and the slope is undefined.

9. A company has total cost C = $0.4x + 15{,}000$, where x is the number of sales and the fixed cost is \$15,000. If the company sells \$30,000 worth of goods, what is the total cost? Sketch the graph of the equation.

Multiple-Choice Exercise Set 1-2

1. Which of the following equations is in general form?
   (a) $y = 2x - 3$ (b) $x = 2y$ (c) $y = 4x$ (d) $x = 5y$ (e) none of these
2. Which of the following equations is <u>not</u> in general form?
   (a) $2x - y = 0$ (b) $4x + y = 6$ (c) $3x - y = 7$ (d) $y = 3x - 2$ (e) $2x + 2y = 9$
3. The y-intercept of the line $2x - y = 6$ is
   (a) –6 (b) 6 (c) 3 (d) –3 (e) none of these
4. The x-intercept of the line $2x - y = 6$ is
   (a) –6 (b) 6 (c) 3 (d) –3 (e) none of these
5. The slope of the line $y = -4x + 2$ is
   (a) 4 (b) 2 (c) –4 (d) 0 (e) none of these
6. The slope of the line $4x + y = 2$ is
   (a) 4 (b) 2 (c) –4 (d) 0 (e) none of these
7. The slope of the line $4x - y = 2$ is
   (a) 4 (b) 2 (c) –4 (d) 0 (e) none of these
8. Which point does <u>not</u> lie on the line $y = 2x - 3$?
   (a) (–1, –1) (b) (1, –1) (c) (4, 5) (d) (0, –3) (e) none of these
9. Which point lies on the line $3x - y = 6$?
   (a) (1, 9) (b) (0, –6) (c) (–2, 0) (d) (2, 6) (e) none of these
10. What is the slope of the line through the points (1, 2) and (2, 5)?
    (a) 0 (b) 1 (c) –3 (d) –4 (e) none of these
11. What is the slope of the line through the points (1, –1) and (2, –1)?
    (a) 0 (b) 1 (c) –3 (d) –4 (e) none of these
12. What is the slope of the line through the points (0, 2) and (1/2, 0)?
    (a) 0 (b) 1 (c) –3 (d) –4 (e) none of these
13. What is the slope of the line with y-intercept 4 and x-intercept 2?
    (a) 2 (b) –2 (c) 1/2 (d) –1/2 (e) none of these
14. What is the slope of the line through the origin and the point (1, –2)?
    (a) 1 (b) –1 (c) 2 (d) –2 (e) none of these

15. Let P_1, P_2, and P_3 be three points on a line with slope m. Which of the following statements is not true?
(a) The slope between P_1 and P_2 is equal to m.
(b) The slope between P_1 and P_2 is equal to the slope between P_2 and P_3.
(c) The slope between P_1 and P_2 is less than the slope between P_2 and P_3.
(d) The slope between P_1 and P_3 is equal to m.
(e) The slope between P_2 and P_3 is equal to m.

16. If a graph has the table of points below then is must be a line.
(a) True (b) False

17. If a graph has the table of points below then is must be a line.
(a) True (b) False

18. If a line has a negative slope then the line is
(a) falling (b) rising (c) horizontal (d) vertical (e) none of these

19. If lines L_1 and L_2 are parallel and the slope of L_1 is equal to 5 then the slope of L_2 is equal to
(a) 1/5 (b) 1/2 (c) –5 (d) -1/5 (e) none of these

20. The slope of a line that is parallel to $y = 3x - 7$ is equal to
(a) –7 (b) 7 (c) –3 (d) 3 (e) none of these

21. The slope of the line $x = 3$ is
(a) 0 (b) undefined (c) –3 (d) 3 (e) none of these

22. The equation of the horizontal line that passes through the point (2, 0) is
(a) $x = 0$ (b) $x = 2$ (c) $y = 0$ (d) $y = 2$ (e) none of these

23. If the x-intercept of a line is equal to the y-intercept of the line, then the slope of the line
(a) is positive (b) is negative (c) equals 1 (d) equals –1 (e) none of these

24. An equation of a line that passes through (2, –1) and has slope 3 is
(a) $y - 2 = 3(x - 1)$ (b) $y - 2 = 3(x + 1)$
(c) $y + 1 = 3(x - 2)$ (d) $y - 1 = 3(x + 2)$
(e) none of these

25. The slope of the line $y = 2(x - 5) + 4$ is equal to
(a) 4 (b) 2 (c) –5 (d) 5 (e) none of these

1-3 INTERSECTION OF LINES

Many applications of mathematics involve more than one linear equation at the same time. In this section we describe several methods of solving systems of two linear equations. The next section applies these techniques to real life situations, which the Case Study develops in greater detail.

The Solution of a System

Consider the two linear equations in general form:

$$\begin{aligned} x + y &= 7 \\ x - y &= 3 \end{aligned}$$

Taken together they form a **system of two linear equations in two variables**. A **solution** is an ordered pair (x, y) of numbers that satisfy both equations at the same time Consequently, the ordered pair (5, 2) is a solution because

$$\begin{aligned} 5 + 2 &= 7 \\ 5 - 2 &= 3 \end{aligned}$$

However, the pair (6, 1) is not a solution because it does not satisfy the second equation, even though it does satisfy the first equation.

The solution (5, 2) was obtained by inspection. The solution is fairly obvious, especially if the equations are first translated into words:

the numbers add up to 7
the difference between the numbers is 3

Since not all systems have such obvious solutions, it is desirable to develop alternate methods. Another way of solving the system is to draw the graphs of both equations on the same coordinate system. Since both graphs are lines, the solution of the system is the point of intersection (5, 2) of the two lines because it satisfies both equations. (See Figure 1–21.) Example 1 illustrates the graphing method.

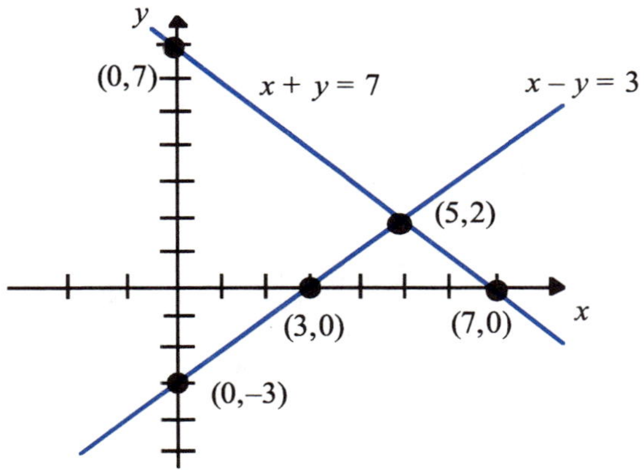

Figure 1–21

EXAMPLE 1

Solve the following system of linear equations by graphing.

$$\begin{aligned} x + 3y &= 5 \\ x - y &= 1 \end{aligned}$$

The graph of each equation is a line. The two lines are drawn on the same coordinate system in Figure 1–22. It appears from the figure that the lines intersect at the point (2 ,1). To verify this, substitute $x = 2$ and $y = 1$ into the equations

$$\begin{aligned} 2 + 3\cdot 1 &= 5 \\ 2 - 1 &= 1 \end{aligned}$$

Since the two equations are satisfied, (2, 1) is the solution of the system.

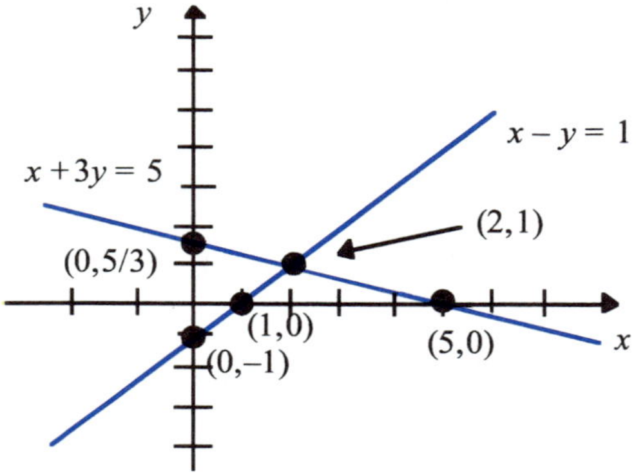

Figure 1–22

The graphical method is valuable for approximating a solution to a system of two equations, but often it is not precise enough to produce exact solutions so we need to develop yet another method.

Elimination

The method of elimination is very efficient in solving systems of linear equations. It consists of eliminating one variable and then solving for the other. We illustrate it on the following system of linear equations:

$$\begin{aligned} 2x + 3y &= 4 \\ 3x - 2y &= 6 \end{aligned}$$

Choose a variable to eliminate. Sometimes it is easier to eliminate one of the variables than the other, but here it does not matter so we arbitrarily choose y. Multiply the first equation by 2 and the second by 3 yielding

$$\begin{aligned} 4x + 6y &= 8 \\ 9x - 6y &= 18 \end{aligned}$$

The numbers 2 and 3 were chosen to multiply the equations in order to make the coefficients of y are opposites of each other. Now, to eliminate y, add the two equations:

$$\begin{array}{rcrcl} 4x & + & 6y & = & 8 \\ 9x & - & 6y & = & 18 \\ \hline 13x & & & = & 26 \end{array}$$

A solution to this equation is also a solution to the original system of equations. Solve for x:

$$x = 2$$

Substitute $x = 2$ into either one of the original equations to get y. From the first equation,

$$\begin{aligned} 2x + 3y &= 4 \\ 2(2) + 3y &= 4 \\ 3y &= 0 \\ y &= 0 \end{aligned}$$

Since $x = 2$ and $y = 0$, the solution to the system is (2, 0).

The variable y was eliminated by producing the numbers 6 and –6 as coefficients of y, and then adding the two equations. This two-step procedure is the essence of the method of elimination.

Two Steps in Eliminating a Variable

1. Multiply the equations by appropriate numbers to produce coefficients of a variable such that one is the negative of the other.
2. Add the two resulting equations.

EXAMPLE 2

Solve the following system of linear equations by elimination:

$$\begin{array}{rcrcl} 3x & - & 4y & = & 1 \\ 4x & - & 2y & = & 3 \end{array}$$

Eliminating y can be accomplished with only one multiplication, namely multiplying the second equation by –2. After multiplying, add the equations:

$$\begin{array}{rcrcl} 3x & - & 4y & = & 1 \\ -8x & + & 4y & = & -6 \\ \hline -5x & & & = & -5 \end{array}$$

Consequently, $x = 1$. Now, substitute $x = 1$ into the first equation of the system to get $y = 1/2$. Therefore, the solution is (1, 1/2).

The next example applies elimination to a system in which one of the variables has the same coefficient in both equations.

EXAMPLE 3

Solve this system of linear equations.

$$\begin{aligned} 3x + 2y &= 2 \\ 3x - y &= 5 \end{aligned}$$

Solution. Multiply the second equation by −1, and add it to the first equation.

$$\begin{array}{r} 3x + 2y = 2 \\ \underline{-3x + \quad y = -5} \\ 3y = -3 \end{array}$$

Solve for y to get $y = -1$. Next, substitute $y = -1$ into the first equation to get $x = 4/3$. Therefore, the solution is $(4/3, -1)$.

Substitution

The method of substitution involves solving one equation for one variable and then substituting that variable into the other equation. It is most useful when a coefficient of one of the variables is equal to 1. This method reduces the system to one equation in one variable, just like the method of elimination.

EXAMPLE 4

Solve the following system of linear equations.

$$\begin{aligned} 2x + 4y &= 1 \\ x - 5y &= -2 \end{aligned}$$

Solution. The coefficient of x in the second equation is 1, so solve the second equation for x, obtaining

$$x = 5y - 2$$

Substitute this expression into the first equation to produce one equation in one variable.

$$\begin{aligned} 2x + 4y &= 1 \\ 2(5y - 2) + 4y &= 1 \\ 10y - 4 + 4y &= 1 \\ 14y - 4 &= 1 \\ y &= \frac{5}{14} \end{aligned}$$

To find x, substitute $y = 5/14$ into $x = 5y - 2$ to get

$$x = 5y - 2 = 5(5/14) - 2 = -3/14.$$

Therefore the solution is $(-3/14, 5/14)$.

Systems Without Unique Solutions

Every system that we have encountered so far contains exactly one solution. This is not always the case, however, as can be seen by examining the geometric meaning of a system of linear equations in two unknowns. The graph of each equation is a line, so the system consists of two lines, l_1 and l_2. Figure 1–23 shows that such a system can have exactly one solution, no solutions, or infinitely many solutions.

A system with no solutions is called **inconsistent**. Figure 1–23(b) shows that the graph of such a system consists of two parallel lines. Example 5 shows how to detect such a system algebraically.

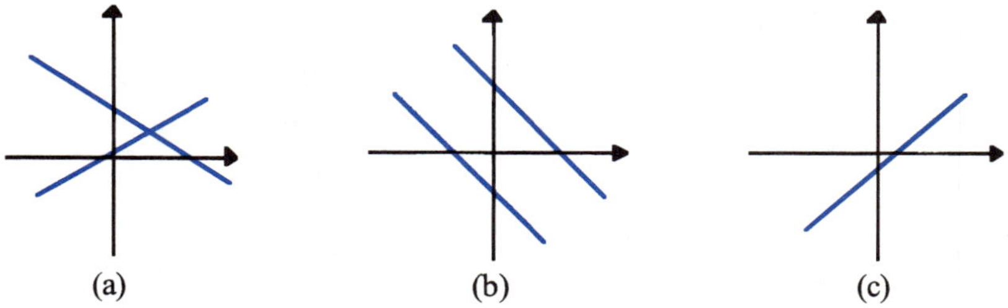

Figure 1–23

EXAMPLE 5

Solve the following system of linear equations.

$$\begin{aligned} x - y &= 1 \\ 2x - 2y &= 3 \end{aligned}$$

Solution. To solve this system by elimination, multiply the first equation by –2 and add the resulting equation to the second equation. (Alternately, substitution could be used by solving the first equation for either variable.)

$$\begin{aligned} -2x + 2y &= -2 \\ \underline{2x - 2y} &\underline{= 3} \\ 0 &= 1 \end{aligned}$$

If an ordered pair (x, y) were to satisfy the system, then 0 would have to equal 1, so a contradiction would result. Therefore the original system cannot have any solutions.

This fact can be seen geometrically by writing each equation in slope-intercept form and by examining their slopes and y-intercepts.

$$\begin{aligned} y &= x - 1 \\ y &= x - (3/2) \end{aligned}$$

These lines have equal slopes but different y-intercepts, so the equations represent two different parallel lines.

A system with infinitely many solutions is called **dependent**. Dependent systems occur when the two equations represent the same line. [See Figure 1–23(c).] Example 6 shows how to detect a dependent system algebraically.

EXAMPLE 6

Solve the following system of linear equations.

$$\begin{aligned} x - y &= 1 \\ 2x - 2y &= 2 \end{aligned}$$

Solution. To solve this system by elimination, multiply the first equation by –2 and add it to the second equation. (Alternately, substitution could be used.)

$$\begin{aligned} -2x + 2y &= -2 \\ \underline{2x - 2y} &\underline{= 2} \\ 0 &= 0 \end{aligned}$$

This equality is true for *all* ordered pairs (x, y) that satisfy the first equation. The reason is that the second equation has the same set of solutions as the first.

Geometrically, what initially appears to be two lines in Example 6 actually turns out to be one line. This is analogous to one person having two different names, say, Christine and Chris, or Robert and Rob. One line, two equations; one person, two names.

Breakeven Analysis

In financial management, the breakeven point is the level of operations where neither profits nor losses are incurred. Above this point, each unit sold contributes to profits; below it, each unit sold contributes to losses. Therefore the breakeven point is the intersection of the sales revenue and the total cost. The revenue is derived from the sales of the units. The total cost usually involves a fixed cost (like salaries and rental expenses) and a variable cost (like sales commissions and advertising). Example 7 illustrates these terms.

EXAMPLE 7

A candy company specializes in making mints. Each box of mints sells for \$2 and has a variable cost of \$1.20. The fixed cost is \$40,000. Determine the company's breakeven point.

Solution. The sales revenue is derived from the \$2 price for each box of mints. Let x be the number of boxes sold. If R is the sales revenue then $R = 2x$. The variable cost is \$1.20 per box, so it is $1.2x$. The total cost C, which includes the \$40,000 fixed cost, is given by $C = 1.2x + 40{,}000$.

Therefore the breakeven point occurs when R = C. Therefore the equation for total cost can be written $R = 1.2x + 40{,}000$. The breakeven point is found by solving the system of linear equations

$$\begin{aligned} R &= 2x \\ R &= 1.2x + 40{,}000 \end{aligned}$$

The solution is $x = 50{,}000$ and $R = 100{,}000$ so the breakeven point occurs when 50,000 boxes of mints are sold, at which point the revenue is \$100,000.

The company breaks even when it sells 50,000 boxes of mints. The graphs of $R = 2x$ and $C = 1.2x + 40{,}000$ are displayed in Figure 1–24. The breakeven point (50,000, 100,000) is the point of intersection of these two lines. If the company sells more than 50,000 boxes it makes a profit, while if it sells less than 50,000 boxes it suffers a loss.

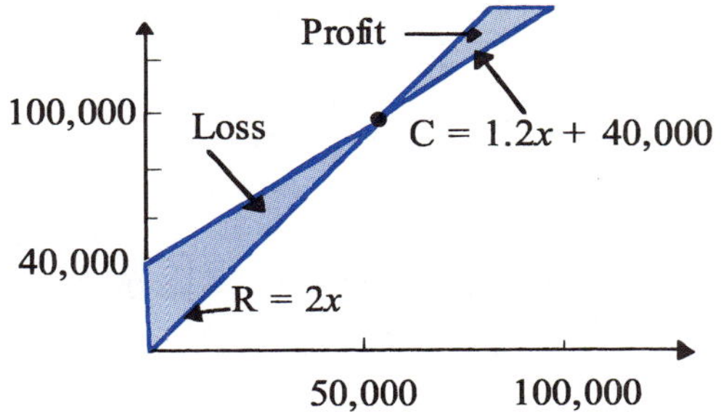

Figure 1–24

EXERCISE SET 1-3

1. $x + y = 5$
   $x + 2y = 7$

2. $x + y = 2$
   $x - y = 2$

In problems 3 and 4 solve the system by graphing.

3. $2x + y = 4$
   $x - y = 2$

4. $x + 3y = 1$
   $2x - y = 9$

In problems 5 to 22 solve the system by elimination.

5. $2x + y = 5$
   $3x + 4y = 10$

6. $2x + 5y = 2$
   $3x - 2y = 3$

7. $5x + 2y = 3$
   $x - 3y = 4$

8. $8x - 3y = 1$
   $3x + 4y = 9$

9. $2x - 3y = 7$
   $x - 2y = 4$

10. $-4x + 2y = -3$
    $3x - 2y = 3$

11. $x + y = 5$
    $40x + 60y = 240$

12. $x + y = 5$
    $40x + 60y = 200$

13. $x + y = 12$
    $6x + 3y = 66$

14. $x + y = 12$
    $6x + 3y = 42$

15. $80x + 100y = 10{,}000$
    $x + y = 110$

16. $100x + 80y = 10{,}000$
    $x + y = 110$

17. $2x + y = 80$
$3x + 2y = 140$

18. $3x + y = 100$
$x + 2y = 100$

19. $2x + 3y = 40$
$100x + 200y = 2500$

20. $0.5x + 0.2y = 9$
$0.1x + 0.3y = 7$

21. $0.5x + 0.4y = 45$
$0.3x + 0.6y = 45$

22. $0.2x + 0.3y = 11$
$0.5x + 0.6y = 26$

In problems 23 and 24 solve the system by substitution.

23. $x + y = -2$
$x - y = 4$

24. $2x - 3y = 2$
$3x - y = 3$

In problems 25 to 28 determine whether the system is inconsistent or dependent.

25. $3x - y = 1$
$6x - 2y = 5$

26. $8x + y = 2$
$8x + y = 3$

27. $2x - 4y = 2$
$3x - 6y = 3$

28. $12x - 6y = 3$
$8x - 4y = 2$

In problems 29 and 30 determine the breakeven point of a candy company which specializes in making mints.

29. Each box of mints sells for \$3 and has a variable cost of \$1.50. The fixed cost is \$30,000.

30. Each box of mints sells for \$5 and has a variable cost of \$2.00. The fixed cost is \$30,000.

In problems 31 to 46 solve the system by any method. If there is not exactly one solution, state whether there are no solutions or infinitely many solutions.

31. $2x - y = 3$
$4x + y = 1$

32. $-x + 2y = 3$
$2x + 6y = 1$

33. $4x + y = 18$
$4x - 3y = 10$

34. $2x + y = 6$
$x - 2y = 3$

35. $x + y = 120$
$100x + 40y = 6000$

36. $45x + 9y = 45$
$10x + 6y = 20$

37. $2x + 3y = 4$
$3x - 2y = 6$

38. $x + 2y = 2$
$x - y = 5$

39. $25x - 50y = 1500$
$30x - 15y = 2700$

40. $1.2x - 3.5y = 23$
$8.5x + 7y = 1.5$

41. $x + 1.5y = 0$
$y = 8.5x$

42. $15x - 110y = 4000$
$10x + 55y = 1000$

43. $x + y = 5$
$2x + 2y = 8$

44. $x - y = 3$
$3x - 3y = 9$

45. $2x + 5y = 40$
$2x + y = 24$

46. $2x + 4y = 40$
$50x + 200y = 1600$

47. A travel agency is planning charter flights to Hawaii for exactly 1400 passengers aboard two types of jets. Each 747 jet holds 400 passengers and each 707 holds 150 passengers. The rental fee is \$15,000 for each 747 and \$9000 for each 707. If exactly \$66,000 must be used for renting the jets, how many of each type should be ordered?

48. A dietitian is preparing a meal of chicken and corn that must provide exactly 63 grams (g) of protein and 21 milligrams (mg) of iron. Each serving of chicken contains 45 g of protein and 9 mg of iron; each serving of corn contains 9 g of protein and 6 mg of iron. How many servings of each ingredient should be prepared?

49. Suppose that \$5000 is invested in two securities: school bonds and public utility bonds. School bonds pay 6% interest and public utility bonds pay 8% interest. If an interest of \$370 is required, how much money should be invested in each security?

50. Suppose that regular gasoline costs \$1 per gallon and unleaded gasoline costs \$1.20 per gallon. How many gallons of each gasoline should be blended to produce 400 gallons of mixture which is worth \$1.05 per gallon?

51. In the 1980 census Ohio was the sixth highest state in population while Florida was seventh. Ten years later Ohio had dropped to seventh while Florida had jumped to fourth. The equations for the populations in these states in the 1980's are given below, where x is the number of years after 1980.

$$F = 325{,}703.8x + 9{,}746{,}324$$
$$O = 8{,}969.5x + 10{,}797{,}638$$

In what month of what year did Florida's population overcome Ohio's population?

52. In 1980 Alaska was the least populated state and Wyoming was next. The equations for the populations in these states in the 1980's are given below, where x is the number of years after 1980.

$$A = 15{,}009.6x + 401{,}851$$
$$W = -1{,}358.2x + 469{,}557$$

In what month of what year did Alaska lose the distinction of being the least populated state?

Referenced Exercise Set 1-3

Problems 1 and 2 are adapted from an article on the ability of the Big Three American automakers to reduce their breakeven point substantially to become more competitive with foreign competition.[13] The variable x represents the number of cars sold (in millions of cars) and y represents the sales revenue and the total cost (in millions of dollars).

1. (a) Determine the breakeven point if the sales revenue is $y = 1.5x$ and the total cost is $y = x + 6.1$.
   (b) Determine the breakeven point if the sales revenue is $y = 2x$ and the total cost is $y = 0.5x + 13.35$.

---

[13] Adapted from Lynn Adkins, "Detroit Gets Lean and Mean," *Dun's Business Month*, January, 1983, pp. 56-58.

(c) The solution in (a) was the breakeven point for the Big Three in 1979. The solution in (b) was the breakeven point in 1983. By how many cars did the Big Three drop the breakeven point in four years?

2. (a) Determine the breakeven point if the sales revenue is $y = 0.5x$ and the total cost is $y = 0.25x + 0.6$.
   (b) Determine the breakeven point if the sales revenue is $y = 0.8x$ and the total cost is $y = 0.6x + 0.28$.
   (c) The solution in (a) was the breakeven point for Chrysler in 1981. The solution in (b) was the breakeven point in 1983. By how many cars did Chrysler drop the breakeven point in two years?

3. Figure 1–25, due to Peter Crawford, head financial analyst for Citicorp Information Services of New York, NY, shows the relationship between short-term interest rates and the volume of funds raised in credit markets.[14] (a) At what rate of interest did the supply of credit equal the demand for credit in 1985? (b) At what rate of interest did the supply of credit equal the demand for credit in 1983?

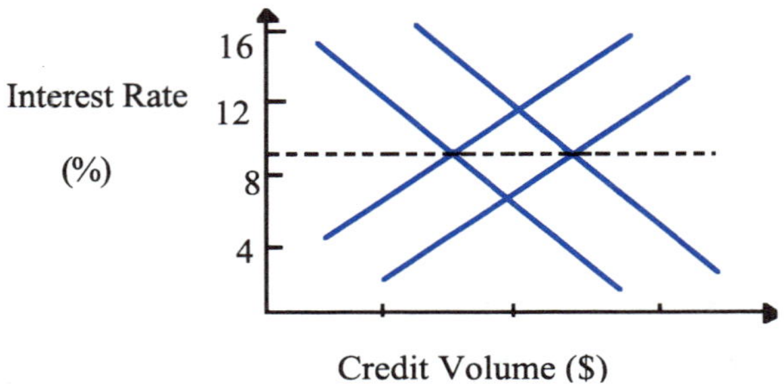

Figure 1–25

4. In 1939 L. F. Richardson proposed a mathematical model of an arms race that is still valid today.[15] The model is based upon equations from calculus called differential equations, but we have adapted it to a system of linear equations. Let X and y be two countries with arms expenditures x and y, respectively. Denote each change in expenditure by Δx and Δy. Then

$$\Delta x = ay - bx + c \quad (a > 0, b > 0)$$
$$\Delta y = dx - ey + f \quad (d > 0, e > 0)$$

The terms ay and dx are called *spurring terms*, the terms bx and ey are called *breaking terms*, and the constants c and f are called *grievance terms*.
A state of equilibrium occurs when there is no change in expenditures, that is, when $\Delta x = \Delta y = 0$. Find the equilibrium point for a = 0.01, b = 0.02, c = 2, d = 0.0125, e = 0.1, and f = 6.25.

---

[14] Peter Crawford, "Interest Rate Outlook: Focus on Credit Supply," *Business Economics*, July 1985, pp. 34-39.

[15] L. F. Richardson, "Generalized foreign politics," *British Journal of Psychology*, Monograph Supplement XXIII, Cambridge University Press, 1939. A BASIC program for solving this system is given in the article by Leonard M. Wapner, "Modeling with Difference Equations: Two Examples," *The Mathematics Teacher*, February 1984, pp. 136-140.

Cumulative Exercise Set 1-3

1. Let $x + 2y = 5$. What is the value of x when $y = 3$?

2. The total cost C (in dollars) for a company is $C = .4x + \$15{,}000$, where x is the amount of sales. If the company sells \$68,500 worth of goods, what is the total cost?

3. Sketch the graph of the line $100x + 40y = 6000$.

4. What is the equation of the vertical line whose x-intercept is -3?

5. Find the slope-intercept form of the line that passes through the points $(-1, 4)$ and $(3, -8)$.

6. What is the slope of the line $3x - 6y = 8$?

7. What is the equation of the line parallel to the line $y = 3x + 2$ and passing through the point $(0, -7)$?

8. Write the slope-intercept form of the line that passes through the points $(0, -1)$ and $(1, -1)$.

9. Solve the following system of linear equations by any method.
   $3x - 2y = -16$
   $2x + 4y = 16$

10. Find the point of intersection of the lines $4x + 2y = 10$ and $y = x + 1$.

11. Determine whether the following system of linear equations is inconsistent or dependent.
    $6x - 3y = 1$
    $8x - 4y = 0$

12. BikeLine manufacturers 10-speed and 12-speed bicycles. Each 10-speed bike requires 2 minutes to assemble the frame and 8 minutes to install the gears. Each 12-speed bike requires 5 minutes to assemble the frame and 19 minutes to install the gears. How many bicycles of each speed can be made per day if exactly 60 minutes must be used for assembling frames and 232 minutes for installing gears?

13. Draw a line having points in the first, second, and third quadrants.
    (a) Is the slope positive or negative?
    (b) Is the y-intercept positive or negative?
    (c) Is the x-intercept positive or negative?

14. A farmer estimates that his seasonal farming costs are \$9700 plus an additional \$130 per acre of corn that he plants.
    (a) Write an equation for the cost C in terms of acres of corn planted.
    (b) Sketch the graph of the equation.
    (c) What is the cost of planting 10 acres of corn?

15. A daytime telephone call from New York City to Los Angeles costs \$0.23 for the first minute and \$.24 for each additional minute. For nights and weekends the rate is a flat \$.13 per minute.
    (a) Write an equation for the cost D of a daytime call in terms of minutes x.
    (b) Write an equation for the amount of money saved by calling during a night or weekend instead of calling during the daytime.
    (c) Interpret the meaning of the slope of the line in part (b).

Multiple-Choice Exercise Set 1-3

1. Which of the following pairs is a solution of the given system of linear equations?
   $x + y = 10$
   $x - y = 4$
   (a) (6, 4) (b) (4, 6) (c) (8, 2) (d) (7, 3) (e) none of these

2. Which of the following pairs is not a solution of the given system of linear equations?
   $2x - 4y = 0$
   $x - 2y = 0$
   (a) (6, 4) (b) (12, 6) (c) (4, 2) (d) (6, 3) (e) none of these

3. Which equation describes the statement "the difference between the numbers is 2?"
   (a) $x + y = 2$ (b) $2x - y = 0$ (c) $x - 2y = 2$ (d) $x - 2y = 0$ (e) none of these

4. Which statement can be described by the equation $x = 2y$?
   (a) a quantity x is half as big as another, y
   (b) a quantity x is twice as big as another, y
   (c) the difference between x and y is 2
   (d) the difference between y and x is 2
   (e) none of these

5. Sketch the graphs of the linear equations below on the same coordinate system. At how many points do the lines meet?
   $y = 2x + 1$
   $y = 2x + 4$
   (a) none (b) one (c) two (d) three (e) infinitely many

6. What equation is obtained after multiplying $2x - y = 6$ by -1?
   (a) $2x + y = 6$ (b) $2x - y = -6$ (c) $-2x - y = 6$
   (d) $-2x + y = -6$ (e) none of these

7. If a system of two linear equations reduces to the equation $4y = 12$ then the solution is
   (a) (4, 12) (b) (12, 4) (c) (3, 0) (d) (0, 3) (e) cannot be specified

8. If a system of two linear equations reduces to the equations $4y = 12$ and $x - 3y = 10$ then the solution is
   (a) (4, 3) (b) (19, 3) (c) (3, 1) (d) (1, 3) (e) cannot be specified

9. Eliminating x from the system
   $2x + y = 9$
   $x - 2y = 2$
   yields $y = 1$. What is the value of x?
   (a) 4 (b) 5 (c) 0 (d) 7 (e) cannot be specified

10. What operation will help to eliminate y from this system of linear equations?
    $2x + 5y = 12$
    $4x - y = 2$
    (a) multiply the top equation by -2
    (b) multiply the top equation by -5
    (c) multiply the bottom equation by $-1/2$
    (d) multiply the bottom equation by 5
    (e) none of these

11. Which method is the most efficient for solving this system of equations?
(a) inspection (b) graphing (c) elimination
(d) substitution (e) none of these

12. Which method is the most efficient for solving the system of equations below?
$2x + 4y = 6$
$3x - 5y = -13$
(a) inspection (b) graphing (c) elimination
(d) substitution (e) none of these

13. Which method is the most efficient for solving the system of equations below?
$2x + 4y = 6$
$3x - y = -5$
(a) inspection (b) graphing (c) elimination
(d) substitution (e) none of these

14. If two lines have different slopes then the number of solutions of the corresponding system of equations must be
(a) 0 (b) 1 (c) 2 (d) infinite (e) none of these

15. If two lines have slope 5 and different y-intercepts then the number of solutions of the corresponding system of equations must be
(a) 0 (b) 1 (c) 2 (d) infinite (e) none of these

16. If two lines have slope 5 and the same y-intercepts then the number of solutions of the corresponding system of equations must be
(a) 0 (b) 1 (c) 2 (d) infinite (e) none of these

17. If two lines have slope 4 and pass through the origin then the number of solutions of the corresponding system of equations must be
(a) 0 (b) 1 (c) 2 (d) infinite (e) none of these

18. If the cost equation is $C = 2x + 900$ and the revenue equation is $R = 5x$ then the breakeven point occurs when $x =$
(a) 3 (b) 30 (c) 300 (d) 3000 (e) none of these

19. If two equations for revenue are $R = 4x$ and $R = 1.2x + 56{,}000$ then $x =$
(a) 2 (b) 20 (c) 200 (d) 2000 (e) none of these

1-4 LEAST SQUARES

Many applications of mathematics are concerned with the relationship between one quantity x and another quantity y. For instance, x can be a person's height and y can be that person's shoe size. If you record the height and shoe size of ten people and then plot the ten ordered pairs of points (x, y) on a coordinate system, you most likely will find that the points do not lie on a straight line. However, they will appear to be distributed along a line, in the sense that a line can be drawn to pass very near to all of them. A set of points is said to be *spaced linearly* when the points seem to lie very near to a straight line. Of all the lines that come close to a linearly spaced set of points, the one that comes closest is called the **line of best fit**. The purpose of the method of least squares is to produce the equation of the line of best fit.

This section presents the method for constructing the equation of the line of best fit. The material is applied to recent voting trends in presidential elections in the United States.

The Line of Best Fit

Table 1–3 lists five points in the plane. Regard them as ordered pairs (x, y) and graph them on a coordinate system. (See Figure 1–26.) The points do not lie on one line, yet they are spaced linearly. The method of least squares produces the line of best fit for these points, where by "best fit" we mean that the sum of squares of the vertical distances from the points to the line are minimal. Since distances are measured in terms of squares, and least is synonymous with minimal, the method is called **least squares**.

x	y
2	1
3	6
4	9
5	13
6	15

The method of least squares involves two equations in the unknowns m (slope) and b (y-intercept). The Greek letter sigma, Σ, denotes "sum."

$$n \cdot b + (\Sigma x)m = \Sigma y \quad \textbf{(LS1)}$$
$$(\Sigma x)b + (\Sigma x^2)m = \Sigma xy \quad \textbf{(LS2)}$$

where

n = the number of points
Σx = the sum of the x values
Σy = the sum of the y values
Σx^2 = the sum of the squares x^2
Σxy = the sum of the products xy

We refer to the equations as LS1 and LS2. Their derivation involves calculus so we illustrate them without deriving them.

Since there are five points in Table 1–3, set $n = 5$. To obtain the four sums, expand Table 1–3 by two more columns. Place a column for x^2 to the left of the x column, and a column for xy to the right of the y column. This is shown in Table 1–4, where the row of sums yields $\Sigma x = 20$, $\Sigma y = 44$, $\Sigma x^2 = 90$, and $\Sigma xy = 211$.

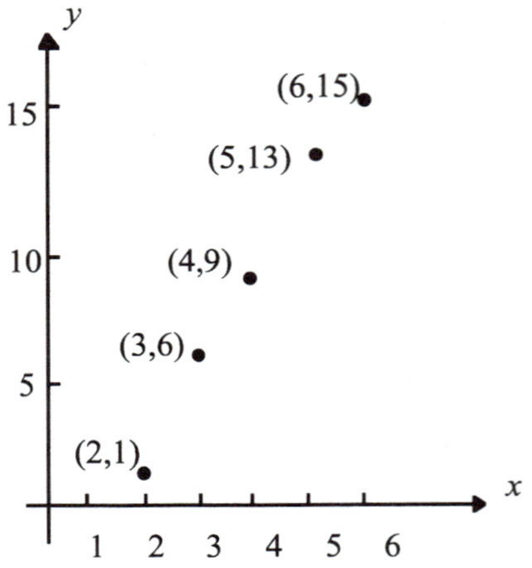

Figure 1–26

	x^2	x	y	xy
	4	2	1	2
	9	3	6	18
	16	4	9	36
	25	5	13	65
	36	6	15	90
Sums	90	20	44	211

Table 1–4

Therefore the least squares equations are:

$$5b + 20m = 44$$
$$20b + 90m = 211$$

This system can be solved by elimination by multiplying the top equation by –4 and then adding the two equations.

$$\begin{array}{rcr} -20b - 80m & = & -176 \\ \underline{20b + 90m} & \underline{=} & \underline{211} \\ 10m & = & 35 \end{array}$$

This produces the slope of the line of best fit: $m = 35/10 = 3.5$. Substitute $m = 3.5$ into the equation

$$5b + 20m = 44$$

Then

$$5b + 20(3.5) = 44$$
$$5b = 44 - 20(3.5) = 44 - 70 = -26$$
$$b = -26/5 = -5.2$$

Consequently, the y-intercept is $b = -5.2$. The line of best fit, $y = mx + b$, is uniquely determined by its slope $m = 3.5$ and y-intercept $b = -5.2$, so

$$y = 3.5x - 5.2$$

The line and the five points are graphed in Figure 1–27.

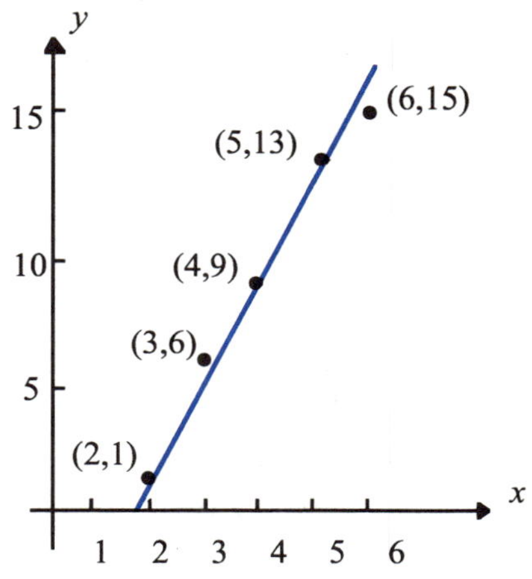

Figure 1–27

The Method of Least Squares

For a given set of data points (x, y), the **line of best fit** is

$$y = mx + b$$

where m and b are obtained by solving the system of **least squares equations**

$$n \cdot b + (\Sigma x)m = \Sigma y \qquad \text{(LS1)}$$
$$(\Sigma x)b + (\Sigma x^2)m = \Sigma xy \qquad \text{(LS2)}$$

and where

n = the number of points
Σx = the sum of the x values
Σy = the sum of the y values
Σx^2 = the sum of the squares x^2
Σxy = the sum of the products xy

EXAMPLE 1

Find the line of best fit for the points (1, 1), (2, 2), (4, 3), (5, 6).

Solution. Construct a vertical table of x values and y values. Place a column of the squares x^2 to the left of the x values, and a column of the products xy to the right of the y values.

	x^2	x	y	xy
	1	1	1	1
	4	2	2	4
	16	4	3	12
	25	5	6	30
Sums	46	12	12	47

There are four points so $n = 4$. The row of sums shows that

$$\Sigma x^2 = 46, \Sigma x = 12, \Sigma y = 12, \Sigma xy = 47$$

Therefore the least squares equations are

$$4b + 12m = 12 \qquad \text{(LS1)}$$
$$12b + 46m = 47 \qquad \text{(LS2)}$$

Multiply LS1 by –3 and then add it to LS2. The resulting equation is $10m = 11$, so $m = 1.1$. Substituting $m = 1.1$ into LS1 yields $b = -0.3$, so the line of best fit is given by the equation $y = 1.1x - 0.3$. This line is graphed in Figure 1–28.

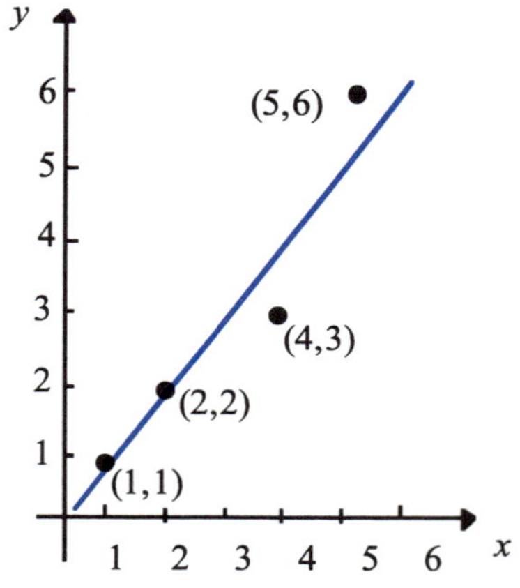

Figure 1–28

Voting Trends

Example 2 applies the method of least squares to presidential elections. We will then discuss one of the drawbacks to basing predictions of future outcomes upon results from the past.

EXAMPLE 2

Table 1–5 lists the percentage of eligible voters who voted in each of the presidential elections from 1960 to 1980. Find the equation of the line of best fit.

Year	Percentage
1960	62.8
1964	61.9
1968	60.9
1972	55.5
1976	54.4
1980	52.3

Table 1–5

Solution. Let x denote the year of the presidential election and let y denote the percentage of voters who voted. To simplify calculations, write years without the prefix 19, so that $x = 60$ stands for 1960. Expand Table 1–5 to include the two additional columns.

	x^2	x	y	xy
	3600	60	62.9	3768.0
	4096	60	62.8	3961.6
	4624	64	61.9	4141.2
	5184	72	55.5	3996.0
	5776	76	54.4	3996.0
	6400	80	52.3	4184.0
Sum	29,680	420	347.8	24,185.2

The system of least squares equations is

$$6b + 420m = 347.8 \qquad \text{(LS1)}$$
$$420b + 29680m = 24185.2 \qquad \text{(LS2)}$$

Multiply LS1 by –70, then add it to LS2.

$$\begin{aligned} -420b - 29400m &= -24346.0 \\ \underline{420b + 29680m} &= \underline{24185.2} \\ 280m &= -160.8 \end{aligned}$$

The slope is $m = -160.8/280 = -1608/2800 = -201/350$. Substituting $m = -201/350$ into LS1 yields the y-intercept b.

$$6b + 420(-201/350) = 347.8$$

$$6b = 347.8 + \left(\frac{84,420}{350}\right)$$

$$6b = 589$$

$$b = \frac{589}{6}$$

Thus the line of best fit is $y = -(201/350)x + (589/6)$.

The primary importance of the line of best fit is for projecting future values. To obtain the projection for the year 1984 substitute $x = 84$ and solve for y.

$$y = -(201/350)x + (589/6) = -(201/350)(84) + (589/6) \approx 49.93$$

Therefore the projection was that slightly less than 50% of the voters would vote in the 1984 election. However, 53.3% voted. Does this reveal a limitation on basing projections upon historical evidence, or is the 1984 figure an exception to the rule of continuing decline?

EXERCISE SET 1-4

In problems 1 to 14 find the equation of the line of best fit for the given points.

1. (1, 3), (2, 1), (3, 4), (4, 3)
2. (1, 3), (2, 7), (4, 9), (5, 12)
3. (1, 3), (2, 4), (3, 1), (4, 2)
4. (1, 12), (2, 9), (4, 7), (5, 3)
5. (–2, –2), (–1, 2), (1, 4), (2, 7)
6. (–1, 0), (0, –2), (1, 1), (2, 0)
7. (2, 15), (3, 13), (4, 9), (5, 6), (6, 1)
8. (–1, –5), (0, 0), (1, 3), (2, 7), (3, 9)
9. (0.3, 0.4), (0.4, 0.3), (0.1, 0.2), (0.3, 0.1)
10. (0.1, 0.5), (0.9, 0.4), (0.7, 0.2), (0.3, 0.1)
11. (–3, 0), (–2, 2.5), (–1, 4.5), (0, 6), (1, 7.5), (2, 10), (3, 13)
12. (–3, –4), (–2, –4), (–1, –3), (0, –3), (1, –2.5), (2, –2), (3, –1)

13. (–3, 1), (–2, 0), (–1, 0), (0, –0.5), (1, –1), (2, –1.5), (3, –2)

14. (–3, 9), (–2, 4.5), (–1, 0), (0, –3.5), (1, –8), (2, –12), (3, –16)

Problems 15 to 18 refer to figures compiled by the Department of Health and Human Services. The average five-year-old boy is 43 inches tall, the average ten-year-old boy is 54 inches tall, and the average fifteen-year-old boy is 67 inches tall.

15. (a) Plot these values on a coordinate system with age along the x-axis and height along the y-axis.
    (b) Find the equation for the line of best fit that links age to height.

16. Use the equation for the line of best fit in Problem 15 to approximate the height of an average eight-year-old boy?

17. Use the equation for the line of best fit in Problem 15 to approximate the height of an average seventeen-year-old boy?

18. Use the equation for the line of best fit in Problem 15 to approximate the height of an average twelve-year-old boy?

In problems 19 to 22 determine whether the points in the figure are spaced linearly.

19.

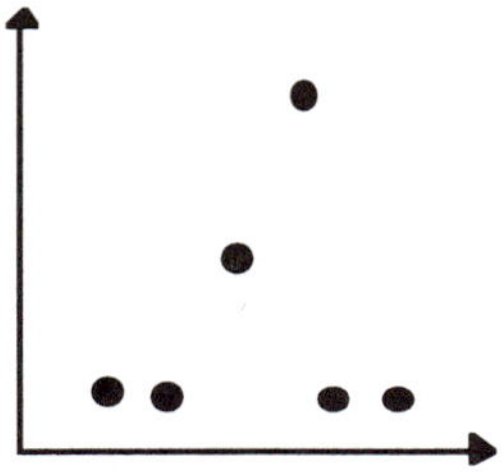

20.

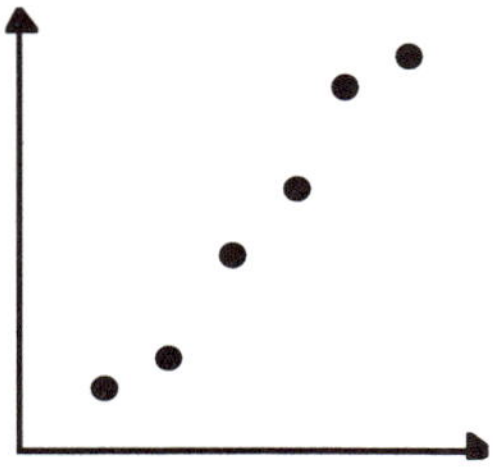

21.

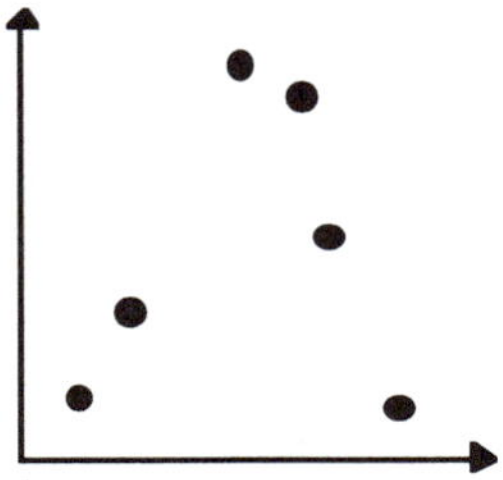

22.

Referenced Exercise Set 1-4

1. Samuel Messick, an official of the Educational Testing Service, admitted that "coaching" for the SAT's could be effective.[16] He revealed that a typical student will increase the SAT-Math score by 10 points after 8 hours of coaching, 20 points after 19 hours, 30 points after 45 hours, and 40 points after 107 hours.
   (a) Plot these points with hours on the x-axis and points on the y-axis.
   (b) Are the points spaced linearly?
   (c) What are the two least squares equations?
   (d) What is the equation for the line of best fit?
   (e) Use part (d) to project how many points the score will be raised with 200 hours of coaching.
   (f) Use part (d) to project how many hours of coaching are required to raise the score 100 points.

2. Table 1–6 displays the difference between U.S. and Europe shoe sizes.[17] Use the table to find the equation of the line of best fit between shoe sizes:
   (a) for women
   (b) for men.

Shoe Sizes

Women		Men	
U.S.	Europe	U.S.	Europe
4 1/2	35	9 1/2	43
6	37	10	44
6 1/2	37.5	10 1/2	44.5
7	38	11	45
7 1/2	38.5	11 1/2	46
8	39	12	46.5
8 1/2	40	13 1/2	48
9	40.5	14	49
9 1/2	41		
10	42		

Table 1–6

3. Table 1–7 is brought to you in part by Delta Air Lines.[18] Draw a coordinate system with the altitude (ft) as the x-axis and the temperature (degrees F) as the y-axis. Plot the points to see if they are spaced linearly. Then compute the line of best fit.

4. Use Problem 3 to project
   (a) the temperature at 40,000 ft.
   (b) the temperature at 4,000 ft.
   (c) the altitude at 32° F
   (d) the altitude at –30° F

---

[16]Edward B. Fiske, "Kicking and Boosting the S.A.T. Scores," *New York Times*, December 23, 1980, p. C4

[17]L. Diane Miller and Jim Miller, "Metric Week – The Capitol Way," *Mathematics Teacher*, September 1989, pp. 454-458.

[18]Courtesy of Delta Air Lines.

How Cold Is Up?

There you are, bareheaded and with no gloves and in the lightest of summer clothes, eating a cold salad at 69° F below (–56° C) zero!

It's routine on Delta Jets, summer or any other season, for you to travel in the upper levels of air far from surface heat. It's a case of the higher, the lower. Perhaps you'd like to know exactly how the temperature changes with altitude, how it can be 104° F (40° C) in the sun of the parking lot and well below zero just four miles from the lot, straight up!

The seasons influence lower level temperatures, of course, but on a "standard" day (59° F, 15° C) at sea level, the mercury would drop this way as you went up:

Feet	Fahrenheit	Meters	Celsius
1,000	56°	300	13°
5,000	41°	1,500	5°
10,000	23°	3,000	–5°
15,000	5°	4,500	–15°
20,000	–15°	6,000	–26°
30,000	–47°	9,000	–44°
36,087	–69°	10,826	–56°

Table 1–7

5. In financial management, an area called "trend analysis" deals with the construction of forecasting models to predict a firm's cash flow (expenditures and receipts).[19] Figure 1–29 is taken from a study of a particular firm's cash receipts. The answers to parts (a) and (b) can be obtained directly from the figure.
   (a) What kind of mathematical method is being used for the model?
   (b) What is the equation of the line of best fit?
   (c) How many receipts does this model forecast after $t = 20$ time periods?
   (d) How many time periods does this model forecast for 33 receipts?

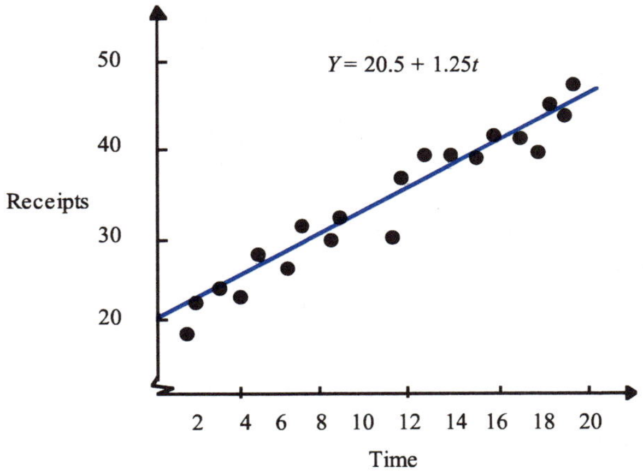

Figure 1–29

[19]James L. Pappas and George P. Huber, "Probabilistic Short-term Financial Planning," *Financial Management*, Autumn 1973, pp. 36-44.

6. Figure 1–30 depicts the relationship between fat intake and death from breast cancer for the inhabitants of many countries around the world.[20] Do they appear to be spaced linearly?

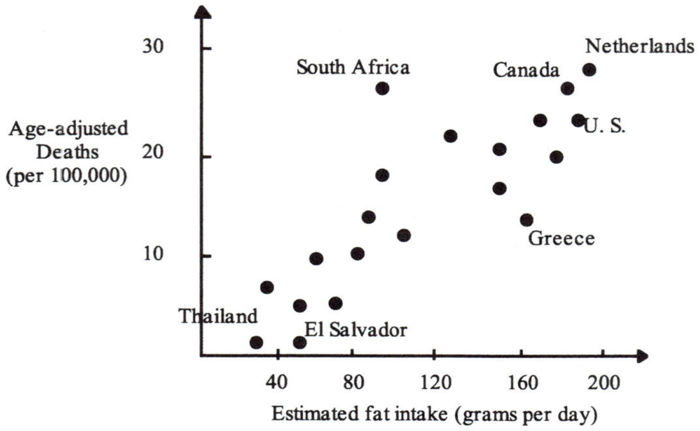

Figure 1–30

Problems 7 and 8 refer to Table 1–8, which lists the percentage of doctorates awarded in mathematics to U.S. citizens over a 23-year period.[21]

7. (a) Find the equation of the line of best fit for the percentage of women who received doctorates. Set $x = 1973$ for the year 1972-1973.
   (b) Use the line to project the percentage in the year 2000-2001.

8. (a) Does the total number of doctorates awarded in mathematics to U.S. citizens seem to be spaced linearly over this period?
   (b) Find the equation of the line of best fit for the total number of U.S. citizens who received doctorates. Set $x = 1973$ for the year 1972-1973.
   (c) Use the line to project the total number in the year 2000-2001.

---

[20]Leonard A. Cohen, "Diet and cancer," *Scientific American*, November 1987, pp. 42-48.

[21]The figures in Table 1–8 are taken from two articles: Edward A. Connors, "Report on the 1987 Survey of New Doctorates," *Notices of the American Mathematical Society*, November 1987, pp. 1081-1086, and John D. Fulton, "AMS-IMS-MAA Annual Survey," *Notices of the American Mathematical Society*, December 1995, pp. 1504-1511.

Mathematics Doctorates, U.S. Citizens by Gender

Year	U.S. Citizens	Male	Female	% Female
1972-1973	774	696	78	10
1973-1974	677	618	59	9
1974-1975	741	658	83	11
1975-1976	722	636	86	12
1976-1977	689	602	87	13
1977-1978	634	545	89	14
1978-1979	596	503	93	16
1979-1980	578	491	87	15
1980-1981	567	465	102	18
1981-1982	519	431	88	17
1982-1983	455	366	89	20
1983-1984	433	346	87	20
1984-1985	396	315	81	20
1985-1986	386	304	82	21
1986-1987	362	289	73	20
1987-1988	363	287	76	21
1988-1989	411	313	98	24
1989-1990	401	312	89	22
1990-1991	461	349	112	24
1991-1992	430	327	103	24
1992-1993	526	381	145	28
1993-1994	469	345	124	26
1994-1995	567	426	141	25

Table 1–8

Cumulative Exercise Set 1-4

1. Let $3x - 2y = 8$. What is the value of y when $x = -2$?
2. Sketch the graph of the equation $2x - 5y = 0$.
3. Find an equation of the line that passes through the origin and has slope $m = 3$.
4. Find the slope of the line passing through the points $(-1, 4)$ and $(1, -1)$.
5. What is the y-intercept of the line $3x - 6y = 8$?
6. Write an equation of the line whose table is given below.

x	y
1	−1
2	−3

7. Solve the system of linear equations:

$$\begin{aligned} x - y &= 4 \\ x - 2y &= 12 \end{aligned}$$

8. Find the point of intersection of the lines $y = 3x + 5$ and $y = 5x + 3$.

9. Determine the equation of the line of best fit for the points
   (1, 5.2), (1.5, 4.7), (1.8, 4.4), (2, 3.8), (2.2, 3.6)

10. Is the slope of a line positive or negative when:
    (a) the x-intercept is positive and the y-intercept is negative?
    (b) the x-intercept is negative and the y-intercept is positive?
    (c) both intercepts are positive?

11. A taxi in Chicago charges \$3 plus \$.25 for each tenth of a mile.
    (a) Write an equation for the cost C in terms of miles.
    (b) Sketch the graph of the equation.
    (c) What is the cost of a 2.6-mile ride?

Multiple-Choice Exercise Set 1-4

1. Any set of three points on the graph of $y = x^2$ is linearly spaced.
   (a) True (b) False

2. Any set of three points on the graph of $2x + 3y = 6$ is linearly spaced.
   (a) True (b) False

3. The line that comes closest to a set of linearly spaced points is called the line of
   (a) least fit (b) minimal fit (c) greatest fit (d) maximal fit (e) none of these

4. The phrase "best fit" refers to the minimal
   (a) sum of squares (b) square of sums (c) absolute value (d) distance (e) none of these

5. The notation for the sum of a set of numbers is
   (a) σ (b) Σ (c) μ (d) Δ (e) none of these

6. How many equations constitute the Least Squares Equations?
   (a) 0 (b) 1 (c) 2 (d) 3 (e) infinitely many

Problems 7 to 10 refer to the table below.

x	y
0	5
1	4
2	3
3	1

7. $\Sigma x =$
   (a) 0 (b) 5 (c) 6 (d) 13 (e) none of these

8. $\Sigma x^2 =$
   (a) 36 (b) 14 (c) 25 (d) 16 (e) none of these

9. $\Sigma xy =$
   (a) 13 (b) 18 (c) 0 (d) 19 (e) none of these

10. The first Least Squares Equation is $kb + 6m = 13$ for $k =$
    (a) 13 (b) 18 (c) 0 (d) 19 (e) none of these

Problems 11 to 14 refer to the following Least Squares System of Equations

$$2m + 5b = 38$$
$$5m + 4b = 27$$

11. The number of points is
(a) 2 (b) 4 (c) 5 (d) 27 (e) 38

12. $\Sigma y =$
(a) 2 (b) 4 (c) 5 (d) 27 (e) 38

13. $m =$
(a) 10 (b) 8 (c) 0 (d) –1 (e) none of these

14. $b =$
(a) 10 (b) 8 (c) 0 (d) –1 (e) none of these

15. If a line of best fit has $m = 7$ and $b = -2$, what is the projection for y when $x = 0$?
(a) 7 (b) –2 (c) –5 (d) 5 (e) none of these

CASE STUDY
THE WOMEN'S 100-METER DASH

The women's 100-meter dash was won in flamboyant style by Florence "Flo-Jo" Griffith-Joyner at the 1988 Olympics in Seoul, Korea. Her winning time was 10.54 seconds. We show how to predict the winning time for this running event in subsequent Olympics based on the winning times since it was first contested as an Olympic event in 1928. We illustrate the method by predicting the winning time for the women's 100-meter dash in the 1992 Olympics based on all winning times through 1988. That prediction is examined in light of the actual time in the 1992 Olympics.

Line of Best Fit

Elizabeth Robinson won the women's 100-meter dash in the Olympics in 1928 in a time of 12.2 seconds. The event has been contested in all Olympic Games ever since. Table 1–9 lists the gold medal winners and their times (in seconds.) The years 1940 and 1944 do not appear because no games were held during World War II. All Olympic data are taken from *The World Almanac and Book of Facts* 1993, Pharos Books, New York.

Year	Gold Medal Winner	Time
1928	Elizabeth Robinson, U.S.	12.2
1932	Stella Walsh, Poland	11.9
1936	Helen Stephens, U.S.	11.5
1948	Francina Blankers-Koen, Netherlands	11.9
1952	Marjorie Jackson, Australia	11.5
1956	Betty Cuthbert, Australia	11.5
1960	Wilma Rudolf, U.S.	11.0
1964	Wyomia Tyus, U.S.	11.4
1968	Wyomia Tyus, U.S.	11.0
1972	Renate Stecher, Germany	11.07
1976	Annegret Richter, Germany	11.08
1980	Lyudmila Kondratyeva, Russia	11.6
1984	Evelyn Ashford, U.S.	10.97
1988	Florence Griffith-Joyner, U.S.	10.54

Table 1–9

Table 1–9 shows that the winning times have been lowered regularly but not steadily (monotonically.) One might argue that the slow time in 1980 was influenced more by politics than sports. Take a close look at the right side of Table 1–9. Of course the times get lower and lower. Is it possible to say more? Is there a pattern to the decrease in times that is obscured by the apparent chaos of facts?

One way to answer these questions is to take a geometric approach. Figure 1 shows the graph of the winning times for the women's 100 meter dash in the Olympics when the

x-axis is the year and the y-axis is the time in seconds. For instance, when x = 1988 the value y = 10.54 reflects Florence Griffith-Joyner's time of 10.54 seconds.

Predicting the Future

Figure 1–31 reveals a remarkable pattern: the points are spaced linearly. This means that the winning performance for the women's 100-meter dash in the Olympics is approximately linear, and if this is true, then the future of the event can be predicted with some degree of confidence. It means that we can project the winning performance in the Olympics in the year 2000. And in 2004. And even in the year 2400. In fact, we can project when the 100-meter dash will be run in a time under ten seconds. Beyond that, we can project when it will be run in 0 seconds flat.

Zero seconds flat! No way. Let us investigate these intriguing possibilities.

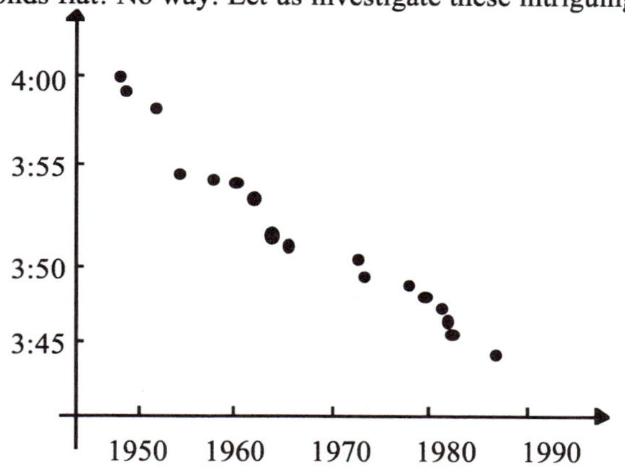

Figure 1–31

EXAMPLE 1

Determine the line of best fit for the women's 100-meter dash in the Olympics based on all Olympic games from 1928 through 1988.

Solution. Expand Table 1–9 by adding the two appropriate columns and inserting the bottom row of sums. Label the new table Table 1–10. The top row of Table 1–10 shows that the winning time in the first Olympics in 1928 was 12.2 seconds.

The Olympics were held 14 times from 1928 to 1988, so n = 14. The sums of the columns are

$\Sigma x^2 = 53802928$
$\Sigma x = 27444$
$\Sigma y = 159.16$
$\Sigma xy = 311907.32$

x^2	x	y	xy
3717184	1928	12.2	23521.60

3732624	1932	11.9	22990.80
3748096	1936	11.5	22264.00
3794704	1948	11.9	23181.20
3810304	1952	11.5	22448.00
3825936	1956	11.5	22494.00
3841600	1960	11.0	21560.00
3857296	1964	11.4	22389.60
3873024	1968	11.0	21648.00
3888784	1972	11.07	21830.04
3904576	1976	11.08	21894.08
3920400	1980	11.6	22968.00
3936256	1984	10.97	21764.48
3952144	1988	10.54	20953.52
3802928	27444	159.16	311907.32

Table 1–10

The system of Least Squares Equations then becomes

$$14b + 27444m = 159.16$$
$$27444b + 53802928m = 311907.32$$

This is not an easy system to solve by hand. The solution, obtained after an exhausting elimination and rounded to four decimal places, is

$$b = 48.4781 \quad \& \quad m = -0.01893$$

Therefore the equation of the line of best fit is

$$y = -0.01893x + 48.4781$$

Consider the meaning of the slope $m = -0.01893$ in Example 1. The negative sign means that the record is being lowered, that the winning time of the event is decreasing. The number 0.01893 measures how much the winning time improves each year. This translates into roughly 0.075 seconds every four years, which is how often the Olympics are held.

One of the basic aims of all of science is to predict the future. Example 2 shows how to use the equation of the line of best fit to obtain projections for the future.

EXAMPLE 2

Use the line of best fit based on data for all Olympic games from 1928 to 1988 to project the winning time in 1992, 1996, and 2000.

Solution. To find the exact times that the line of best fit projects for the future, substitute the desired year into the equation $y = -0.01893x + 48.4781$. For the year 1992 set $x = 1992$. Then $y = -0.01893(1992) + 48.4781 = 10.76$ so the winning time in the 1992 Olympic games is 10.76 seconds. To find the projection for the winning time in the year 1996 set $x = 1996$. Then $y = 10.69$ so the projected winning time is 10.69 seconds. Similarly the projected winning time for the year 2000 is 10.62.

Example 2 shows that the least squares method predicts a winning time of 10.76 seconds for the 1992 Olympics. How accurate was this prediction?

Gail Devers of the United States won the women's 100-meter dash at the 1992 Olympics at Barcelona in a time of 10.82 seconds. Thus the predicted time of 10.76 is off by six one-hundredth of a second.

This result is quite impressive, but you should beware of this model. Actually, human events cannot always be predicted with such accuracy. To provide some evidence for this skepticism we look into the distant future by predicting the winning time for the women's 100-meter dash in some futuristic Olympics.

EXAMPLE 3

Use the line of best fit based on data for all Olympic games from 1928 to 1988 to project the winning time in the year 2400.

Solution. Substitute $x = 2400$ into the equation $y = -0.01893x + 48.4781$. Then the winning time is given by $y = -0.01893(2400) + 48.4781 = 3.05$. Therefore the winning time in the Olympic games in the year 2400 will be about three seconds.

Stating the result of Example 3 in other terms, the speed of running 100 meters in 3 seconds translates into 75 miles per hour. Can a human ever running that fast?

Projecting Years

In all the examples so far we have projected the winning time in a given year. Sometimes, however, it is necessary to determine when a performance will reach a certain level; for instance, when the women's 100m dash will be run in the Olympics under 10 seconds. Example 4 demonstrates how to proceed.

EXAMPLE 4

Use the line of best fit based on data for all Olympic games from 1928 to 1988 to project the first Olympics when the event will be run under 10 seconds.

Solution. We saw in Example 1 that the equation of the line of best fit is given by $y = -0.01893x + 48.4781$. Substitute $y = 10$ into this equation and solve for x.

$$y = -0.01893x + 48.4781$$

$$10 = -0.01893x + 48.4781$$

$$0.01893x = 38.4781$$

$$x = \frac{38.4781}{0.01893} \approx 2032$$

Therefore, according to the given model, the women's 100-meter dash will be run under 10 seconds for the first Olympics in the year 2032.

The result of Example 4 seems reasonable. Following the process described in that example we find, after substituting $y = 5$ into the equation of the line of best fit and solving for x, that the women's 100-meter dash will be run under 5 seconds for the first Olympics in the year 2296. Probably none of us will witness this event.

Now extend this method to an inhuman feat. Substituting $y = 0$ into the line of best fit yields $x = 2560$, so the 100-meter dash will be run in 0 seconds flat in the 2560 Olympics. This is patently absurd. Let us discuss the ramifications.

The least squares method can be carried out on any set of data, but it is most effective when the points appear to lie along a straight line. If the points are not spaced linearly then the method of least squares will not provide reasonable projections, so curves other than lines have to be used. And that is what will eventually occur with the 100-meter dash. Since that event was first contested in 1928 progress has been remarkably linear. At some time in the future the winning time will not continue to follow a linear path, and then we will have to apply another curve to model the situation.

CASE STUDY EXERCISE SET

1. Use the equation of the line of best fit $y = -0.0189x + 48.4781$ for the women's 100-meter dash based on all Olympics through 1988 to predict the year of the first Olympic Games in which the winning time is better than Flo-Jo's world record of 10.54 seconds.

2. Use Table 1–9 to construct a table of the winning times for the 100-meter dash in the Olympics from 1960 through 1992.
   (a) Write the two Least Squares Equations.
   (b) What is the equation of the line of best fit?
   (c) What time does this model project for 1996?
   (d) How does the projection by this model compare with that in Exercise 1(b)?

3. Use Table 1–9 to construct a table of the winning times for the 100-meter dash in all Olympics from 1948 through 1992.
   (a) Write the two Least Squares Equations.
   (b) What is the equation of the line of best fit?
   (c) What time does this model project for 1996?
   (d) How does the projection by this model compare with that in Exercise 1(b)?

4. Use Table 1–9 to construct a table of the winning times for the 100-meter dash in all Olympics from 1948 through 1992.
   (a) Write the two Least Squares Equations.
   (b) What is the equation of the line of best fit?
   (c) What time does this model project for 1996?
   (d) How does the projection by this model compare with that in Exercise 1(b)?

5. Use Table 1–9 to construct a table of the winning times for the 100-meter dash in all Olympics from 1948 through 1992 except 1980, whose winning time was influenced by politics.
   (a) Write the two Least Squares Equations.
   (b) What is the equation of the line of best fit?
   (c) What time does this model project for 1996?
   (d) How does the projection by this model compare with that in Exercise 1(b)?

1-5 WORD PROBLEMS

In the next chapter we study word problems that are expressed in terms of linear equations and linear inequalities. In this section we look at some examples of this type of word problem and cover some of the basic strategies used to translate them into mathematical language.

To express a word problem in mathematical language there are several points of the problem that must be identified.

1. The variables must be labeled. Assign a letter to each quantity and be careful about the units that are used.
2. Each equation is of the form $Ax + By = C$ where A, B and C are positive numbers.
3. In each equation the three terms, Ax, By and C, must have the same units, so determine the units of each term and check that they match.

To illustrate these points we translate the following word problem.

> *Two types of busses are to be leased, the smaller type of bus seats* 40 *people and the larger type seats* 60 *people; how many busses of each type should be leased if there is to be a total of* 240 *seats and a total of* 5 *busses leased?*

The decision of how many busses of each type to lease is determined by the two limitations in the problem: there are to be 240 *seats* available and there are to be 5 *busses* leased. This means that there are two equations in the translation of the word problem into mathematical language. Each equation is called a **constraint** because it sets a limit or establishes a boundary to the problem.

The first step in the translation is to label the variables. The variables are determined from the phrase

"how many busses of each type"

Let the variables be x and y. Assign x to represent "the number of the smaller type of busses to be leased" and let y represent "the number of the larger type of busses to be leased." This assignment is arbitrary in that any letters can be used to represent the two variables. It is good form in solving word problems to make the assignment of the variable labels in a formal way as follows so you can refer back to your assignment as you solve the problem:

Let x = the number of the smaller type of busses to be leased
Let y = the number of the larger type of busses to be leased

To translate into a mathematical expression the phrase "a total of 5 busses are to be leased," first translate the phrase "total number of busses" which is the sum of the number of smaller busses, x, plus the number of larger busses, y. Thus the phrase, "total number of busses," is translated as

$$x + y$$

Since the total number of busses, $x + y$, is to be 5, we equate the two expressions to get the equation

$$x + y = 5$$

The units of each of the three terms, x, y and 5, are "number of busses," so the units check. This equation is the first constraint in the translation of the word problem. It is the constraint on the total number of busses leased.

Next we translate the constraint on the number of seats available by considering the phrase

"the smaller type of bus seats 40 people"

If one smaller type of bus is leased there are $40(1) = 40$ seats available; if two smaller type busses are leased there are $40(2) = 80$ seats available; and if three smaller type busses are leased there are $40(3) = 120$ seats available. In general, if x smaller type busses are leased there are $40(x)$, which we express as $40x$, total seats available in the smaller busses that are leased. In other words, since the smaller type of bus seats 40 people, $40x$ represents the total number of seats in all the smaller busses leased. To check the units think of the phrase "$40x$" as "40 times x" which is:

40	*TIMES*	x	*EQUALS*	$40x$
"number of seats in one smaller bus"		"number of smaller busses leased"		"total number of seats in all smaller busses"

Thus the product $40x$ represents the *total number of seats in all the smaller busses leased.*

Likewise the term "$60y$" represents the total number of seats in the larger busses leased. Check the units in the same way. Thus the product $60y$ is the *total number of seats in all the larger busses.*

The problem requires us to find an expression for the total number of seats in all the busses. If the smaller busses have $40x$ seats available and the larger busses have $60y$ seats available, then the total number of seats available is the sum of these two expressions,

$$40x + 60y$$

The problem requires this total to be 240 seats, so we need $40x + 60y$ to equal 240. This is the second equation or constraint in the problem:

$$40x + 60y = 240$$

The translation of the word problem is therefore the system of two equations in two variables:

$$\begin{aligned} x + y &= 5 \\ 40x + 60y &= 240 \end{aligned}$$

The solution of the problem is the solution of this system of equations. Multiplying the first equation by -40 yields

$$\begin{aligned} -40x - 40y &= -200 \\ 40x + 60y &= 240 \end{aligned}$$

Adding the equations yields

$$20y = 40$$

which simplifies to

$$y = 2$$

Substituting $y = 2$ into the first equation yields $x = 3$. Hence the solution is $x = 3$ and $y = 2$. In the next chapter we will graph the lines so we can express the solution as a point, (3,2). This means that the two constraints are satisfied when three smaller busses and two larger busses are leased.

The next three examples are similar to the types of word problems encountered in the following chapter. Each problem contains two variables and two constraints on the variables that can be translated into algebraic statements.

EXAMPLE 1

A window manufacturer produces windows in two styles, regular and deluxe. It costs \$80 to make each regular window and it costs \$100 to make each deluxe window. The manufacturer will spend \$10,000 to make windows. How many windows of each type should be produced each day if the total number of windows produced per day is 110.
(a) Translate the cost constraint into a linear equation.
(b) Translate the total number of windows constraint into a linear equation.
(c) Find the point of intersection of the lines in parts (a) and (b).

Solution. First label the variables.

Let x = the number of the regular windows
Let y = the number of the deluxe windows

(a) The algebraic statement $80x$ represents the cost to produce the regular windows and the algebraic statement $100y$ represents the cost to produce the deluxe windows. The total cost is \$10,000 so the cost constraint is

$$80x + 100y = 10{,}000$$

(b) The total number of windows produced is 110 which is the sum of the number of regular windows x and the number of deluxe windows y. Thus the total-number-of-windows constraint is

$$x + y = 110$$

(c) The two equations form a system of two equations in two variables.

$$\begin{aligned} 80x + 100y &= 10{,}000 \\ x + y &= 110 \end{aligned}$$

To eliminate x multiply the second equation by -80

$$\begin{aligned} 80x + 100y &= 10{,}000 \\ -80x - 80y &= -8{,}800 \end{aligned}$$

and add the equations to get

$$20y = 1{,}200$$

or

$$y = 60$$

Substitution of $y = 60$ into $x + y = 110$ yields

$$x = 50$$

Therefore the solution is (50,60).

EXAMPLE 2

A travel agency has sold 800 tickets for a Hawaiian cruise that includes air fare. There are two types of airplanes for the charter flights, type 1 can carry 50 passengers and type 2 can carry 100 passengers. How many airplanes of each type should be used if the total number of airplanes is to be 10?

(a) Translate the number of seats constraint into a linear equation.
(b) Translate the total number of airplanes constraint into a linear equation.
(c) Find the point of intersection of the lines in parts (a) and (b).

Solution. First label the variables.

Let x = the number of type 1 airplane
Let y = the number of type 2 airplane

(a) The expression $50x$ represents the number of seats available in type 1 airplanes and the expression $100y$ represents the number of seats available in type 2 airplanes. The total number of people requiring seats is 800 so the number of seats constraint is

$$50x + 100y = 800$$

(b) The total number of airplanes is to be 10 which is the sum of the number of type 1 airplanes x and the number of type 2 airplanes y. Thus the total number of airplanes constraint is

$$x + y = 10$$

(c) The two equations form a system of two equations in two variables.

$$\begin{aligned} 50x + 100y &= 800 \\ x + y &= 10 \end{aligned}$$

To eliminate x multiply the second equation by –80

$$\begin{aligned} 50x + 100y &= 800 \\ -50x - 50y &= -500 \end{aligned}$$

and add the equations to get

$$50y = 300$$

or

$$y = 6$$

Substitution of $y = 6$ into $x + y = 10$ yields

$$x = 4$$

Therefore the solution is (4, 6).

EXAMPLE 3

A farmer is going to plant 12 acres using two crops, peas and corn. The cost per acre per week for maintenance is \$6 for peas and \$3 for corn. The total cost for maintenance is to be \$66 per week. How many acres of each crop should be planted?

Solution. First label the variables.

Let x = the number of acres planted with peas
Let y = the number of acres planted with corn

The total number of acres planted is $x + y$ and this is to be 12, so the first constraint on the total number of acres to be planted is

$$x + y = 12$$

The statement $6x$ represents the cost of maintenance for peas and the statement $3y$ represents the cost of maintenance for corn. The total cost for maintenance is to be \$66 so the second constraint on cost of maintenance is

$$6x + 3y = 66$$

The word problem is translated into the following system of equations:

$$\begin{aligned} x + y &= 12 \\ 6x + 3y &= 66 \end{aligned}$$

To eliminate y multiply the first equation by –3 to get

$$\begin{aligned} -3x - 3y &= -36 \\ 6x + 3y &= 66 \end{aligned}$$

Adding the equations yields

$$3x = 30$$

which simplifies to

$$x = 10$$

Substituting $x = 10$ into the first equation yields $y = 2$. Hence the solution is (10,2).

Inequalities

In the next chapter constraints will be expressed as inequalities instead of equations. For example, if the farmer in Example 3 wanted to include the possibility of planting 12 or more acres for peas and corn, the problem could be expressed as follows.

> A farmer is going to plant at least 12 acres using two crops – peas and corn. The cost per acre per week for maintenance is \$6 for peas and \$3 for corn. The total cost for maintenance is not to exceed \$66 per week.

The phrase "at least 12 acres" is translated as

$$x + y \geq 12$$

Therefore it is possible that the solution will be to plant more than 12 acres, but the farmer will not plant fewer than 12 acres. In a similar way, the phrase "is not to exceed \$66" is translated as

$$6x + 3y \leq 66$$

Consequently, the farmer's solution of how many acres of each crop to plant might entail spending less than \$66 per week. However, this constraint means that the farmer will not spend more than \$66 per week.

When the word problem is expressed using inequalities another expression is needed for the problem to make sense. For example, the farmer might want to maximize the total profit of the two crops or minimize the total cost of leasing the land. This type of problem forms the basis of the word problems in the next chapter.

The next two examples are designed to illustrate how to translate phrases into inequalities.

EXAMPLE 4

A computer manufacturer produces two types of computers, desktop computers and laptop computers. It costs \$800 to make each desktop computer and it costs \$1000 to make each laptop computer. The manufacturer will spend at most \$10,000 per month to produce computers. The manufacturer will produce at least 150 computers per month.
(a) Translate the cost constraint into a linear inequality.
(b) Translate the total number of computers produced constraint into a linear inequality.

Solution. First label the variables.

Let x = the number of desktop computers
Let y = the number of laptop computers

(a) The statement $800x$ represents the cost to produce x desktop computers and the statement $1000y$ represents the cost to produce y laptop computers. The total cost is not to exceed \$10,000, meaning the cost could be less than \$10,000 or equal to \$10,000, but it cannot be larger. This is translated by "less than or equals 10,000." The cost constraint is

$$800x + 1000y \leq 10{,}000$$

(b) The phrase "the manufacturer will produce at least 150 computers per month" means that 150 or more computers will be produced. This means that the total number of computers produced, $x + y$, is greater than or equal to 150. Thus the total number of computers produced constraint is

$$x + y \geq 150$$

EXAMPLE 5

A travel agency has sold 600 tickets for a trip that includes air fare. There are two types of airplanes for the charter flights, type 1 can carry 50 passengers and type 2 can carry 100 passengers. The type 1 airplanes utilize a crew of 5 attendants and the type 2 airplanes utilize a crew of 6 attendants. There is a pool of 60 attendants available for the trip.
(a) Translate the number of seats constraint into a linear inequality.
(b) Translate the number of attendants available constraint into a linear inequality.

Solution. First label the variables.

Let x = the number of type 1 airplane
Let y = the number of type 2 airplane

(a) The statement $50x$ represents the number of seats available in type 1 airplanes and the statement $100y$ represents the number of seats available in type 2 airplanes. The total number of people requiring seats is 600, and so the number of seats must be at least equal to 600. The number of seats cannot be less than 600 because then some people would not have a seat. Thus the number of seats constraint is

$$50x + 100y \geq 600$$

(b) The number of attendants utilized in x type 1 airplanes is $5x$ and the number of attendants in y type 2 airplanes is $6y$. There are at most 60 attendants available so the number of attendants constraint is

$$5x + 6y \leq 60$$

EXERCISE SET 1-5

In problems 1 to 10 translate the given phrase into an equation where the variables are labeled as follows:

Let x = the number of small planes
Let y = the number of large planes

1. The number of planes to be leased is 10.
2. The number of planes to be leased is 15.
3. The cost to lease each small plane is \$6000 and the cost to lease each large plane is \$8000 and the total cost is to be \$70,000.
4. The cost to lease each small plane is \$5000 and the cost to lease each large plane is \$7500 and the total cost is to be \$60,000.
5. Each small plane has 50 seats and each large plane has 100 seats and a total of 750 seats are required.
6. Each small plane has 60 seats and each large plane has 150 seats and a total of 900 seats are required.
7. Each small plane requires a crew of 3 attendants and each large plane requires a crew of 5 attendants and there are 40 attendants available.
8. Each small plane requires a crew of 2 attendants and each large plane requires a crew of 6 attendants and there are 40 attendants available.
9. Each small plane has a cargo capacity of 2000 cubic feet and each large plane has a cargo capacity of 3000 cubic feet and 25,000 cubic feet of cargo space is to be used.
10. Each small plane has a cargo capacity of 2500 cubic feet and each large plane has a cargo capacity of 3500 cubic feet and 30,000 cubic feet of cargo space is to be used.

In problems 11 to 24 (a) translate the word problem into mathematical language by translating both constraints into linear equations and then (b) find the point of intersection of the lines in part (a).

11. A clothing manufacturer makes two types of suits, wool and a blend. The cost for manufacturing each wool suit is \$150 and the cost of each blend suit is \$100. If the manufacturer makes 100 suits per week and budgets \$12,000 per week, how many of each type should be made?

12. A clothing manufacturer makes two types of suits, wool and a blend. The cost for manufacturing each wool suit is \$120 and the cost of each blend suit is \$80. If the manufacturer makes 60 suits per week and budgets \$5200 per week, how many of each type should be made?

13. A clothing manufacturer makes two types of gloves, leather and cotton. The cost for manufacturing each pair of leather gloves is \$10 and the cost of each pair of cotton gloves is \$6. If the manufacturer makes 500 pairs of gloves per week and budgets \$4000 per week, how many of each type should be made?

14. A clothing manufacturer makes two types of gloves, leather and cotton. The cost for manufacturing each pair of leather gloves is \$8 and the cost of each pair of cotton gloves is \$4. If the manufacturer makes 1000 pairs of gloves per week and budgets \$7000 per week, how many of each type should be made?

15. A shirt manufacturer cuts each shirt on machine I and then sews the shirt on machine II. On machine I each long sleeve shirt requires 2 minutes of time and each short sleeve shirt requires 1 minute of time. On machine II each long sleeve shirt requires 3 minutes of time and each short sleeve shirt requires 2 minutes of time. On a given shift machine I operates for 80 minutes and machine II operates 140 minutes. How many of each type of shirt should be made?

16. A shirt manufacturer cuts each shirt on machine I and then sews the shirt on machine II. On machine I each long sleeve shirt requires 3 minutes of time and each short sleeve shirt requires 2 minutes of time. On machine II each long sleeve shirt requires 2 minutes of time and each short sleeve shirt requires 1 minute of time. On a given shift machine I operates for 160 minutes and machine II operates 60 minutes. How many of each type of shirt should be made?

17. A furniture manufacturer has a retail store and also sells furniture under a different brand name to a large chain store. A delivery truck takes 2 hours to make a round trip to the retail store and 3 hours to make a round trip to the chain store. The cost of a round trip to the retail store is \$100 and the cost of a round trip to the chain store is \$200. How many shipments per week should be scheduled to each store if the manufacturer uses 48 hours per week delivery time and a total delivery cost of \$2500?

18. A furniture manufacturer has a retail store and also sells furniture under a different brand name to a large chain store. A delivery truck takes 3 hours to make a round trip to the retail store and 5 hours to make a round trip to the chain store. The cost of a round trip to the retail store is \$200 and the cost of a round trip to the chain store is \$400. How many shipments per week should be scheduled to each store if the manufacturer uses 40 hours per week delivery time and a total delivery cost of \$3600?

19. A candy manufacturer mixes pretzels and nuts in a candy mix. Two types of mix are made, regular and deluxe. Each pound of regular mix contains 0.5 pound of pretzels

and 0.3 pound of nuts, while each pound of deluxe mix contains 0.4 pound of pretzels and 0.6 pound of nuts. There are 96 pounds of pretzels and 120 pounds of nuts in stock. How many pounds of each type of mix should be made?

20. A candy manufacturer mixes pretzels and nuts in a candy mix. Two types of mix are made, regular and deluxe. Each pound of regular mix contains 0.4 pound of pretzels and 0.4 pound of nuts, while each pound of deluxe mix contains 0.3 pound of pretzels and 0.6 pound of nuts. There is 70 pounds of pretzels and 100 pounds of nuts in stock. How many pounds of each type of mix should be made?

21. A company is going to lease two types of cars, compacts and mid-sized for its sales staff. The company will lease 100 cars. The cost of each compact is $200 per month and the cost of each mid-sized car is $350 per month. The company has decided to allocate $26,000 per month for the lease of the cars. How many of each type should be leased?

22. A company is going to lease two types of cars, compacts and mid-sized for its sales staff. The company will lease 50 cars. The cost of each compact is $250 per month and the cost of each mid-sized car is $400 per month. The company has decided to allocate $17,000 per month for the lease of the cars. How many of each type should be leased?

23. Four hundred delegates to a conference will be housed in two types of rooms, singles and doubles. The maintenance cost of each single room is $25 per day and the maintenance cost of each double room is $30 per day. How many of each type of room is needed if the total maintenance cost is to be $8000 per day?

24. Six hundred delegates to a conference will be housed in two types of rooms, singles and doubles. The maintenance cost of each single room is $30 per day and the maintenance cost of each double room is $50 per day. How many of each type of room is needed if the total maintenance cost is to be $17,000 per day?

In problems 25 to 34 translate the given phrase into an inequality where the variables are labeled as follows:

Let x = the number of color printers
Let y = the number of laser printers

25. The number of printers to be leased is at least 100.

26. The number of printers to be leased is at most 125.

27. The number of printers to be leased is to be no more than 200.

28. 50 or more printers will be leased.

29. The cost to lease each color printer is $300, the cost to lease each laser printer is $400, and the total cost is not to exceed $30,000.

30. The cost to lease each color printer is $250, the cost to lease each laser printer is $350, and the total cost will be at least $10,000.

31. Each color printer has a maintenance cost of $5 per month, each laser printer has a maintenance cost of $10, and the total maintenance cost is to be no more than $1000 per month.

32. Each color printer has a maintenance cost of $7 per month, each laser printer has a maintenance cost of $8, and the total maintenance cost is to be no more than $1500 per month.

33. Each color printer can print 2 pages per minute and each laser printer can print 5 pages per minute. The printers must be able to print at least 600 pages per minute.

34. Each color printer can print 3 pages per minute and each laser printer can print 6 pages per minute. The printers must be able to print at least 1000 pages per minute.

In problems 35 to 40 translate the word problem into mathematical language by translating both constraints into linear inequalities.

35. A clothing manufacturer makes two types of suits, wool and a blend. The cost for manufacturing each wool suit is $150 and the cost of each blend suit is $100. The manufacturer makes at least 100 suits per week and budgets no more than $12,000 per week.

36. A clothing manufacturer makes two types of gloves, leather and cotton. The cost for manufacturing each pair of leather gloves is $10 and the cost of each pair of cotton gloves is $6. The manufacturer makes at least 500 pairs of gloves per week and budgets no more than $4000 per week.

37. A shirt manufacturer cuts each shirt on machine I and then sews the shirt on machine II. On machine I each long sleeve shirt requires 2 minutes of time and each short sleeve shirt requires 1 minute of time. On machine II each long sleeve shirt requires 3 minutes of time and each short sleeve shirt requires 2 minutes of time. On a given shift machine I operates for at least 80 minutes and machine II operates for no more than 140 minutes.

38. A furniture manufacturer has a retail store and also sells furniture under a different brand name to a large chain store. A delivery truck takes 2 hours to make a round trip to the retail store and 3 hours to make a round trip to the chain store. The cost of a round trip to the retail store is $100 and the cost of a round trip to the chain store is $200. The manufacturer is limited to no more than 48 hours per week delivery time and a total delivery cost of no more than $2500?

39. A travel agent has sold 400 tickets to the Super Bowl. The weekend package includes air fare, and they have a choice of two types of airplanes for the charter flights. Type 1 can carry 50 passengers and type 2 can carry 100 passengers. The consortium is allowed to lease no more than 6 planes.

40. A farmer plants two crops, wheat and corn. The cost per acre for fertilizer is $4 for wheat and $6 for corn. No more than 12 acres will be planted with wheat and corn. No more than $160 is to be spent for fertilizer.

Cumulative Exercise Set 1-5

1. Let $4x - 3y = 24$. Find the intercepts of the linear equation.

2. Find the equation of the line that passes through the point (1,2) and is parallel to the line $2x - y = 10$.

3. Find the equation of the line that passes through the points (1,2) and (2,5).

4. Find the point of intersection of the system of equations $30x + 40y = 1200$ and $x + y = 70$.

5. Find the point of intersection of the system of equations $50x + 40y = 3000$ and $x + y = 70$.

6. Determine the line of best fit for the points
   (1, 1.5), (2,1.5), (3,4), (4,4.5)

Problems 7 to 10 refer to the following word problem.

A farmer plants two crops, wheat and corn. The cost per acre for fertilizer is \$2 for wheat and \$3 for corn. No more than 15 acres will be planted with wheat and corn. No more than \$35 is to be spent for fertilizer. The profit per acre for wheat is predicted to be \$40 and the profit per acre for corn is predicted to be \$35. Total profit must be at least \$575.

7. Find the cost constraint.

8. Find the number of acres constraint.

9. Find the profit constraint.

10. Find the point of intersection of the cost constraint and the number of acres constraint.

Multiple-Choice Problem Set 1-5

Problems 1 to 4 refer to the following word problem.

A company wants to purchase 25 cars. The cost for each large car is \$20,000, the cost of each compact car is \$12,000, and \$420,000 will be used for the purchase. How many of each type of car should be purchased?

1. One constraint is
   (a) $32{,}000 = 20{,}000x + 12{,}000y$
   (b) $x + y = 420{,}000$
   (c) $x + y = 20{,}000$
   (d) $420{,}000 = 20{,}000x + 12{,}000y$
   (e) $12{,}000x + 20{,}000y = 25$

2. One constraint is
   (a) $x + y = 25$
   (b) $20{,}000 = 12{,}000x + 20{,}000y$
   (c) $x + y = 420{,}000$
   (d) $x + y = 20{,}000$
   (e) $420{,}000 = 1200x + 2000y$

3. The system of two constraints is
   (a) $420{,}000 = 20{,}000x + 12{,}000y$ and $x + y = 420{,}000$,
   (b) $x + y = 420{,}000$ and $25 = 20{,}000x + 12{,}000y$
   (c) $x + y = 20{,}000$ and $420{,}000 = 20{,}000x + 12{,}000y$
   (d) $x + y = 25$ and $420{,}000 = 20{,}000x + 12{,}000y$
   (e) $12{,}000x + 20{,}000y = 25$ and $x + y = 20{,}000$

4. The point of intersection of the constraints is
   (a) (20, 5) (b) (15, 10) (c) (25, 0) (d) (5, 20) (e) (10, 15)

Problems 5 to 8 refer to the following word problem.

A shirt manufacturer cuts each shirt on machine I and then sews the shirt on machine II. On machine I each long sleeve shirt requires 4 minutes of time and each short sleeve shirt requires 2 minutes of time. On machine II each long sleeve shirt requires 3 minutes of time and each short sleeve shirt requires 3 minutes of time. On a given shift machine I operates for 50 minutes and machine II operates 45 minutes. How many of each type of shirt should be made?

5. One constraint is
(a) $4x + 3y = 50$
(b) $4x + 2y = 45$
(c) $4x + 2y = 50$
(d) $3x + 2y = 45$
(e) $2x + 3y = 45$

6. One constraint is
(a) $3x + 3y = 50$
(b) $4x + 2y = 45$
(c) $4x + 2y = 8$
(d) $4x + 3y = 50$
(e) $3x + 3y = 45$

7. The system of two constraints is
(a) $3x + 3y = 50$ and $4x + 2y = 45$
(b) $4x + 2y = 45$ and $3x + 3y = 50$
(c) $4x + 3y = 50$ and $3x + 3y = 45$
(d) $4x + 2y = 50$ and $3x + 3y = 45$
(e) $4x + 2y = 45$ and $3x + 3y = 45$

8. The point of intersection of the constraints is
(a) (4, 2) (b) (3, 3) (c) (45, 50) (d) (10, 10) (e) (10, 5)

Problems 9 to 12 refer to the following word problem.

A consortium of travel agents has sold 1000 tickets to the Super Bowl. The weekend package includes air fare, and they have a choice of two types of planes for the flight. Type 1 planes can carry 100 passengers while type 2 planes can carry 200 passengers. Each type 1 flight costs $6000 and each type 2 flight costs $8000. Each type 1 plane requires 4 flight attendants and each type 2 plane requires 8 flight attendants. There are at most 40 flight attendants available. They can lease no more than 8 planes.

9. One constraint is
(a) $100x + 200y \geq 10$
(b) $x + y \leq 8$
(c) $6000x + 8000y \leq 14{,}000$
(d) $C = 6000x + 8000y$
(e) $100x + 200y = 1000$

10. One constraint is
(a) $100x + 200y \leq 1000$
(b) $x + y = 8$
(c) $6000x + 8000y \leq 14{,}000$
(d) $14{,}000 = 6000x + 8000y$
(e) $100x + 200y \geq 1000$

11. One constraint is
(a) $4x + 8y \leq 1000$
(b) $4x + 8y = 12$
(c) $4x + 8y \leq 40$
(d) $40 = 4x + 8y$
(e) $100x + 200y \geq 40$

12. The point of intersection of the cost constraint and the number of planes constraint is
(a) (8, 0) (b) (0, 8) (c) (4, 4) (d) (6, 2) (e) (2, 6)

CHAPTER REVIEW

Key Terms

1-1 The Cartesian Coordinate System

Real Line
Cartesian Coordinate System
Unit Distance
x-axis; y-axis
x-coordinate; y-coordinate
Coordinates
Origin
Quadrants
Graph
The Plotting Method
Linear Equation

1-2 Lines

General Form
x-intercept; y-intercept
Slope
Slope-intercept Form
Parallel Lines
Horizontal & Vertical Lines
Point-slope Form
Linear Depreciation

1-3 Intersection of Lines

System of Equations
Solution of a System
Dependent & Inconsistent Systems
Breakeven Point

1-4 Least Squares

Line of Best Fit
Least Squares Equations

1-5 Word Problems

Constraints
Inequalities

Summary of Important Concepts

- The slope of a line is $m = \dfrac{y_2 - y_1}{x_2 - x_1}$, unless $x_2 = x_1$
- Horizontal lines have form $y = c$; their slope is 0.
- Vertical lines have form $x = k$; their slope is undefined.
- Three principal forms for the equation of a line are:

Name of Form	Equation
general	$Ax + By = C$
slope-intercept	$y = mx + b$
point-slope	$y - y_1 = m(x - x_1)$

- The Method of Least Squares

  For a given set of data points (x, y) the line of best fit is $y = mx + b$ where m and b are obtained by solving the system of Least Squares Equations

$$n \cdot b + (\Sigma x)m = \Sigma y$$
$$(\Sigma x)b + (\Sigma x^2)m = \Sigma xy$$

REVIEW EXERCISE SET

1. Find the values of y when $x = -2, -1, 0, 1$, and 2 for the equation $y = 3 - 2x$. Plot the five points and sketch the graph of the equation.
2. Find the slope of the line passing through the points $(-2, 3)$ and $(1, -3)$.

In problems 3 to 5 sketch the graph of the stated line.

3. $y = -x - 5$
4. $y - 2 = 3(x + 1)$
5. $2x - 4y = 8$

6. Find the slope-intercept form of the line which is parallel to the line $x + y = 0$ and whose x-intercept is equal to -2.
7. Write an equation of the line whose y-intercept is -3 and slope is 3.
8. Find the equation of the horizontal line passing through the point $(-2, -5)$.
9. Write the equation for one day's car rental cost in terms of miles driven if the charges are \$35 per day and 32 cents for each mile the car is driven. Let C denote the rental cost and x the miles driven.

In problems 10 and 11 determine whether the system has a solution. If it does, write the solution. If it does not, tell whether it is inconsistent or dependent.

10. $\begin{aligned} 15x - 5y &= 20 \\ -12x + 4y &= -16 \end{aligned}$

11. $\begin{aligned} x + 2y &= 1 \\ x - y &= -3 \end{aligned}$

12. Determine the point of intersection of the lines $y = 2x - 1$ and $y = -x + 5$.

In problems 13 and 14 determine how many solutions the system has. If it has one, state it; if there are more than one, state two solutions.

13. $\begin{aligned} x - y &= 2 \\ 2x - 2y &= 4 \end{aligned}$

14. $\begin{aligned} 2x + 4y &= -6 \\ -3x - 5y &= 12 \end{aligned}$

15. Find the equation of the line of best fit for the points $(-1, 5)$, $(0, 3)$, $(1, 6)$, and $(2, 5)$.

2 LINEAR PROGRAMMING

Chapter Overview

Linear programming is a method of solving problems in which a particular quantity is to be minimized or maximized. The problem could be to maximize profit, minimize cost, minimize caloric intake or a similar optimization problem. The quantity to be optimized is limited in some ways. Profit is dependent on production capacity while cost depends upon personnel. The solution of these problems involves linear inequalities and linear equations. In Section 1 linear programming problems are translated into mathematics. Then in Section 2 we use the skills developed in Chapter 1 to solve the linear programming problems using a geometric approach.

Case Study Preview

In 1947 George Dantzig was a young mathematician looking for a method of solving very complicated optimization problems. The problems he considered were real-world problems originating in the military, business and economics. He was employed by the military but it was his friends who were economists and mathematicians, not to mention the engineers who were working on building a new type of machine called a "computer", who led him on the path to success.

Many of the problems facing Dantzig involved systems of equations with 50 or even 100 equations in as many unknowns. The method he created, called the simplex method became an instant hit. One of the first problems he applied the method to was a transportation problem involving the large steel industry. These companies had built a virtual monopoly by claiming that the transportation cost of steel was too complicated to compute, so all steel companies, large and small, had to use the same simplified formula which favored the large companies. The simplex method showed them how to compute the cost, thereby allowing smaller companies to use their proximity to the user to cut cost and remain competitive.

After Section 1 we present a case study based on the Berlin airlift, one of the first and most significant applications of linear programming. The problem was to supply Berlin and break the Russian blockade after World War II.

The second case study after Section 2 shows how Federal Express Corporation used linear programming to answer a major question: whether to shift their operations from small air craft to larger capacity ones.

2-1 Linear Programming Problems

A linear programming problem involves finding a largest or smallest value of a variable given that certain restrictions are met. For instance, a firm wants to maximize profit subject to limitations on production capacity and cost.

In this section we present three examples of typical linear programming problems. They are analogous to the ones which originated George Dantzig's research in this area. The aim is to develop a method for translating a given problem into mathematical terms. In the next section we will show how to solve the problems once the translation has been carried out.

We first state a typical problem and then use it to illustrate the first few steps in expressing linear programming problems in mathematical terms. It is a type of problem that is frequently found in the business world.

An Example: The Manufacturing Problem

Manufacturing Problem. A clothing manufacturer makes two types of suits - wool and a blend of wool and polyester. The wool suit sells to the retail stores for \$160 each and the blend for \$90 each. The cost for manufacturing each wool suit is \$100 and the cost of each blend suit is \$40. If the manufacturer makes no more than 120 suits a week and budgets no more than \$6000 per week, how many of each type should be made to maximize profit?

To express a linear programming problem in mathematical language, there are certain parts of the problem that must be identified.

1. The variables must be defined. Assign a letter to each quantity and be careful about the units that are used. The question that is asked will quite often dictate how the variables will be defined.
2. Determine the equation that governs the quantity to be maximized or minimized. This equation is called the **objective function**.
3. Each problem contains restrictions that are called **constraints**. They are translated into linear inequalities. For each constraint the terms must have the same units.

Begin by rereading the problem once or twice. Often it is convenient to organize the data in a table. Let us demonstrate the technique by translating the manufacturing problem.

EXAMPLE 1

Translate the manufacturing problem into mathematical language.

Solution. First the variables must be defined. There are two types of suits – wool and blend – and the problem entails finding the number of each type of suit that maximizes profit. Therefore the two variables must be:

x = the number of wool suits manufactured per week
y = the number of blend suits manufactured per week

One constraint is that the total number of suits manufactured cannot exceed 120. This fact translates into the inequality

$$x + y < 120$$

Generally the coefficients of the variables are entered into the table, but for this constraint it seems to be clearer if the variables themselves, x and y, are put into the table. Hence x and y are entered in the first row together with the limiting value, 120. The next constraint is cost so we enter the cost of each type of suit, \$100 and \$40 respectively, together with the limiting value, \$6000. From the last sentence in the problem, the objective function entails maximizing profit, so one of the rows in the table consists of profit. The cost of each wool suit is \$100 and the revenue is \$160 so the profit of each wool suit is \$60. The profit on each blend suit is \$90 – \$40 = \$50. These values are entered in the last row of the table. The problem can be summarized Table 2–1.

	Wool suits	Blend suits	Limiting value
Number manufactured	x	y	120
Cost (\$)	100	40	6000
Profit (\$)	60	50	

Table 2–1

The second row of Table 2–1 represents cost, and since no more than \$6000 is to be spent per week, it is interpreted as

$$100x + 40y \le 6000$$

Notice that each term represents cost in dollars. 6000 is the budgeted amount that can be spent and is given in dollars. Since 100 represents cost per wool suit in dollars and x is number of wool suits, $100x$ is

(cost in dollars/wool suit)(number of wool suits) = cost in dollars

Similarly, $40y$ is

(cost in dollars/blend suit)(number of blend suits) = cost in dollars

Two additional constraints are implicitly included in the problem. It makes no sense to allow x or y to take on negative values, and thus we must include the two inequalities $x \ge 0$ and $y \ge 0$ in the list of constraints. These constraints are easy to overlook because the problem does not explicitly mention them, but they are a part of almost every linear programming problem.

The objective function is obtained from the third row of the table. In this row there is no limiting value because we are to find the largest value that profit can be under the given constraints. Letting P represent profit, the third row is interpreted as the equation

$$P = 60x + 50y$$

The problem is stated in mathematical language by the following:

Maximize	$P = 60x + 50y$	*Objective function*
Subject to	$100x + 40y \le 6000$	
	$x + y \le 120$	*Constraints*
	$x \ge 0, \quad y \ge 0$	

The objective function is separated from the constraints because they play different roles in the solution to the linear programming problem.

Definition of a Linear Programming Problem

A linear programming problem originates as a word problem. When it is translated into mathematics it is expressed in a similar way to the problem in Example 1. While each problem has specific characteristics, it is possible to describe in general the formulation of linear programming problems in two variables. In the next chapter we will study linear programming problems in more than two variables.

DEFINITION

A **linear programming problem in two variables** is a problem that entails maximizing or minimizing a linear expression $z = ax + by$, called the **objective function**, subject to several **constraints**. The constraints are expressed as linear inequalities such as $cx + dy \leq k$ (or $cx + dy \geq k$). The object is to find values of x and y that satisfy the constraints and yield the largest (or smallest) value of z.

Key Steps in Translating the Problem

Example 1 illustrates the key steps in translating a linear programming problem into mathematical language. We record them for easy reference.

Key Steps for Translating a Linear Programming Problem into Mathematical Language

1. Read and reread the problem to assess what is known and what is to be determined.
2. Organize the data, preferably into a table.
3. Identify unknown quantities and assign variables to them.
4. Determine the objective function.
5. Translate the constraints into linear inequalities.

Steps 2, 3 and 4 were described earlier in more detail. Notice that their order was different. This is to emphasize that the objective function can be determined before or after the constraints, just be sure to keep it separate and express it as an equation rather than an inequality like the constraints.

Remember that most linear programming problems are complicated and require a second or third reading to get started. Also, when organizing the data in a table, check to insure that the numbers in each row and column agree with its description.

In the remainder of this section we present two more examples of linear programming problems. One is from the health field and the other is a transportation problem.

EXAMPLE 2

A hospital nutritionist prepares a menu for patients on low-fat diets. The primary ingredients are chicken and corn. The meal must contain enough iron and protein to meet at least the U.S. recommended daily allowance (RDA) for each nutrient. The RDA for protein is 45 grams and for iron it is 20 milligrams. Each 3-once serving of chicken contains 45 grams of protein, 10 milligrams of iron and 4 grams of fat. Each 1-cup serving of corn contains 9 grams of protein, 6 milligrams of iron and 2 grams of fat. How much chicken and corn should be included in the meal in order to minimize the amount of fat? Translate the problem into mathematical language.

Solution. There are two types of food, chicken and corn. The limiting values are at least 45 grams of protein and at least 20 milligrams of iron. Hence the units of the constraints are protein and iron, respectively. They become the rows of the table while chicken and corn are the columns. The objective function is to minimize the amount of fat. It becomes the third row.

	Chicken (3-oz serving)	Corn (1-cup serving)	Limiting value
Protein (g)	45	9	45
Iron (mg)	10	6	20
Fat (g)	4	2	

Table 2–2

The variables are

x = the number of servings of chicken
y = the number of servings of corn

To determine the objective function let F represent the amount of fat in the meal (in grams). From the third row in Table 2–2 it becomes

$$F = 4x + 2y$$

The first two rows are the constraints. In each case the limiting value is prefaced by the term "at least" which translates into "≥." Thus the two constraints are $45x + 9y \geq 45$ and $10x + 6y \geq 20$. As before, it makes no sense to let x and y take on negative values, so the two additional constraints, $x \geq 0$ and $y \geq 0$ are included. The problem can be stated as follows:

Maximize	$F = 4x + 2y$	*Objective function*
Subject to	$45x + 9y \geq 45$	
	$10x + 6y \geq 20$	*Constraints*
	$x \geq 0, \quad y \geq 0$	

The next example is a common type of problem facing many businesses. It is called a transportation problem. We have chosen to illustrate this type of problem by maximizing profit subject to time and cost. In actual practice sometimes a firm needs to minimize time or cost subject to other limitations.

EXAMPLE 3

A bicycle manufacturer owns a retail outlet and, in addition, sells some bikes to a large chain store. One truck makes all deliveries from the manufacturing warehouse to the retail outlet and to the chain store distribution center. It takes 2 hours to make a round trip to the retail outlet and 4 hours to make a round trip to the distribution center. The cost of a round trip to the retail outlet is \$50 and the cost of a round trip to the distribution center is \$200. The profit of each truckload sold through the retail outlet is \$210, and the profit is \$400 for each truckload sold to the chain store. How many shipments per week should be scheduled to each store in order to maximize profit if the manufacturer is limited to no more than 40 hours of travel time per week and a cost of \$1600? Translate the problem into mathematical language.

Solution. From the question stated in the problem, the variables can be determined. They are given by

x = the number of trips per week to the retail outlet
y = the number of trips per week to the chain store

In addition, the question states that we are maximizing profit, so the objective function deals with profit, given by the variable P, and is given by the formula

$$P = 210x + 400y$$

Since the constraints involve time and cost they are the labels of the rows. The data are then put into Table 2–3.

	Retail Outlet	Chain Store	Limiting Value
Round trip time (hrs)	2	4	40
Cost of a round trip (\$)	50	200	1600
Profit (\$)	210	400	

Table 2–3

The first two rows represent the constraints on time and cost. Since each uses the phrase "no more than," the inequalities use ≤. They are $2x + 4y \le 40$ and $50x + 200y \le 1600$. The problem can be stated as follows:

Maximize	$P = 210x + 400y$	*Objective function*
Subject to	$2x + 4y \le 40$	
	$50x + 200y \le 1600$	*Constraints*
	$x \ge 0,\ \ y \ge 0$	

The method of solving linear programming problems involves two distinct operations. First, the word problem must be expressed in mathematics and then the mathematical formulation must be solved. This section concentrates on the first skill, translating the word problem into mathematics. In the next section the method of solving the mathematical formulation of the linear programming problem will be studied.

When doing the exercises it is important to distinguish the objective function from the constraints. This is because the solution in the next section treats them separately. The first step in the solution is to graph the system of linear inequalities formed by the constraints. This graph produces a region in the plane. It is called the region of feasible solutions because the points in the region are the only ones that are allowed to be substituted into the objective function. Any point not in this region is not a possible solution. Then the points of intersection of the boundaries of the region are located. These points are called corner points. The corner points are then substituted into the objective function to find the solution. In the next section we will see why this method produces the solution.

EXERCISE SET 2-1

In problems 1 to 36, translate the linear programming problem into mathematical language.

1. A clothing manufacturer makes two types of outer-wear, a wool coat and a polyester all-weather coat. The wool coat sells to the retail stores for \$190 each and the polyester coat for \$130 each. The cost for manufacturing each wool coat is \$150 and the cost of each polyester coat is \$100. If the manufacturer makes no more than 100 coats a week and budgets no more than \$12,000 per week, how many of each type should be made to maximize profit?

2. A clothing manufacturer makes two types of jogging attire - regular and designer. The attire sells to the retail stores for \$10 each for the regular suits and \$40 each for the designer suits. The cost for manufacturing each regular suit is \$5 and it is \$10 for a designer suit. If the manufacturer makes no more than 200 suits a week and budgets no more than \$1500 per week, how many of each type should be made to maximize profit?

3. A hospital nutritionist prepares a menu for patients on low-fat diets. The primary ingredients are chicken and corn. The meal must contain at least 45 grams of protein and at least 15 milligrams of iron. Each 3-once serving of chicken contains 45 grams of protein, 9 milligrams of iron and 4 grams of fat. Each 1-cup serving of corn contains 9 grams of protein, 6 milligrams of iron and 2 grams of fat. How much chicken and corn should be included in the meal in order to minimize the amount of fat?

4. Work Problem 3 with the constraint that the nutritionist wants the meal to contain at least the full U.S. RDA for protein but requires only that at least 1/2 the U.S. RDA for iron be met.

5. The computer center director at a large university wants to staff consulting stations with two types of shifts: type 1 will contain 2 senior programmers and 1 student worker and type 2 will consist of 1 senior programmer and 5 student workers. The director wants to use no more than 36 individuals. There will be at least 24 hours to be filled during the week, with a type 1 shift serving for 3 hours and a type 2 shift serving for 2 hours. The cost of a type 1 shift is \$44 per hour, and The cost of a type 2 shift is \$40 per hour. Determine the number of shifts of each type in order to minimize cost.

6. Work Problem 5 if a type 2 shift is to consist of 1 senior programmer and 7 student workers, so that the cost of a type 2 shift is \$48 per hour, and all other figures remain the same.

7. A bicycle manufacturer has a retail outlet and, in addition, sells some bikes to a large chain store. One truck makes all deliveries, and it takes 3 hours to make a round trip to the retail store and 4 hours to make a round trip to the chain store. The cost of a round trip to the retail store is \$100 and the cost of a round trip to the chain store is \$200. The profit of each truckload sold through the retail store is \$500; it is \$800 for each truckload sold to the chain store. How many shipments per week should be scheduled to each store in order to maximize profit if the manufacturer is limited to no more than 48 hours per week and a cost of \$2000?

8. Work Problem 7 if the profit on each truckload sold through the retail store is \$600 and the profit is \$800 for a truckload sold to the chain store.

9. A meat manufacturer mixes beef and pork in sausage links. Two types of links are made, regular and "all-beef". Each pound of regular sausage meat contains 0.3 pound of beef and 0.2 pound of pork, while each pound of all-beef sausage meat contains 0.4 pound of beef and 0.1 pound of pork. There are 120 pounds of beef and 60 pounds of pork in stock. If the profit for regular links is 50 cents per pound and the profit for all-beef links is 70 cents per pound, how many pounds of each type link should be made to maximize profit?

10. A consortium of travel agents has sold 1200 tickets to the Super Bowl. The weekend package includes air fare, and they have a choice of two types of airplanes for the charter flights. Type 1 can carry 100 passengers and type 2 can carry 150 passengers. Each flight of a type 1 aircraft will cost $9000 and each type 2 flight will cost $15,000. The consortium is allowed to lease no more than 10 planes. How many airplanes of each type should be leased in order to minimize cost?

11. A furniture manufacturer has a retail store and also sells furniture under a different brand name to a large chain store. A delivery truck takes 4 hours to make a round trip to the retail store and 5 hours to make a round trip to the chain store. The cost of a round trip to the retail store is $200 and the cost of a round trip to the chain store is $500. The manufacturer will use the delivery truck at least 40 hours per week in delivery time and will hold the delivery cost to no more than $3000 per week? How many shipments per week should be scheduled to each store to maximize profit it the profit on each shipment to the retail store is $2000 and the profit on each shipment to the chain store is $6000?

12. Work Problem 11 if the profit on each shipment to the retail store is $4000 and the profit on each shipment to the chain store is $6000.

13. A club has sold 480 tickets for a bus tour of Washington, DC. Two types of busses are to be leased, the smaller type of bus seats 40 people and the larger type seats 60 people. No more than 10 busses can be leased. How many busses of each type should be leased to minimize cost if the cost of the trip if the cost of a smaller type of bus is $500 and the cost of a larger type of bus is $600?

14. A club has sold 480 tickets for a bus tour of Washington, DC. Two types of busses are to be leased, the smaller type of bus seats 40 people and the larger type seats 60 people. No more than 10 busses can be leased. How many busses of each type should be leased to maximize profit of the trip if the profit for each smaller type of bus is $500 and the profit of a larger type of bus is $600?

15. Officials of the state high school basketball tournament must plan accommodations for at least 600 players. They have two types of rooms available, dormitory rooms that sleep three people and motel rooms that sleep two people. For meals, they have budgeted $10 daily per person for those in dorms and $20 daily per person for those in motels, and they must not spend more than $4000 in meal money per day. If the daily cost of a room is $20 for a dormitory room and $30 for a motel room, how many rooms of each kind must they schedule in order to minimize cost?

16. Work Problem 15 with the meal money constraint changed to a limiting value of $2400.

17. Four hundred delegates to a conference will be housed in two types of rooms, singles and doubles. The maintenance cost of each single room is $20 per day and the maintenance cost of each double room is $50 per day. The total maintenance cost cannot exceed $9000 per day. How many of each type of room should be

booked to minimize total cost if the total cost for each delegate in a single room is \$100 and the total cost of each delegate in a double room is \$120?

18. Work Problem 17 if the total cost for each delegate in a single room is \$100 and the total cost of each delegate in a double room is \$220.

19. Work Problem 17 with the number of delegates changed to 360.

20. Work Problem 17 with the number of delegates changed to 460.

21. A candy manufacturer mixes pretzels and nuts in a candy mix. Two types of mix are made, regular and deluxe. Each pound of regular mix contains 0.5 pound of pretzels and 0.4 pound of nuts, while each pound of deluxe mix contains 0.3 pound of pretzels and 0.6 pound of nuts. There are 960 pounds of pretzels and 1200 pounds of nuts in stock. How many pounds of each type of mix should be made to maximize profit if the profit on each pound of regular mix is \$4 and the profit on each pound of deluxe mix is \$5?

22. A candy manufacturer makes two types of chocolate bars, one with nuts and one with raisins. The cost to manufacture each candy bar with nuts is \$1 and the cost to manufacture each candy bar with raisins is \$1.2. The cost of packaging each candy bar with nuts is \$0.10 and the cost of packaging each candy bar with raisins is \$0.15. The manufacturer allocates no mote than \$300 per day to manufacture the candy bars and spends no more than \$36 per day to package the candy bars. How many of each type of candy bar should be made to maximize profit if the profit on each candy bar with nuts is \$3 and the profit on each candy bar with raisins is \$4?

23. A computer manufacturer produces two types of computers, desktop computers and laptop computers. It costs \$800 to make each desktop computer and it costs \$1000 to make each laptop computer. The manufacturer will spend at most \$16,000 per month to produce computers. The manufacturing capacity of the company is 180 computers produced per month. How many computers of each type should be produced to maximize profit if the price of each desktop computer is \$1800 and the profit for each laptop computer is \$2100?

24. A manufacturer of class rings makes two types of rings, silver and gold. The manufacturer must make at least 3000 rings. The cost of each silver ring is \$10 and the cost of each gold ring is \$15. How many of each type of ring should the manufacturer make to maximize profit if the profit on each silver ring is \$40 and the profit on each gold ring is \$70 and the manufacturer can allocate no more than \$36,000 in manufacturing cost?

25. An automobile leasing company buys two types of cars, compacts and mid-sized. The company wants to purchase at most 1000 cars. The cost of each compact is \$8000 and of each mid-sized car is \$10,000. The company has decided to allocate no more than \$9 million for the purchase of the cars. If the anticipated return from the sale of the cars at the end of the company's 3-year leasing period is \$1000 for a compact and \$1200 for a mid-sized car, how many of each type should be purchased in order to maximize return?

26. Work Problem 25 if the company increases the amount of money allocated for the purchase of the cars to \$12 million dollars.

27. A window manufacturer produces windows in two styles, regular and thermopane. It costs \$100 to make each regular window, which the manufacturer sells for \$150. It costs \$120 to make each thermopane window and each sells for \$175. The daily

production capacity is 110 windows and the daily cost cannot exceed $12,000. How many windows of each type should be made per day to maximize profit?

28. Work Problem 27 if the company replaces the daily production capacity constraint by assuming that the company will produce at least 90 windows per day.

29. A heating oil delivery firm has two processing plants, plant A and plant B. Plant A processes daily 150 barrels of high-grade oil for commercial use and 50 barrels of low-grade for residential use. Plant B processes daily 100 barrels of high-grade oil and 50 barrels of low-grade oil. The daily cost of operation is $20,000 for plant A and $15,000 for plant B. An order is placed for 1000 barrels of high-grade oil and 450 barrels for low-grade oil. Find the number of days that each plant should operate to fill the order and minimize the cost?

30. An automobile leasing company has warehouses in Los Angeles and San Francisco. A company in Las Vegas sends an order to lease 25 cars and a firm in Fresno orders 15 cars. The Los Angeles warehouse has 30 cars and the San Francisco warehouse has 20 cars. It costs $100 to ship a car from Los Angeles to Las Vegas, $90 to ship a car from Los Angeles to Fresno, $150 to ship a car from San Francisco to Las Vegas, and $80 to ship a car from San Francisco to Fresno. Find the number of cars to be shipped from each warehouse to each city in order to minimize cost.

31. A consortium of travel agents has sold 1800 tickets to the Super Bowl. The weekend package includes air fare, and they have a choice of two types of airplanes for the charter flights. Type 1 can carry 100 passengers and type 2 can carry 300 passengers. Each flight of a type 1 aircraft will cost $9000 and each type 2 flight will cost $15,000. The consortium is allowed to lease no more than 12 planes. Each type 1 flight requires 2 stewards and each type 2 flight requires 3 stewards. No more than 30 stewards are available. How many airplanes of each type should be leased in order to minimize cost?

32. A manufacturer makes two types of sausage meat. Each pound of type 1 sausage meat contains 3 ounces of beef, 1 ounce of pork and 2 ounces of spices. Each pound of type 2 sausage meat contains 6 ounces of beef, 3 ounces of pork and 1 ounce of spices. There is 225 pounds of beef, 100 pounds of pork and 200 pounds of spices in stock. If the profit for type 1 sausage meat is 90 cents per pound and the profit for type 2 sausage meat is 70 cents per pound, how many pounds of each type should be made to maximize profit?

33. A farmer plants two crops, wheat and corn. The profit per acre for wheat is $600 and for corn it is $900. The cost per acre for fertilizer is $8 for wheat and $16 for corn. The cost per acre for maintenance is $6 for wheat and $3 for corn. No more than 12 acres will be planted with wheat and corn. No more than $160 is to be spent for fertilizer and no more than $66 is to be spent for maintenance. How many acres of each crop should be planted to maximize profit?

34. A farmer plants two crops, wheat and corn. The profit per acre for wheat is $500 and for corn it is $600. The cost per acre for maintenance is $6 for wheat and $3 for corn. Time spent per acre per week for maintenance is one hour for wheat and two hours for corn. No more than 12 acres will be planted with wheat and corn. No more than $66 is to be spent for maintenance. Time spent for maintenance per week must be at least 20 hours. How many acres of each crop should be planted to profit?

35. A lamp manufacturer makes two types of lamps: deluxe and regular. The cost of manufacturing each type of lamp is $60 for a deluxe model and $30 for a regular model. It costs $10 to ship each deluxe model and $15 to ship each regular model.

The maximum weekly costs are $3600 for manufacturing and $1200 for shipping. No more than 100 lamps can be manufactured per week. How many lamps of each type should be made to maximize profit if the profit is $140 on each deluxe model $90 on each regular model?

36. A consulting firm has two types of consulting teams to handle different types of business: type I consists of 2 architects and 2 salespersons; type II consists of 1 architect and 3 salespersons. There are 12 architects available and 24 salespersons available to make up the teams. They must form at least 10 teams. If the average weekly revenue from each type of team is $4000 for type I and $5000 from type II, how many teams of each type should be made to maximize profit?

In problems 37 to 40 match Tables A to D with the given word problem.

			Limiting Value
	2	1	20
	4	5	30
Objective Function	80	70	

Table A

			Limiting Value
	4	1	70
	5	2	80
Objective Function	20	30	

Table B

			Limiting Value
	4	1	70
	5	2	80
Objective Function	20	30	

Table C

			Limiting Value
	4	1	70
	5	2	80
Objective Function	20	30	

Table D

37. A manufacturing firm makes steel tubing. Each tube is shaped and cut on machine I and then is buffed on machine II. Each large tube requires 4 minutes of time on machine I and 5 minutes of time on machine II. Each small tube requires 1 minute of time on machine I and 2 minutes of time on machine II. On a given shift machine I can operate a no more than 70 minutes and machine II can operate no more than 80 minutes. If the profit on each large tube is $30 and it is $20 on each small tube, how many of each type should be made to maximize profit?

38. A manufacturing firm makes steel tubing. Each tube is shaped and cut on machine I and then is buffed on machine II. Each large tube requires 4 minutes of time on machine I and 1 minute of time on machine II. Each small tube requires 5 minutes of time on machine I and 2 minutes of time on machine II. On a given shift machine I can operate a no more than 30 minutes and machine II can operate no more than

20 minutes. If the profit on each large tube is $70 and it is $80 on each small tube, how many of each type should be made to maximize profit?

39. A manufacturer makes shirts. Each shirt is cut on machine I and then is sewn on machine II. On machine I each long sleeve shirt requires 2 minutes of time and each short sleeve shirt requires 1 minute of time. On machine II each long sleeve shirt requires 4 minutes of time and each short sleeve shirt requires 5 minutes of time. On a given shift machine I can operate a no more than 20 minutes and machine II can operate no more than 30 minutes. If the profit on each short sleeve shirt is $70 and it is $80 on each long sleeve shirt, how many of each type should be made to maximize profit?

40. A manufacturer makes shirts. Each shirt is cut on machine I and then is sewn on machine II. On machine I each long sleeve shirt requires 2 minutes of time and each short sleeve shirt requires 4 minutes of time. On machine II each long sleeve shirt requires 1 minute of time and each short sleeve shirt requires 5 minutes of time. On a given shift machine I can operate a no more than 80 minutes and machine II can operate no more than 70 minutes. If the profit on each short sleeve shirt is $30 and it is $20 on each long sleeve shirt, how many of each type should be made to maximize profit?

Referenced Exercise Set 2-1

1. In 1967 the Missouri Air Conservation Commission adopted air pollution regulations for metropolitan St. Louis that created heated controversy. Coal burning accounted for more than 30% of the pollutants and a major coal producer sought an injunction to prohibit enforcement of the regulations, claiming, among other things, that it was not economically feasible. A great deal of study ensued, thereby establishing precedent on the legality of pollution standards. A study by Robert Kohn applied linear programming to determine the economic impact of the regulations on the coal industry.[1] One problem investigated by Kohn dealt with two methods of controlling the pollutant ammonia nitrate. The two control methods were the use of a wet scrubber and a cyclone collector. The regulation required that there be a reduction of at least 165 million pounds of sulfur dioxide. Another constraint was that if the two controls operated at peak efficiency, the largest reduction of particulate would be 230 million lb. For each ton of ammonia nitrate that was controlled by the wet scrubber, there was a reduction of sulfur dioxide by 9.2 pounds and particulates by 7.6 lb. For each ton of ammonia nitrate that was controlled by the cyclone collector, there was a reduction of sulfur dioxide by 4.6 pounds and particulates by 3.1 lb. If the cost per ton was $0.07 to operate the wet scrubber $0.03 to operate the cyclone collector, find the number of tons of ammonia nitrate production controlled by each type of control method in order to minimize cost.

2. The allocation of space is often a complicated problem for large enterprises. Large school systems and universities especially have a major task assigning classroom

---

[1] Kohn, Robert, "Application of linear programming to a controversy on air pollution control," *Management Science*, 17, 1971, pp. 609-621.

space. Gosselin and Truchon devised a linear programming model to help solve this type of allocation of space problem.[2] The procedure has been applied at various locals, including the Universite Laval. As an example of their procedure, consider a segment of a university that uses rooms with 20 seats, called preferred rooms, and rooms with 40 seats, called regular rooms. The total number of these rooms available is 50. There are 1600 students to accommodate. The objective is to minimize the number of times a request for a particular type of room is denied. This is called a penalty. Suppose 10% of requests for preferred rooms are usually denied and 5% of requests for regular rooms are usually denied. Find the number of each type of room that should be allocated to this segment of the university to minimize the penalty function.

3. In his book on linear programming, G. H. Symonds describes one of the first practical applications of the subject, which centered on the solution of oil refinery production problems. The following problem is adapted from this book.[3] An oil refinery produces at most 2000 gallons of fuel oil per day. Two types are produced, commercial fuel oil and residential fuel oil. Blending requirements are that the number of gallons of commercial fuel oil be at least three times as many as the number of gallons of residential fuel oil. If the profit per gallon is 34 cents for commercial fuel oil and 22 cents for residential fuel oil, how many gallons of each type should be produced to maximize profit?

Multiple-Choice Exercise Set 2-1

Problems 1 to 4 refer to the following linear programming problem.
A company wants to purchase at most 25 computers. The cost for each IBM is $1500 and for MAC is $2000 and no more than $40,000 will be used for the purchase. If the expected profit is $2000 for each IBM and $3000 for each MAC how many of each type of computer should be purchased to maximize profit?

1. The objective function is
   (a) $P = 1200x + 1800y$
   (b) $x + y \le 25$
   (c) $1500x + 2000y \le 40{,}000$
   (d) $P = 2000x + 3000y$
   (e) $P = 1500x + 2000y$

2. One constraint is
   (a) $P = 1200x + 1800y$
   (b) $x + y \le 25$
   (c) $1000x + 2000y = 40{,}000$
   (d) $P = 2000x + 3000y$
   (e) $P = 1000x + 2000y$

3. One constraint is
   (a) $P = 1200x + 1800y$
   (b) $x + y \ge 25$
   (c) $1000x + 2000y \le 40{,}000$
   (d) $x + y = 25$
   (e) $1000x + 2000y \ge 40{,}000$

---

[2] Gosselin, K. and Truchon, M., "Allocation of classrooms by linear programming," *Journal of Operations Research Society*, 37, 1986. pp. 561-569.

[3] Symonds, G. H., *Linear Programming: The Solution of Refinery Problems*, Esso Standard Oil Co., New York, 1955.

4. The limiting values are
   (a) 1200 and 1800
   (b) 25 and 40000
   (c) 2500 and 4000000
   (d) x and y
   (e) 1000 and 2000

Problems 5 to 8 refer to the following linear programming problem.

A consortium of travel agents has sold 1600 tickets to the Super Bowl. The weekend package includes air fare, and they have a choice of two types of planes for the flight. Type 1 can carry 100 passengers while type 2 can carry 200 passengers. Each type 1 flight costs $6000 and each type 2 flight costs $8000. They can lease no more than 10 planes. How many planes of each type should be used to minimize cost?

5. The objective function is
   (a) $C = 100x + 200y$
   (b) $x + y \leq 10$
   (c) $6000x + 8000y \leq 14{,}000$
   (d) $C = 6000x + 8000y$
   (e) $100x + 200y = 14{,}000$

6. One constraint is
   (a) $100x + 200y \geq 10$
   (b) $x + y \leq 10$
   (c) $6000x + 8000y \leq 14{,}000$
   (d) $C = 6000x + 8000y$
   (e) $100x + 200y = 1600$

7. One constraint is
   (a) $100x + 200y \leq 1600$
   (b) $x + y = 10$
   (c) $6000x + 8000y \leq 14{,}000$
   (d) $C = 6000x + 8000y$
   (e) $100x + 200y \geq 1600$

8. The limiting values are
   (a) 100 and 200
   (b) 10 and 14,000
   (c) 10 and 1600
   (d) x and y
   (e) 6000 and 8000

Problems 9 to 14 refer to the following linear programming problem.

An automobile leasing company buys large and small cars. The company wants to purchase at least 25 cars. The cost of each large car is $12000 and the cost of each small car is $8000. It will allocate no more than $4 million for the purchase of the cars. If the profit is $1200 for each large car and is $1800 for each small car, how many of each type should be purchased to maximize profit?

9. The title of the objective function is
   (a) profit
   (b) cost
   (c) large cars
   (d) small cars
   (e) automobiles

10. The titles of the constraints are
    (a) profit and size
    (b) cost and number of cars
    (c) large and small cars
    (d) profit and cost
    (e) cost and size

11. The objective function is
    (a) $P = 1200x + 1800y$
    (b) $x + y \geq 2500$,
    (c) $12{,}000x + 8000y = 4{,}000{,}000$
    (d) $P = 2500x + 1800y$,
    (e) $P = 12{,}000x + 8000y$

12. One constraint is
    (a) P = $1200x + 1800y$
    (b) $x + y \geq 2500$
    (c) $12{,}000x + 8000y = 4{,}000{,}000$
    (d) P = $2500x + 1800y$
    (e) P = $12{,}000x + 8000y$

13. One constraint is
    (a) $x + y \leq 25$
    (b) $x + y \geq 2500$
    (c) $12{,}000x + 8000y \leq 4{,}000{,}000$
    (d) P = $2500x + 1800y$
    (e) P = $12{,}000x + 8000y$

14. The limiting values are
    (a) 1200 and 1800
    (b) 2500 and 8000
    (c) 2500 and 4000000
    (d) x and y
    (e) 12000 and 8000

15. One constraint is <u>not</u>
    (a) $12{,}000x + 1800y \leq 4{,}000{,}000$
    (b) $x + y \geq 2500$
    (c) $x \geq 0$
    (d) $12x + 8y \leq 4000$
    (e) P = $1200x + 1800y$

CASE STUDY FEDERAL EXPRESS CORPORATION[4]

Seldom does a research paper in a college Political Science course amount to more than simply a grade for the writer. But Mr. Frederick W. Smith's 1964 paper, written while he was a student at Yale University, describing a new type of business venture, was put into practice in 1972. It was then that Mr. Smith founded a company he called Federal Express Corporation. His concept was to airlift small packages overnight with door-to-door pickup and delivery.

Just to open for business Mr. Smith needed a great deal of capital and planning. All airplanes, delivery trucks, offices, personnel and government licenses had to be in place. In the eight years since writing his political science paper Mr. Smith firmed up his ideas while running a very successful airplane company. After leaving the Marines he purchased a company that serviced airplanes and gradually changed the business to buying and selling used aircraft. Over a two-year period Mr. Smith increased sales from $1 million to $9 million and turned its profit situation from a $40,000 loss to a $250,000 profit.

To verify the profit potential of his overnight delivery concept Mr. Smith hired two business consultant firms. Both were enthusiastic. In fact, the president of one firm, Mr. Arthur Bass, joined Mr. Smith as a Senior Vice-President for Planning in 1972. It was Mr. Bass who decided to use analytical methods to help in making the complex decisions Federal Express would have to face in the near future. According to Mr. Bass, it was his job to translate Mr. Smith's far-reaching ideas into reality, or to turn "Mr. Smith's view of Federal Express in 1990 into a profit in the 1970's." It was this view to the future, their ability to visualize and anticipate future trends, that provided them with almost limitless potential and distinguished them from the competition. In 1978, reflecting on the meteoric growth of Federal Express, Mr. Smith said,

> *In five years we have gone through phases of growth which other companies normally experience over a period of twelve to fifteen years. From June 1977 to June 1978 alone our revenues have jumped from $108 million to $160 million. In order to understand where we are, we have to look not only at the past but also at the future.*

The past achievements of Federal Express and potential growth in a market they defined was summarized in an article that appeared in *The South Magazine*, March-April, 1975, which stated that:

> *If anyone wants a refresher course in the free enterprise system, in the basic premise that creativity, dedication and sweat adds up to better, in the assumption that a work force can be inspired by the challenge of its mission, and that free give-and-take competition equals progress, then go to Federal Express. It is undoubtedly one of America's great adventures.*

The key ingredient in Federal Express' strategy was that in a developed economy such as in the U.S., with advanced technology and high capital investment, the need for

---

[4] Adapted from John C. Camillus, "Federal Express Corporation: History and Concept of Business", *Business Policy and Strategy Management*, by William Glueck and Laurence Jauck, McGraw-Hill Book Co., New York, 1980.

swift transportation of parts and supplies is inevitable and frequent. While the needs of any individual customer might be unpredictable, from a statistical point of view, the timing and transportation requirements between large population centers would be fairly constant.

Federal Express' management believed that they could meet these demands and win 100% of the market if they could satisfy six primary requirements of their customers. They were: (1) coverage - they had to service all major population centers; (2) timing - they had to deliver before noon of the following day; (3) reliability - they would guarantee delivery; (4) cost - equitable rates were needed; (5) billing - it must be timely and accurate; (6) service - door-to-door pickup and delivery was essential.

Federal Express' most important consideration was the type of aircraft to use. The initial decision was guided by the Civil Aeronautics Board (CAB) ruling that distinguished between "air taxi" operators and cargo firms. The former could fly anywhere in the U.S. without the need to obtain a certificate of "public necessity and convenience". Operators not falling into this classification had to obtain a certificate for each city-pair route, of which Federal Express had 5000. No new certificate had been granted in over 20 years, as existing carriers' objections had to be overruled, with the burden of proof resting on the applicant. Federal Express' strategy to start up overnight with a multitude of city-pair routes made it impossible to classify their business as air cargo. They had to fall into the air taxi classification. This meant that the capacity of their planes was limited to 7500 pounds. They chose the French-built Dassault Fanjet Falcon 20, and later added some Learjets.

These served as their only aircraft until 1978 when Federal Express' Washington lobbying resulted in an amendment to the CAB ruling which essentially exempted cargo carriers from CAB control. This meant Federal Express could purchase larger aircraft. The obstacle was that their delivery system, based on the swift economy of small aircraft, was working efficiently and profitably. Would even a partial shift to large planes enhance operating efficiency? And if so, what mix would be most desirable?

United Airlines was retiring 20 Boeing 727's and Federal Express' management decided to analyze the feasibility of purchasing some or all of them. A comparison of the two types of planes yielded the following data.

	Boeing 727	Falcon 20
Direct operating cost	\$1400/hour	\$500/hour
Payload	42,000 pounds	6,000 pounds

Hourly operating cost was limited to \$35,000 and total payload had to be at least 672,000 pounds. These facts translate into two constraints but a third constraint was that only 20 727's were available. The objective was to maximize the number of aircraft given these constraints.

To solve the problem let

x = the number of 727's
y = the number of Falcon 20's

Then the problem becomes:

Maximize $N = x + y$ subject to

$$1400x + 500y \le 35,000$$
$$42,000x + 6000y \ge 672,000$$
$$x \le 20$$
$$x \ge 0,\ y \ge 0$$

In the next section we will see how to find the solution to this type of problem. The solution is (10,42). That is, to maximize the number of aircraft under the given constraints Federal Express should purchase 10 727's and 42 Falcon 20's. (See Figure 2–1.)

In 1978, following the removal of CAB regulation, Federal Express purchased 10 Boeing 727's from United Airlines. The management of Federal Express was sure that the choice of the Fanjet Falcons and the Learjets played a vital role in the firm's early success. But as competition grew, both from new private companies and the U.S. Post Office, the larger aircraft would provide more flexibility and a wider range of cargo capability. In fact, as they were purchasing the 727's, Federal Express management was studying the feasibility of using 747's.

By the way, Mr. Smith received a grade of C on his paper.

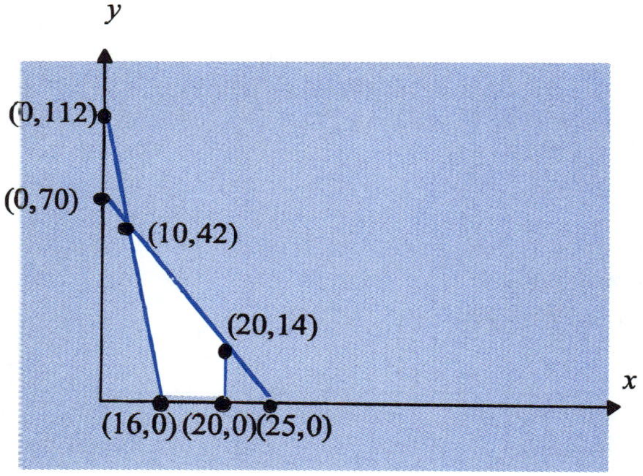

Figure 2–1

2-2 GRAPHICAL SOLUTIONS TO LINEAR PROGRAMMING PROBLEMS

Once a linear programming problem in two variables has been translated into mathematical language, then the graphical solution can be applied. In this section we will demonstrate this geometric approach to the solution. It is effective in handling problems in two unknowns, like those in Section 2-1.

The Manufacturing Problem (Revisited)

Let us demonstrate the steps of the graphical method by looking at the manufacturing problem of Section 2-1. A clothing manufacturer made two types of suits. The problem was to maximize profit subject to constraints of cost and capacity. We start with the mathematical formulation of the problem.

Maximize $P = 60x + 50y$ subject to

$$100x + 40y \le 6000$$
$$x + y \le 120$$
$$x \ge 0,\ y \ge 0$$

The task is to find the values of x and y that simultaneously satisfy the inequalities and yield the largest value of P. The method is separated into three steps.

1) The first step is to find the points that satisfy the inequalities by graphing the inequalities. This graph is called the solution of the system of inequalities.
2) The second step is to find the **corner points** of the graph of the inequalities The corner points are the points of intersection of the boundaries of the region of solutions of the inequalities.
3) The third step is to find which corner point yields the largest number when substituted into the objective function. This point is called the **optimal solution**.

We will solve the manufacturing problem and then show why the method works.

1. *Graph the region defined by the inequalities.*
We first make two general points about graphing inequalities in linear programming problems
   i) Almost every linear programming problem contains the constraints, $x \ge 0$ and $y \ge 0$. The inequality $x \ge 0$ means that every point must have a nonzero first coordinate and the inequality $y \ge 0$ means that every point must have a nonzero second coordinate. This means that the only points that we will consider in a linear programming problem are those in the first quadrant, those on the positively directed x- and y- axes, and the origin.
   ii) Every other constraint will be of the form $Ax + By \ge C$ or $Ax + By \le C$, where A, B and C are positive numbers. As we will demonstrate after this example, the points that satisfy the constraint $Ax + By \ge C$ are the points above the line

$Ax + By = 0$, and the points that satisfy the constraint $Ax + By \le C$ are the points that lie below the line $Ax + By = C$.

Replace the inequality signs with equals signs, and then graph the lines in the first quadrant by plotting the intercepts and drawing a line through the two points. For each line shade the region that is not part of the solution. Then the region that is left unshaded is the solution of the inequalities.

2. *Find the corner points.*
The boundary of the solution of the unshaded region consists of the line segments of the four lines $x + y = 120$, $100x + 40y = 6000$, $x = 0$, and $y = 0$. From Figure 2–2 the boundary consists of four line segments, and thus there are four points of intersection. Three of them, (0, 0), (60, 0) and (0, 120) were determined when we graphed the lines. The fourth is the point of intersection of $x + y = 120$ and $100x + 40y = 6000$. By the methods of Section 1-3, multiply the first equation by –40 and add the two equations. The point of intersection is (20, 100). Hence the four corner points are (0, 0), (60, 0), (0, 120) and (20, 100).

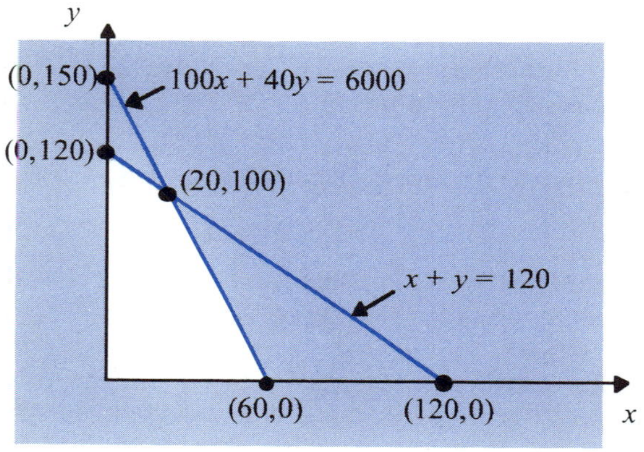

$$x + y \le 120$$
$$100x + 40y \le 6000$$
$$x \ge 0,\ y \ge 0$$

Figure 2–2

3. *Evaluate the objective function at the corner points.*
Enter the computations in the following table.

Corner Point	Value of $60x + 50y$	= P
(0, 0)	$60 \cdot 0 + 50 \cdot 0$	= 0
(60, 0)	$60 \cdot 60 + 50 \cdot 0$	= 3600
(0, 120)	$60 \cdot 0 + 50 \cdot 120$	= 6000
(20, 100)	$60 \cdot 20 + 50 \cdot 100$	= 6200

Table 2–4

The largest value of P in Table 2–4 is 6200. Therefore the solution is given by the corner point (20, 100), which means that $x = 20$ and $y = 100$. Referring to the meaning of the description of the manufacturing problem, the largest profit is $6200 when the manufacturer makes $x = 20$ wool suits and $y = 100$ blend suits.

The Graphical Method of Solution

Before presenting the general method we need some terminology. The set of points that are solutions of the constraints, the unshaded region in Figure 1, is called the **region of feasible solutions**. The feasible solutions are the only points in the plane that have a chance to be solutions of the linear programming problem. The vertices of the boundaries are called corner points and the optimal solution is the point (or points) that yields the largest or smallest value of the objective function, depending upon whether it is to be maximized or minimized, respectively.

It is possible that a linear programming problem does not have a solution. For instance, if the region of feasible solutions is unbounded then an objective function such as $z = x + y$ cannot be maximized. On the other hand, if the region is bounded, then the solution occurs at a corner point. In cases where a solution does exist for an unbounded region, the solution will also occur at a corner point.

The Graphical Method for Solving a Linear Programming Problem

1. Graph the constraints and find the region of feasible solutions.
2. Find the corner points of the region of feasible solutions. They are the points of intersection of the boundary line segments.
3. Evaluate the objective function at each corner point. The optimum solution, if it exists, is that point that yields
   a. the largest value if the objective function is to be maximized
   b. the smallest value if the objective function is to be minimized.

Justification of the Method

There are two natural questions about how the method works. Why does the shading technique work and why is it sufficient to consider only the corner points when substituting numbers into the objective function? Let us first answer the question about the shading.

We stated earlier that the solutions to the inequality $Ax + By \geq C$ for A, B and C positive numbers, are the points above the line $Ax + By = C$. We only consider the case where A, B and C are positive because linear programming problems involve only positive constants. A point that is above the line is directly above a point on the line with the same first coordinate. Call this point on the line (a, b), so that any point directly above this point can be expressed as (a, b + c) for some positive number c. Since (a, b) is on the line it satisfies $Ax + By = C$ meaning that $Aa + Bb = C$. We now claim that the point (a, b + c) satisfies the inequality $Ax + By \geq C$. This is true if when we substitute this point into the inequality and it produces a true statement. Substituting the point into the inequality yields

$$Aa + B(b + c) \geq C$$

Expanding the left-hand side yields

$$Aa + Bb + Bc \geq C$$

Recall that because (a, b) is on the line we have $Aa + Bb = C$ so the latter inequality is equivalent to

$$C + Bc \geq C$$

This inequality is a true statement since Bc is positive because both B and c are positive. This argument shows that any point above the line satisfies the inequality. A very similar argument proves that the points below the line are the solutions to the inequality $Ax + By \leq C$.

The next question about the method of solving linear programming problems centers on how we can be sure that the optimal solution is always one of the corner points. After all, there are an infinite number of points in the region of feasible solutions, each of which could be a point that optimizes the objective function. Why must it be a corner point? Let us answer this question for the manufacturing problem. The essence of the argument can be used to show that the method works for any linear programming problem.

The object of the problem is to find values of x and y that are in the region of feasible solutions and yield the largest value of $P = 60x + 50y$. If P takes on a value, for example $P = 600$, then this equation becomes $600 = 60x + 50y$ which is a linear equation in x and y. The slope-intercept form of this equation is found by solving for y. First subtract $60x$ from each side giving $50y = 600 - 60x$, and then divide each side by 50 yielding the slope-intercept form:

$$y = -\frac{6}{5}x + 12$$

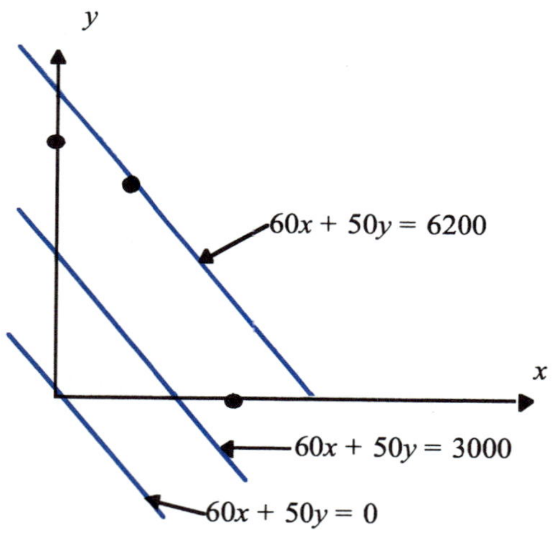

Figure 2–3

This line has slope –6/5 and y-intercept 12. Now let P take on another value, say $P = 6200$. The result is another linear equation in x and y. Its slope-intercept form is

$$y = -\frac{6}{5}x + 124$$

Notice that this line has the same slope of –6/5 as the previous line. In general, if P takes on a value, say $P = P_0$, then the objective function becomes a linear equation in x and y whose slope is –6/5 and whose y-intercept is $P_0/50$. This means that all of these lines are parallel and that they differ only in their y-intercept. Figure 2–3 gives some examples of the lines $60x + 50y = P_0$.

The optimal solution is a point in the region of feasible solutions that is also substituted into the objective function, $P = 60x + 50y$. Graphically, this means we are looking for a point in the region of feasible solutions that is also on one of the lines given by $60x + 50y = P_0$. We need the point that yields the largest value of P_0. Making P_0 as large as possible means choosing the line with the largest y-intercept. Therefore we are searching for the line with the largest y-intercept that also touches the region of feasible solutions. From Figure 2–3 that line is the line that goes through (20, 100). Any line below it has a smaller y-intercept and any line above it does not touch the region of feasible solutions. This is why (20, 100) is the optimal solution.

This argument shows that the line with the largest y-intercept (or smallest if the objective function is to be minimized) that also touches the region of feasible solutions must be a corner point. Thus the corner points are the only ones that must be evaluated; the optimal solution must be one of them.

We have implied that there may be more than one optimum solution. Multiple solutions can only happen if the objective function has the same slope as one of the boundaries. In Figure 2–4 not only are points A and B optimum solutions but all points on the line segment between A and B are also optimum solutions. Since they lie on the same line they yield the same value of the objective function.

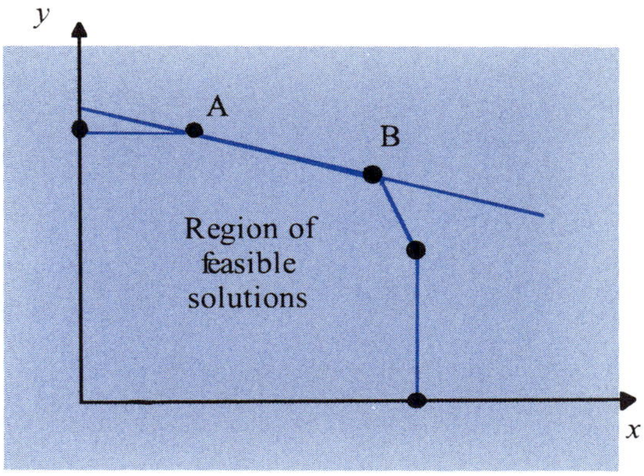

Figure 2–4

Applying the Procedure

We can now apply the graphical method to solve the problems that were translated into mathematical language in Section 2-1.

EXAMPLE 1

Find the optimum solution to the nutrition problem in Example 2 of Section 2-1.

Minimize $F = 4x + 2y$

Subject to $45x + 9y \geq 45$

$10x + 6y \geq 20$

$x \geq 0, y \geq 0$

Solution.

1. Graph the constraints and find the region of feasible solutions. The inequality signs are replaced by equality signs and the intercepts are plotted. The graph is in Figure 2–5; it is the unbounded region above the line segments.

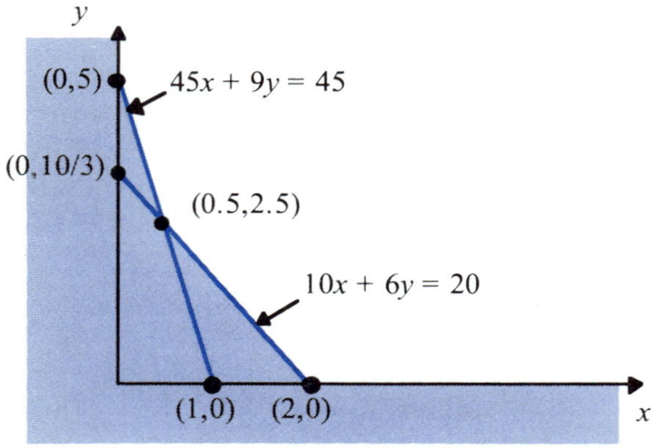

$$45x + 9y \geq 45$$
$$10x + 6y \geq 20$$
$$x \geq 0,\ y \geq 0$$

Figure 2–5

2. The corner points are (2, 0), (0, 5) and (0.5, 2.5), he latter point being the intersection of the two lines $45x + 9y = 45$ and $10x + 6y = 20$.
3. Evaluate the objective function at each corner point. The computations are in the table.

Corner Point	Value of $4x + 2y$	$= F$
(2, 0)	$4 \cdot 2 + 2 \cdot 0$	$= 8$
(0, 5)	$4 \cdot 0 + 2 \cdot 5$	$= 10$
(0.5, 2.5)	$4 \cdot 0.5 + 2 \cdot 2.5$	$= 7$

Table 2–5 The smallest value of F in the last column of Table 2–5 is 7. Therefore the solution is (0.5,2.5), which means that $x = 0.5$ and $y = 2.5$. Referring to the meaning of the description of the diet problem, the optimum meal with the smallest amount of fat has 0.5 serving of chicken (1.5 ounces) and 2.5 servings of corn (2.5 cups).

If the problem in Example 1 called for maximizing the objective function, there would be no solution because the region of feasible solutions is unbounded. The objective function, $F = 4x + 2y$, can be made as large as can be by choosing large enough values of x and y. Since the region is unbounded, such values can always be found in the region. This would be an example of the situation where you must check to see if the problem has a solution before applying the method.

EXAMPLE 2

Find the optimum solution to the transportation problem in Example 3 of Section 2-1, which is

Maximize $P = 210x + 400y$

Subject to $2x + 4y \le 40$

$50x + 200y \le 1600$

$x \ge 0, y \ge 0$

Solution.

1. Graph the constraints and find the region of feasible solutions. (See Figure 2–6.)

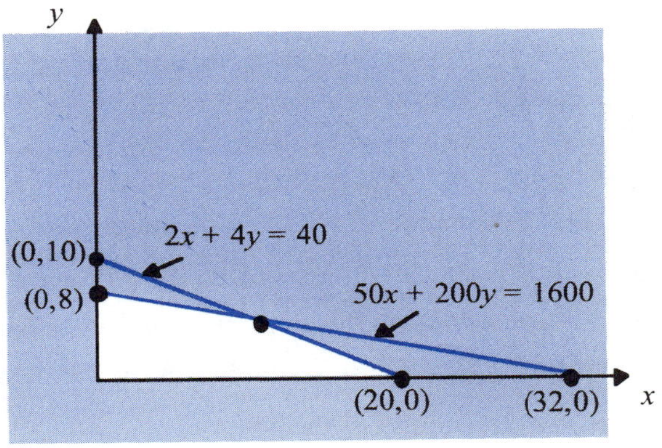

$2x + 4y \le 40$

$50x + 200y \le 1600$

$x \ge 0, y \ge 0$

Figure 2–6

2. The corner points are (20,0), (0,8) and (8,6).
3. Evaluate the objective function at each corner point.

Corner Point	Value of $210x + 400y$	= P
(0, 0)	$210 \cdot 0 + 400 \cdot 0$	=0
(20, 0)	$210 \cdot 20 + 400 \cdot 0$	= 4200
(0, 8)	$210 \cdot 0 + 400 \cdot 8$	= 3200
(8, 6)	$210 \cdot 8 + 400 \cdot 6$	= 4080

The optimum solution is (20,0). The manufacturer should schedule 20 trips to the retail outlet and no trips to the chain store each week to maximize profit.

EXERCISE SET 2-2

In problems 1 to 6 find the graph of the set of feasible solutions and the corner points.

1. $x + 4y \le 12$
   $x + 3y \le 10$
   $x \ge 0, y \ge 0$

2. $2x + y \ge 10$
   $3x + 2y \ge 12$
   $x \ge 0, y \ge 0$

3. $x + 6y \le 18$
   $2x + 3y \ge 12$
   $x \ge 0, y \ge 0$

4. $x + 3y \le 12$
   $2x + 5y \ge 10$
   $x \ge 0, y \ge 0$

5. $2x + 3y \le 10$
   $4x + 7y \ge 28$
   $x \ge 0, y \ge 0$

6. $3x + 2y \ge 18$
   $x + 4y \ge 12$
   $x \ge 0, y \ge 0$

In problems 7 to 12 solve the linear programming problem with the given objective function and the constraints from the indicated problem from this exercise set.

7. Maximize $z = 4x + 10y$ subject to the constraints in Problem 1.
8. Minimize $z = 3x + 4y$ subject to the constraints in Problem 1.
9. Minimize $z = 20x + 18y$ subject to the constraints in Problem 3.
10. Minimize $z = 10x + 12y$ subject to the constraints in Problem 2.
11. Maximize $z = 20x + 18y$ subject to the constraints in Problem 4.
12. Minimize $z = 10x + 12y$ subject to the constraints in Problem 6.

In problems 13 to 30 solve the indicated problem from the exercise set in Section 2-1.

13. Problem 1
14. Problem 2
15. Problem 3
16. Problem 4
17. Problem 5
18. Problem 6
19. Problem 7
20. Problem 8
21. Problem 9
22. Problem 10
23. Problem 11
24. Problem 12
25. Problem 13
26. Problem 14
27. Problem 15
28. Problem 16
29. Problem 17
30. Problem 18

In problems 31 to 38 find the values of x and y that maximize $P = 2x + 3y$ subject to the given constraints.

31. $x + 3y \le 6$
    $x + 4y \ge 4$
    $x \ge 1, y \ge 1$

32. $x + 3y \le 3$
    $x + 2y \le 6$
    $x \ge 1, y \ge 2$

33. $4x + y \geq 8$
$x + y \geq 5$
$x + 3y \geq 9$
$x \geq 0, y \geq 0$

34. $x + y \leq 8$
$x + 2y \leq 10$
$2x + 5y \geq 10$
$x \geq 0, y \geq 0$

35. $x + y \leq 10$
$x + 2y \leq 12$
$2x + y \leq 12$
$x \geq 0, y \geq 0$

36. $x + y \leq 10$
$x + 2y \geq 12$
$2x + y \leq 12$
$x \geq 0, y \geq 0$

37. $x + y \leq 10$
$x + 2y \geq 12$
$2x + y \geq 12$
$x \geq 0, y \geq 0$

38. $x + y \leq 20$
$x + 3y \geq 30$
$y \leq 25$
$x \geq 5, y \geq 0$

39. A firm manufactures two types of lamps, gold-plated and chrome-plated. Each gold-plated lamp requires 2 hours of plating, 1 hour of finishing and 3 hours of assembling. Each chrome-plated lamp requires 2 hours of plating, 2 hours of finishing and 1.5 hours of assembling. The capacity of the plating department is 200 hours, the capacity of the finishing department is 120 hours and the capacity of the assembly department is 180 hours. If the profit of each gold-plated lamp is $50 and the profit of each chrome-plated lamp is $30, find the number of each type of lamp that should be manufactured to maximize profit.

40. A hospital nutritionist prepares a menu consisting of chicken and rice that must contain at least 40 units of protein, at least 90 units of carbohydrates and at least 50 units of iron. Each serving of chicken contains 4 units of protein, 1 unit of carbohydrates, 1 unit of iron and 4 units of fat. Each serving of rice contains 1 units of protein, 1 unit of carbohydrates, 3 units of iron and 5 units of fat. How many servings of chicken and rice should the menu contain to minimize the amount of fat?

41. A meat manufacturer mixes beef and pork in sausage links. Two types of links are made, regular and spicy. Each pound of regular sausage meat contains 0.1 pound of beef, 0.1 pound of pork and 0.1 pound of spices. Each pound of spicy sausage meat contains 0.3 pound of beef, 0.1 pound of pork and 0.2 pound of spices. The manufacturer can use no more than 120 pounds of beef, and 80 pounds of pork but at least 100 pounds of spices must be used. The profit of each pound of regular sausage meat is $.40 while the profit of each pound of spicy sausage meat is $.60. Find the number of pounds of each type of sausage meat that should be made to maximize profit.

42. Solve problem 41 if the manufacturer is also required to manufacture at least 100 pounds of regular sausage meat but no more than 700 pounds of spicy sausage meat.

REFERENCED EXERCISE SET 2-2

1. Many farming strategies require crop rotations. In a model that was tested on a northeastern Oregon farm, researchers studied crop rotations of four crops; corn, potatoes, wheat and alfalfa.[5] In one part of the study, the number of acres planted with corn and the number of acres planted with potatoes was investigated. The total number of acres planted was limited to at most 100. Each acre of corn required 0.9 pounds of nitrogen while each acre of potatoes required 3 pounds of nitrogen. No more than 30 pounds of nitrogen was to be used. Find the number of acres of each type of crop that should be planted to maximize profit if the profit per acre was \$126 for corn and \$1600 for potatoes.

2. In an article exploring new techniques in the application of linear programming to various aspects of planning and management, Robert Bland studied a small brewery whose ale and beer are in high demand but whose production is limited by materials that are in short supply.[6] Ale and beer are made from different proportions of corn, hops and malt. Each barrel of ale requires 5 pounds of corn, 4 oz of hops and 35 pounds of malt. Each barrel of beer requires 15 pounds of corn, 4 oz of hops and 20 pounds of malt. The brewery has on hand 480 pounds of corn, 160 oz of hops and 1190 pounds of malt. If the profit per barrel is \$13 for ale and \$23 for beer, how many barrels of each type should be produced to maximize profit?

3. An integrated crop and intensive beef production firm is a farm enterprise that grows much of its own foodstuffs for its cattle. To operate with a minimum cost goal requires detailed long-term planning. In an article applying linear programming to the solution of this type of problem, J.J. Glen considered various foodstuff rations for the livestock.[7] In one part of the study, two variables were considered, the number of cattle fed primarily enterprise-grow wheat and the number of cattle fed primarily enterprise-grow hay. The number of cattle in the study was limited to 500. Each animal fed primarily enterprise-grow wheat consumed 0.2 hectare of foodstuff while each animal fed primarily enterprise-grow hay consumed 0.7 hectare of foodstuff. The number of hectares for these animals was limited to 140. Find the number of animals that should be given each type of ration to minimize cost if the cost is \$35 to feed an animal the wheat ration diet and the cost is \$30 to feed an animal the hay ration diet.

---

[5] El-Nazer, Talaat, and McCarl, Bruce, "The choice of crop rotation: a modeling approach and case study", *American Journal of Agricultural Economics*, 68, pp. 127-36, 1986.

[6] Bland, Robert G., "The allocation of resources by linear programming", *Scientific American*, June, 1983, pp. 126-144.

[7] Glen J. J., "A linear programming model for an integrated crop and intensive beef production enterprise", *Journal of theOperations Research Society*, 37, 1986, pp. 487-494.

Cumulative Exercise Set 2-2

1. Express the word problem as a system of inequalities, and graph its solution. A manufacturer makes two types of jackets - a wool jacket and a cotton jacket. The wool jacket sells to the retail stores for $250 each and the cotton jacket sells for $125 each. The manufacturer can budget no more than $2000 per week to produce the coats. The manufacturer can make no more than 100 jackets a week.

2. Sketch the graph of the system of inequalities in Problem 1.

In problems 3 to 6 translate the linear programming problem into mathematical language.

3. A tour director has sold 450 tickets for a weekend excursion. which includes a bus trip. The director can lease two types of buses. Type 1 can carry 50 passengers and type 2 can carry 75 passengers. Each type 1 bus will cost $900 and each type 2 bus will cost $1,000. The director will lease no more than 8 buses. How many buses of each type should be leased in order to minimize cost?

4. A door manufacturer produces doors in two styles, regular and ornate. It costs $500 to make each regular door, which the manufacturer sells for $650. It costs $700 to make each ornate door and each sells for $975. The daily production capacity is 100 doors and the daily cost cannot exceed $60,000. How many doors of each type should be made per day to maximize profit?

5. An automobile leasing company buys two types of cars, compacts and mid-sized. The company wants to purchase at most 2000 cars. The cost of each compact is $10,000 and of each mid-sized car is $12,000. The company has decided to allocate no more than $21 million for the purchase of the cars. It must purchase at least 200 of each type of car. If the anticipated return from the sale of the cars at the end of the company's 3-year leasing period is $1000 for a compact and $1200 for a mid-sized car, how many of each type should be purchased in order to maximize return?

6. An oil delivery firm has two processing plants, plant A and plant B. Plant A processes daily 100 barrels of high-grade oil for commercial use and 200 barrels of low-grade for residential use. Plant B processes daily 150 barrels of high-grade oil and 500 barrels of low-grade oil. The daily cost of operation is $20,000 for plant A and $35,000 for plant B. An order is placed for 500 barrels of high-grade oil and 1400 barrels for low-grade oil. Find the number of days that each plant should operate to fill the order and minimize the cost?

In problems 7 to 10 solve the linear programming problem referred to in this cumulative exercise set.

7. Problem 3
8. Problem 4
9. Problem 5
10. Problem 6

MULTIPLE-CHOICE EXERCISE SET 2-2

1. What are the corner points of the system of linear inequalities?

$$x + 2y \geq 12$$
$$x + 4y \leq 20$$
$$x \geq 0, y \geq 0$$

(a) (4,4), (12,0), (20,0) (b) (4,4), (12,0), (0,5), (0,0)
(c) (4,4), (0,5), (0,6) (d) (4,4), (20,0), (0,6)

2. What are the corner points of the system of linear inequalities?

$$x + 2y \geq 12$$
$$x + 4y \geq 20$$
$$x \geq 0, y \geq 0$$

(a) (4,4), (12,0), (0,20) (b) (4,4), (12,0), (20,0)
(c) (4,4), (0,5), (0,6) (d) (4,4), (20,0), (0,6)

3. What are the corner points of the system of linear inequalities?

$$x + 2y \leq 12$$
$$x + 4y \leq 20$$
$$x \geq 0, y \geq 0$$

(a) (4,4), (12,0), (0,20) (b) (4,4), (12,0), (0,20)
(c) (4,4), (0,5), (0,6) (d) (4,4), (20,0), (12,0)

4. What are the corner points of the system of linear inequalities?

$$x + y \geq 7$$
$$2x + 3y \leq 18$$
$$x \geq 0, y \geq 0$$

(a) (3,4), (7,0), (0,6) (b) (3,4), (7,0), (9,0)
(c) (3,4), (0,6), (0,7) (d) (3,4), (0,6), (7,0)

5. What are the corner points of the system of linear inequalities?

$$x + y \leq 7$$
$$2x + 3y \geq 18$$
$$x \geq 0, y \geq 0$$

(a) (3,4), (7,0), (0,6) (b) (3,4), (7,0), (9,0)
(c) (3,4), (0,6), (0,7) (d) (3,4), (0,6), (7,0)

6. Given the linear programming problem:

Maximize $P = 3x + 7y$ subject to

$$x + y \geq 7$$
$$2x + 3y \leq 18$$
$$x \geq 0, y \geq 0,$$

the point that solves the problem is
(a) (3,4) (b) (7,0) (c) (9,0) (d) (0,6) (e) (0,7)

7. Given the linear programming problem:

   Minimize $P = 3x + 7y$ subject to

   $$x + y \geq 7$$
   $$2x + 3y \leq 18$$
   $$x \geq 0,\ y \geq 0,$$

   the point that solves the problem is
   (a) (3,4) (b) (7,0) (c) (9,0) (d) (0,6) (e) (0,7)

8. Given the linear programming problem:

   Maximize $P = 2x + 9y$ subject to

   $$x + y \geq 7$$
   $$2x + 3y \leq 18$$
   $$x \geq 0,\ y \geq 0,$$

   the point that solves the problem is
   (a) (3,4) (b) (7,0) (c) (9,0) (d) (0,6) (e) (0,7)

9. Given the linear programming problem:

   Minimize $P = 3x + 7y$ subject to

   $$x + y \leq 7$$
   $$2x + 3y \geq 18$$
   $$x \geq 0,\ y \geq 0,$$

   the point that solves the problem is
   (a) (3,4) (b) (7,0) (c) (9,0) (d) (0,6) (e) (0,7)

10. Given the linear programming problem:

    Maximize $P = 5x + 9y$ subject to

    $$x + y \leq 7$$
    $$2x + 3y \leq 18$$
    $$x \geq 0,\ y \geq 0,$$

    the point that solves the problem is
    (a) (3,4) (b) (7,0) (c) (9,0) (d) (0,6) (e) (0,7)

CASE STUDY
THE BERLIN AIRLIFT

One of the first concrete problems that George Dantzig's linear programming helped solve was the massive supply of West Berlin shortly after World War II.

Berlin surrendered to the Russian army on May 2, 1945. During the following weeks, the Russians shipped most of the city's industrial goods to Russia. The American, British and French troops did not arrive until July, 1945. Even before the war ended the Allies decided to divide the city into four sectors, each country occupying one. Berlin lay deep in the Russian sector, but the Western powers assumed that the Soviets would allow them free access to the city. However, on June 24, 1948 the Soviets blocked all land and water routes through East Germany to Berlin. They hoped to drive the Western powers from East Germany. The problem facing the Western Allies was how to keep West Berlin supplied during the Russian blockade.

The problem was turned over to the Planning Research Division of the U.S. Air Force. Their staff had been working on similar programming problems, most of which were theoretical. But now they were asked to apply their new methods to a very practical problem. Their solution helped shatter Soviet hopes of using the blockade to win total occupation of Berlin. A gigantic airlift was organized to supply more than 2 million people in West Berlin. It was a large scale program that required intricate planning. To break the blockade, hundreds of American and British planes delivered massive quantities of food, clothing, coal, petroleum and other supplies. At its peak, a plane landed in West Berlin every 45 seconds.

The number of variables in the formulation of the problem exceeded 50. They included: number of planes, crew capacity, runways, supplies in Berlin, supplies in West Germany, money. A simplified version of the problem is: maximize the cargo capacity of the planes subject to the restrictions; (1) The number of planes is limited to 44, (2) The American planes are larger than the British planes and so they need twice the number of personnel per flight, so if we let one "crew" represent the number of flight personnel required for a British plane, and thus two "crews" are required for an American flight, then the restriction is that the number of crews available is 64, and (3) the cost of an American flight is $9000 and the cost of a British flight is $5000 and the total cost per week is $300,000. The cargo capacity of an American plane is 30,000 cu ft and it is 20,000 cu ft for a British plane.

These data can be arranged in the following table.

	American Planes	British Planes	Limiting Value
Number of Planes	x	y	44
Crew Capacity	2	1	64
Cost per Flight	9000	5000	300000
Capacity	30000	20000	

The corresponding linear programming problem is:

Maximize $C = 30000x + 20000y$ subject to

$$x + y \le 44$$
$$2x + y \le 64$$
$$9000x + 5000y \le 300000$$
$$x \ge 0, y \ge 0$$

Though the system involves three inequalities, the solution can be found in exactly the same manner as in those with two constraints. The solution to the problem is given by $x = 20$ and $y = 24$. Therefore, 20 American and 24 British planes should be used to maximize cargo capacity.

CASE STUDY EXERCISE SET

In problems 1 to 4 solve the linear programming problem in the case study for the given change in the objective function.

1. Maximize $C = 20{,}000x + 30{,}000y$
2. Maximize $C = 15{,}000x + 40{,}000y$
3. Maximize $C = 40{,}000x + 15{,}000y$
4. Maximize $C = 45{,}000x + 10{,}000y$

CHAPTER REVIEW

Key Terms

2-1 Linear Programming Problems
Linear Programming
Objective Function
Constraints

2-2 Graphical Solution of Linear Programming Problems
Region Of Feasible Solutions
Corner Points
Optimal Solution

REVIEW EXERCISE SET

1. Express the following word problem as an inequality.
A manufacturer makes two types of stoves, gas and electric. The gas stove sells for \$280 and the electric stove sells for \$330. The manufacturer can budget no more than \$4800 per week to produce the stoves.

2. Express the following word problem as a system of inequalities.
A manufacturer makes two types of lamps - large and small. The small lamp sells for \$20 and the large lamp sells for \$40. The manufacturer can budget no more than \$2400 per week to produce the lamps and can make no more than 100 lamps per week.

In problems 3 to 6 translate the given linear programming problem into mathematical language.

3. A consortium of travel agents has sold 2500 tickets to the worlds fair. The weekend package includes air fare, and they have a choice of two types of airplanes for the charter flights. Type 1 can carry 150 passengers and type 2 can carry 200 passengers. Each flight of a type 1 aircraft will cost \$12,000 and each type 2 flight will cost \$15,000. The consortium is allowed to lease no more than 15 planes. How many airplanes of each type should be leased in order to minimize cost?

4. A door manufacturer produces doors in two styles, regular and deluxe. It costs \$100 to make each regular door, which the manufacturer sells for \$175. It costs \$120 to make each deluxe door and each sells for \$250. The daily production capacity is 110 doors and the daily cost cannot exceed \$12,000. How many doors of each type should be made per day to maximize profit?

5. Officials of a convention must plan accommodations for at least 800 participants. They have two types of rooms available, motel rooms that sleep three people and hotel rooms that sleep two people. For meals, they have budgeted \$20 daily per person for those in motel rooms and \$40 daily per person for those in hotels, and they must not spend more than \$24,000 in meal money per day. If the daily cost of a room is \$40 for a motel room and \$60 for a hotel room, how many rooms of each kind must they schedule in order to minimize cost?

6. A manufacturer makes two types of hot dog meat. Each pound of type 1 hot dog meat contains 3 ounces of beef, 1 ounce of pork and 1 ounce of ham. Each pound of

type 2 hot dog meat contains 4 ounces of beef, 2 ounces of pork and 1 ounce of ham. There is 120 pounds of beef, 50 pounds of pork and 30 pounds of ham in stock. If the profit for type 1 hot dog meat is 30 cents per pound and the profit for type 2 hot dog meat is 40 cents per pound, how many pounds of each type should be made to maximize profit?

7. Find the region of feasible solutions, the corner points and the point that maximizes $z = 4x + 5y$ subject to

$$2x + 3y \geq 12$$
$$3x + y \leq 11$$
$$x \geq 0,\ y \geq 0$$

8. Find the region of feasible solutions, the corner points and the point that minimizes $z = x + 8y$ subject to

$$x + 3y \geq 24$$
$$x + 2y \leq 20$$
$$x \geq 0,\ y \geq 0$$

In problems 9 to 12 solve the linear programming problem for the given problem in this set of exercises.

9. Problem 3.

10. Problem 4.

11. Problem 5.

12. Problem 6.

Tomatoes
$1.25

3 SETS AND COUNTING

Chapter Overview

This chapter lays the foundation for the next two chapters on probability and statistics. It begins with a review of the basic terms and notation from set theory, introducing various sets of real numbers. Since several meanings are attached to the verb "count," a discussion of the mathematical meaning is given. That discussion is followed by a development of two fundamental counting techniques, the Inclusion-Exclusion Principle, and the Multiplication Principle. Venn diagrams and tree diagrams are introduced as aids in visualizing certain types of situations, like the scheme concocted by a gang of crooks to rig the Pennsylvania lottery.

Case Study Preview

Write down the license plate number of your car (or a friend's car if you do not own one.) More than likely the number that you wrote down is not a "number" at all, but a sequence of three letters followed by three digits. The reason for this assertion is that more than half of the states currently use this so-called *standard format*.

California has more passenger vehicles than any other state, so many that the standard format was not able to account for all license plates. A second format was introduced, and the combined formats seemed to account for all of the state's motoring needs. But then California introduced yet a third format. Is this an instance of irrational behavior in the Golden State, or is there a mathematical justification? The Case Study reveals the answer.

3-1 SETS

The little word "set" plays a very big role in mathematics. Set theory was developed to organize, synthesize, and unify all of mathematics. This section introduces some of the basic aspects of set theory. It also defines sets of real numbers that will be used later in the book. We have already used sets intuitively in Chapter 2, where sets of feasible solutions to linear programming problems were encountered.

Notation

There are several ways to describe a **set**. The most direct way is the *descriptive method*, in which the phrase "the set of all ... " precedes a description of the set. The set must be *well defined* in the sense that it is possible to tell whether or not a given object is in the set. Otherwise it is called a *collection*. For instance, "large numbers" are not well-defined because the word "large" is ambiguous, so they form a collection but not a set, whereas "prime numbers" form a set because they are well defined.

EXAMPLE 1

Determine whether each of the following forms a collection or a well-defined set:
(a) good senators
(b) nonnegative numbers

Solution.
(a) Each senator is regarded as good by some people and bad by others. Since such a situation is not well-defined, "good senators" form a collection but not a set.
(b) "Nonnegative numbers" consist of zero and all positive numbers. It is easy to tell whether any element in the universe is one of these numbers, to it is well-defined. Therefore we speak of the set of nonnegative numbers.

Usually capital letters stand for sets. *Throughout this chapter the letter* S *will denote the set of all license plates in standard format*, where **standard format** means three letters followed by three digits.

The objects in a set are called **elements**. The symbol $\in$ means "is an element of" while $\notin$ means "is not an element of." Thus AOK 007 $\in$ S, but 007 AOK $\notin$ S.

Another way to specify a set is to use *braces* to enclose the elements. The notation

$$A = \{ 4, 5, 6 \}$$

means that A is the name of the set whose elements are 4, 5, and 6. It is read "A is equal to the set consisting of 4, 5, and 6." Braces can also define sets implicitly. The set

$$F = \{ 1, 2, 3, \ldots , 50 \}$$

refers to the whole numbers from 1 to 50. Here the ellipsis … means that some elements are deliberately missing but that the reader knows what they are from the context.

The sets A and F are finite. Two examples of infinite sets are the **natural numbers**

$N = \{ 1, 2, 3, \ldots \}$

and the **integers**

$Z = \{ \ldots , -3, -2, -1, 0, 1, 2, 3, \ldots \}$

A third way to describe sets is to use *set-builder notation* which is given in the form { _ : _ _ _ }. This notation is read "the set of all _ such that _ _ _ ." A common set that requires the use of this notation is the set of **rational numbers** Q, sometimes called the set of fractions, which is defined by :

$$Q = \{\frac{a}{b} : a,b \in Z, b \neq 0\}$$

In words, Q consists of all quantities that can be expressed in the form a/b, where a and b are integers such that $b \neq 0$. Below are some examples of rational numbers, with the reason for their rationality shown in parentheses.

$$\frac{8}{5}, \quad 4\ (= \frac{4}{1}), \quad 0\ (= \frac{0}{1}), \quad 1.25\ (= \frac{5}{4}), \quad 0.333K\ \ (= \frac{1}{3})$$

Subsets

Let X and Y be arbitrary sets. Then X is a **subset** of Y if every element in X also belongs to Y. Equivalently, X is a subset of Y if and only if $x \in X$ implies that $x \in Y$. This relationship is denoted by $X \subset Y$. If $X \subset Y$, another common expression is to say that "Y *contains* X." If X is not a subset of Y, we write $X \not\subset Y$. This occurs when some element is in the set X but not in the set Y. Since every natural number is an integer it follows that for the infinite sets N and Z described earlier, $N \subset Z$.

EXAMPLE 2

Let D be the set $D = \{0, 1/2, 1\}$.
(a) Is it true that $D \subset N$?
(b) Is it true that $D \subset Z$?
(c) Is it true that $D \subset Q$?

Solution.
(a) The number 0 belongs to D but not to N, so $D \not\subset N$.
(b) Similarly, $D \not\subset Z$ since $1/2 \in D$ but $1/2 \notin Z$.
(c) Each number in D is rational because $0 = 0/1$, and $1 = 1/1$. Therefore it is true that $D \subset Q$.

Since the set of rational numbers contains all of the natural numbers, all of their negatives, and 0, it follows that Z ⊂ Q. A **Venn diagram** depicts this in Figure 3–1. The innermost circle represents Z and is contained in the larger circle representing Q.

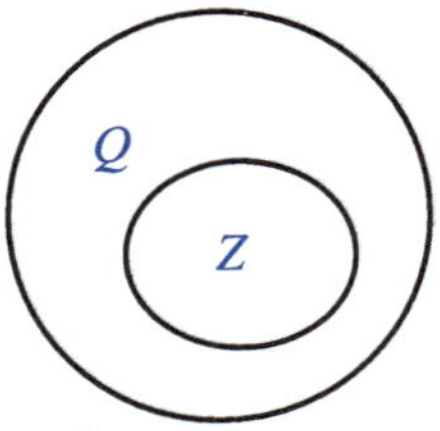

Figure 3–1

Two sets are **equal** if they contain precisely the same elements. Thus the two sets {i, o ,u} and {u, i, o} are equal. Note that when elements are listed in a set, the order in which the elements are written is not important. However, it is customary to write letters in alphabetical order and numbers in numerical order. Word processors use the term *sort* to alphabetize words and numbers. (In the latter case, letters and digits collectively are called *alphanumeric characters*.)

Two sets are not equal if one of the sets contains at least one element that is not contained in the other set. Thus {i, o, u} ≠ {i, o, x}.

Operations on Sets

The **complement** of a set X is the set of elements that do not belong to X. It is denoted by X′, which is read "X prime." The definition raises the issue of how far to look outside X. Consider, for instance, the set Q of rational numbers. Is the proverbial kitchen sink "not in Q?" Yes, it is not in Q. However, we usually limit the search within a set that is either specified or understood from context. That set is called the **universal set** and is denoted by U. The universal set for Q in this book will be the set $\mathfrak{R}$ of real numbers.

The Venn diagram in Figure 3–2 shows a set and its complement. The universal set U is the entire region inside the rectangle, X is the region inside the circle, and X′ is the region that is inside the rectangle U but outside the circle X.

Usually the universal set is understood from context, but sometimes it must be specified. Example 3 addresses this issue.

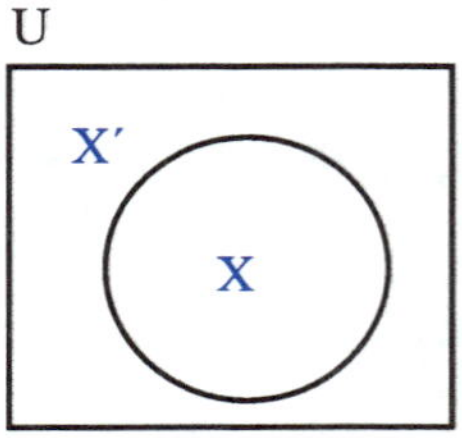

Figure 3–2

EXAMPLE 3

Find A′ for the following choices of a set A and a universal set U.

(a) A = {1, 2, 3}, U = {1, 2, 3, 4, 5, 6}
(b) A = {1, 2, 3}, U = N (N stands for the natural numbers)
(c) A is the set of even positive integers, U = N

Solution.

(a) After removing the numbers 1, 2, and 3 from the set U, you are left with the numbers 4, 5, and 6. These numbers make up the set A′ = {4, 5, 6}.
(b) In a similar way, A′ = {4, 5, 6, 7, ... }.
(c) After removing the even positive integers from the set of all positive integers, you are left with the set of odd positive integers. The descriptive way to write this is: A′ is the set of odd positive integers.

Two additional operations on sets are union and intersection. Let X and Y be arbitrary sets. The **intersection** of X and Y, denoted by $X \cap Y$, is the set of all elements that are in X and Y simultaneously.

EXAMPLE 4

Consider the sets $A = \{0, 1, 2, 3, 4\}$, $B = \{1, 3, 5\}$, and the natural numbers N.
(a) Find $A \cap B$.
(b) Find $A \cap N$.

Solution.
(a) The set $A \cap B$ consists of those numbers that lie in both sets A and B. By observation we conclude that $A \cap B = \{1, 3\}$.
(b) The set $A \cap N$ consists of those numbers in A that are natural numbers. Recall that 0 is not a natural number, but all other numbers in A are natural numbers. Therefore, $A \cap N = \{1, 2, 3, 4\}$.

EXAMPLE 5

Let U be the months in a year, A those months with days in summer, and B those months that end in the letters "ber."
(a) Find $A \cap B$.
(b) Find $(A \cap B)'$.

Solution. The first task for both parts of the Example is to determine the months that lie in each of the sets A and B. Since summer runs from June to September we have

A = {June, July, August, September}

Listing the 12 months reveals that

B = {September, October, November, December}

(a) Since $A \cap B$ consists of those months that lie in A and B simultaneously, it follows by observation that $A \cap B$ = {September}.
(b) The set $(A \cap B)'$ consists of those months that do not lie in the set $A \cap B$. From part (a), $A \cap B$ = {September}. Therefore $(A \cap B)'$ consists of all months except September. This descriptive way of defining $(A \cap B)'$ is preferable to listing all 11 months in $(A \cap B)'$, although that method is correct too.

The **union** of X and Y, denoted by $X \cup Y$, is the set of all elements that are in X or in Y (or in their intersection). This concept frequently causes difficulties. We illustrate it with two examples, which show that the way to form $X \cup Y$ is to first list the elements in X and then add to the list those elements from Y that are not already accounted for.

EXAMPLE 6

Consider the sets $A = \{0, 1, 2, 3, 4\}$, $B = \{1, 3, 5\}$, and N, the natural numbers. Find (a) $A \cup B$ and (b) $A \cup N$.

Solution.

(a) The union of two sets A and B is formed by first listing the elements in A and then adding those elements from B that are not already listed. So first list the numbers 0, 1, 2, 3, and 4. The numbers 1 and 3 are already in the list so they are not repeated, but the number 5 is not in the list so it must be added to the list. Therefore $A \cup B = \{0, 1, 2, 3, 4, 5\}$.

(b) To form the set $A \cup N$ begin by listing 0, 1, 2, 3, and 4. Then throw into this list all natural numbers that are not already in it; that is, throw in all natural numbers that are greater than 4. One way to write this set is $A \cup N = \{0, 1, 2, 3, \dots\}$.

The set $\{0, 1, 2, 3, \dots\}$ is sometimes called the set of *whole numbers*. We will have no occasion to use this terminology in this book, however.

EXAMPLE 7

Let U be the months in a year, A be those months with days in summer, and B those months that end in the letters "ber."

(a) Find $A \cup B$.

(b) Find $A' \cup B'$.

Solution. The sets A and B were formed in Example 5:

A = {June, July, August, September}
B = {September, October, November, December}

(a) Begin to form $A \cup B$ by listing the months in A: June, July, August, and September. Add to this list all months in B except September, which is already listed. Then $A \cup B$ = {June, July, August, September, October, November, December}.

(b) The set $A' \cup B'$ is a union of two sets, A' and B'. We list the elements in those sets separately by forming the complements from the universal set of all 12 months.

A′ = {January, February, March, April, May, October, November, December}
B′ = {January, February, March, April, May, June, July, August}

The union of these two sets is formed by listing the months in A'. The next step is to add to this list those months in B' not already in the list; that is, add June, July, and August to the list. Therefore

$A' \cup B'$ = {January, February, March, April, May, October, November, December, June, July, August}

This way of listing the months in the set $A' \cup B'$ may seem awkward, but it is correct. It is also possible to write the months in order, in which case an equivalent answer is

$A' \cup B'$ = {January, February, March, April, May, June, July, August, October, November, December}

Now we turn from demonstrating the three basic operations on sets (complement, intersection, and union) to using them to define sets of numbers. The number $\sqrt{2}$ cannot be written as the ratio of any pair of integers, so it is called an **irrational number**. (We do not prove this fact in this book, but it is a classic in the history of mathematics.) The set Q' consists of all irrational numbers. Since no number can be both rational and irrational simultaneously, the set $Q \cap Q'$ contains no elements. The set with no elements is called the **empty** (or **null**) **set**, denoted by Ø. Hence $Q \cap Q' = \emptyset$.

Any two sets X and Y that have no elements in common are called **disjoint**. Every set X and its complement X′ are disjoint because $X \cap X' = \emptyset$. In particular, Q and Q' are disjoint.

For any set X and universal set U we have $X \cup X' = U$. What about the union $Q \cup Q'$? This set equal to the set $\Re$ of all **real numbers**. Therefore $\Re$ is the union of the set of rational numbers and the set of irrational numbers, so that $\Re = Q \cup Q'$, where Q and Q' are disjoint sets. The Venn diagram in Figure 3–3 shows the set inclusions:

$$N \subset Z \subset Q \subset \Re$$

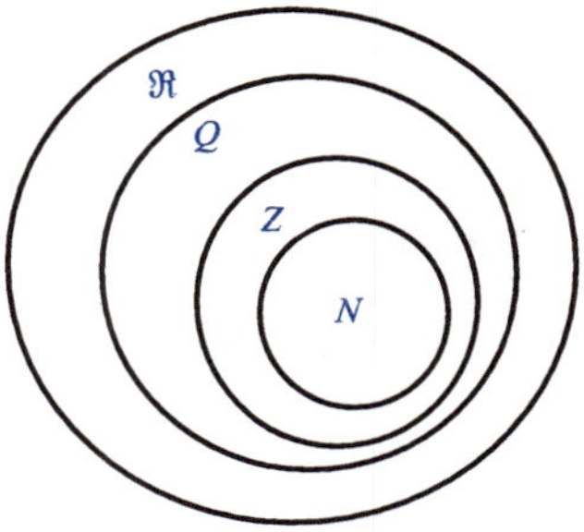

Figure 3–3

Example 8 applies set theory to a kind of problem that is encountered frequently.

EXAMPLE 8

The American Chamber of Commerce Researchers Association conducted a study to see how much it costs to maintain a middle-management lifestyle in 242 cities.[1] Table 3–1 shows some parts of the study for ten major cities. Let G denote those cities whose cost-of-living index for grocery items is less than 100, and let L denote those cities whose cost-of-living index for utilities is less than 100. Determine the elements in each of the sets G, L, $G \cap L$, and $G \cup L$.

Solution. The sets G and L can be read from the third and fifth columns of Table 3–1:

G = {San Diego, Los Angeles, Phoenix}
L = {San Francisco, San Diego, Los Angeles, San Jose, Denver}

The intersection consists of all cities that are in both sets G and L simultaneously, so

$G \cap L$ = {San Diego, Los Angeles}

To form the union of G and L, first list the cities in G:

San Diego, Los Angeles, Phoenix

Add to the list those cities in L that are not already on the list:

San Francisco, San Jose, Denver

The set $G \cup L$ is formed by listing all these cities, so

$G \cup L$ = {San Diego, Los Angeles, Phoenix, San Francisco, San Jose, Denver}

[1] *Source*: American Chamber of Commerce Researchers Association, 1987.

City	All Items	Grocery Items	Housing	Utilities	Transportation	Health Care	Misc. Goods & Services
New York	136.3	107.6	179.8	172.9	117.6	164.3	108.9
San Francisco	127.2	109.9	189.9	59.5	119.9	151.6	113.4
San Diego	116.8	95.5	162.1	82.6	119.7	122.6	105.6
Philadelphia	116.4	106.8	117.8	152.4	113.1	125.2	107.2
Los Angeles	114.1	95.4	145.8	65.2	113.1	130.5	115.9
San Jose, CA	113.0	102.4	158.6	57.7	109.9	130.4	103.2
Dallas	112.0	110.5	119.6	120.7	111.3	108.8	105.2
Denver	107.6	105.8	134.9	75.1	110.1	111.7	98.7
Phoenix	106.3	99.2	118.6	102.5	106.9	119.0	99.6
Columbus, OH	105.2	106.8	98.2	120.3	107.7	104.5	103.1

Cost-of-living index for major items; national average = 100

Source: American Chamber of Commerce Researchers Association.

Table 3–1

EXERCISE SET 3-1

In problems 1 to 4 determine if the description is a well-defined set or a collection.

1. rich countries.
2. hot countries.
3. communist countries.
4. Asian countries.

In problems 5 to 8 list the elements in the set.

5. The set of integers between –6 and 1 inclusive.
6. The set of integers between 0 and 5 inclusive.
7. The set of perfect squares between 1 and 50 inclusive.
8. The set of even integers between –10 and –4 inclusive.

In problems 9 to 12 determine whether $A \subset X$ or $A \not\subset X$.

9. $X = \{1, 2, 3, 4\}$, $A = \{0, 1, 2\}$
10. $X = \{-1, 0, 1, 2, 3, 4\}$, $A = \{0, 1, 4\}$
11. $X = Q$, $A = \{-5, 0, 3/8, 4.67\}$
12. $X = Z$, $A = \{-1/2, 1, 2\}$

In problems 13 to 16 list the elements in X′ for the given choice of universal set U.

13. $X = \{6, 7, 8\}$, $U = \{2, 3, 4, \ldots, 12\}$

14. $X = \{1, 2, 3\}$, $U = \{0, 1, 2, 3, 4, 5, 6, 7\}$

15. $X = \{1, 2, 3, \ldots, 49\}$, $U = N$

16. $X = \{2, 4, 6, 8, 10\}$, $U = N$

In problems 17 to 20 list the elements in the set $X \cup Y$.

17. $X = \{0, 1, 2, 3\}$, $Y = \{0, 2, 4, 6, 8\}$.

18. $X = \{-1, 0, 1\}$, $Y = \{-1, 1\}$.

19. $X = N$, $Y = Z$.

20. $X = Q$, $Y = Z$.

In problems 21 to 24 list the elements in the set $X \cap Y$.

21. $X = \{0, 1, 2, 3\}$, $Y = \{0, 2, 4, 6, 8\}$.

22. $X = \{-1, 0, 1\}$, $Y = \{-1, 1\}$.

23. $X = N$, $Y = Z$.

24. $X = Q$, $Y = Z$.

In problems 25 to 28 determine whether $A \subset X$ or $A \not\subset X$.

25. $X = Q$, $A = \{x \in Z : -3 \le x \le 2\}$.

26. $X = Q$, $A = \{x : -3 \le x \le 2\}$.

27. $X = Q$, $A = \{x : x > 1\}$.

28. $X = Q$, $A = \{x \in Z : x < -4\}$.

In problems 29 and 30 list the elements in the set X' for the given choice of U.

29. $X = \{4, 8, 12, \ldots\}$, $U = \{2, 4, 6, \ldots\}$.

30. $X = \{2, 4, 6, \ldots\}$, $U = \{\ldots, -4, -2, 0, 2, 4, \ldots\}$.

In problems 31 to 34 list the elements in $X \cup Y$.

31. $X = \{x \in Z : x \ge 7\}$, $Y = \{x \in Z : x \ge 1\}$.

32. $X = \{x \in Z : x \le 7\}$, $Y = \{x \in Z : x \ge 1\}$.

33. $X = N$, $Y = \{x \in Z: x \le 0\}$.

34. $X = N$, $Y = \{x \in Z : x \ge 0\}$.

In problems 35 to 38 list the elements in $X \cap Y$.

35. $X = Q$, $Y = \{\sqrt{2}, \sqrt{4}, \sqrt{8}, \sqrt{16}, \ldots\}$.

36. $X = Q'$, $Y = \{\sqrt{2}, \sqrt{4}, \sqrt{8}, \sqrt{16}, \ldots\}$.

37. $X = N$, $Y = \{x : x < 7\}$.

38. $X = \{x \in Z : x \ge -2\}$, $Y = \{x \in Z : x < 5\}$.

39. Describe $F \cap X$ where $F = \{4, 8, 12, 16, \ldots\}$ and $X = \{6, 12, 18, 24, \ldots\}$.

40. Describe $M \cap K$ where M consists of all multiples of m, and K consists of all multiples of k, where m and k are given positive integers.

41. Let U be the set of letters in the alphabet, let V be the set of vowels, and let M be the set of letters in the word "mathematics." List the elements in each set.

 (a) $V \cap M$ (b) $V \cup M$ (c) $V' \cap M$

42. Let U be the set of letters in the alphabet, F = { a, b, c, d}, and V = { a, e, i, o, u}. List the elements in each set.

 (a) $F \cap V$ (b) $F \cup V$ (c) $F \cap V'$

43. Let U = {1, 2, 3, ... , 100}, let E be the set of even numbers in U, and let T be the first 10 numbers in U. List the elements in each set.

 (a) $E \cap T$ (b) $E \cup T$ (c) $E' \cap T$

44. Let U = {1, 2, 3, ... , 20}, T = {3, 6, 9, 12, 15, 18}, and E be the set of even numbers in U List the elements in each set.

 (a) $E \cap T$ (b) $E \cup T$ (c) $E' \cap T$

45. Let $B = \{x \in Z : x^2 < 16\}$ and let $C = \{x : x^2 = 9\}$. Answer each of the following as true or false.

 (a) $-3 \in C$ (b) $\{-2\} \in B$ (c) $C \subset B$

46. List all of the elements in the set $S = \{x < 10 : x \text{ is a positive even integer}\}$.

Referenced Exercise Set 3-1

Problems 1 to 4 refer to Table 3–1 in Example 8.

1. Write the set of cities whose cost of housing or miscellaneous goods and services is less than \$100.

2. Write the set of cities whose cost of housing and utilities exceeds \$100.

3. Let U = the ten cities listed in the chart, H = cities where the cost of housing is more than \$100, and M = cities where the cost of miscellaneous goods and services is more than \$100. Describe in words the fact that $H \cup M = U$.

4. Let U = the ten cities listed in the chart and T = cities where the cost of transportation is more than \$100. Describe in words the fact that U = T.

Problems 5 and 6 refer to Table 3–2, which lists certain countries[2] and the percentage of their population for two age groups as of 1986. Let F denote the three countries with the highest percentages aged 15-29 and let T denote the three countries with the highest percentages aged 30-44.

Age Group	Brazil	Canada	China	Ireland	US	USSR
15-29	28.6	28.9	29.1	24.9	22.4	26.6
30-44	16.4	20.0	17.5	17.2	19.1	19.1

Table 3–2

5. List all elements in (a) F (b) T.
6. List the elements in (a) $F \cap T$ (b) $F \cup T$.
7. Is there any difference between Ø and {Ø}? If so, what is the difference?
8. In music theory "the diatonic universe of seven pitch classes" consists of the notes A, B, C, D, E, F, and G, irrespective of inflection by sharps and flats.[3]
   (a) Use set notation to describe this universal set.
   (b) Use set notation to describe the set in which each note can be sharp or flat (refer to a piano keyboard; that is, G, G-sharp, and G-flat are different notes.)
9. The Office Network, a consortium of real estate firms, conducted a study of rental rates in the central business districts (CBD's) of several cities.[4] Table 3 shows part of the results of the survey.
   (a) Let A be those cities having the 5 highest CBD rates. Write the set A.
   (b) Let B be those cities having the 5 highest rates outside CBD's. Write the set B.
   (c) What would it mean if $A \cap B = A$?
   (d) For this study, is it true that $A = B$?

Typical annual rate, in dollars per square foot, for rent and operating expenses in space that is new or competitive with new space. Rates are for space inside central business district or outside CBD.

City	CBD	Outside
Baltimore	\$23.38	\$17.50
Boston	35.00	15.50
Chicago	28.95	19.56
Houston	12.00	9.50
Los Angeles	34.00	22.00
Miami	29.00	19.00
New York	40.00	35.00
Philadelphia	22.10	14.85
San Francisco	23.65	22.25
Washington	32.00	22.50
Wilmington	18.00	16.10

***Source:* The Office Network**

Table 3–3

---

[2]Source: 1986 Britannica Book of the Year, pp. 826-831.

[3]John Clough, "Aspects of Diatonic Sets," *Journal of Music Theory*, Spring 1979, pp. 45-62.

[4]*Source*: The Office Network, 1988.

Multiple-Choice Exercise Set 3-1

1. Which of the following descriptions forms a set?
(a) good students (b) bad professors (c) large numbers
(d) prime numbers (e) none of these

2. Which of the following descriptions is <u>not</u> a set?
(a) even integers (b) 23 (c) irrational numbers
(d) {n, o, *t*} (e) none of these

3. If the universal set is U = {a, b, c, Y , z} then
(a) $a \in U$ (b) $d \notin U$ (c) $\{a\} \in U$ (d) $\{d\} \not\subset U$ (e) none of these

4. If the universal set is U = {a, b, c, Y , z} then
(a) $a \subset U$ (b) $d \notin U$ (c) $\{a\} \in U$ (d) $\{d\} \not\subset U$ (e) none of these

5. Which number is a natural number?
(a) 0 (b) −2 (c) 1/2 (d) 100 (e) none of these

6. Which number is a natural number?
(a) 0.2 (b) −2 (c) 1/2 (d) 0.333… (e) none of these

7. Which expression is a rational number?
(a) π (b) i (c) 2 (d) $\sqrt{2}$ (e) none of these

8. Which expression is a rational number?
(a) $\pi/2$ (b) 1/0 (c) $\sqrt{3}$ (d) 0.25 (e) none of these

9. Let V = {a, e, i, o, u} and B = {m, a, t, h, e}. Then $V \cap B =$
(a) a, e (b) a, e, i, o ,u, m, t, h (c) {a, e}
(d) {a, e, i, o, u, m, t, h} (e) none of these

10. Let V = {a, e, i, o, u} and B = {m, a, t, h, e}. Then $V \cup B =$
(a) a, e (b) a, e, i, o, u, m, t, h (c) {a, e}
(d) {a, e, i, o, u, m, t, h} (e) none of these

11. If the universal set is U = {a, b, c, … , z} and V = {a, e, i, o, u} then $V' =$
(a) consonants (b) constants (c) {a} (d) {U, V} (e) none of these

12. Let U = {a, b, c, … , z} be a universal set. If V = {a, e, i, o, u} and B = {m, a, t, h, e} then $V \cap B' =$
(a) {a, e} (b) {m, t, h} (c) {V, B} (d) {i, o, u} (e) none of these

13. Which set represents the expression "at least one head is thrown?"
(a) {0, 1} (b) {1, 2, 3, …} (c) {1, 2, 3}
(d) {0, 1, 2} (e) none of these

14. Which set represents the expression "at most two heads are thrown?"
(a) {0, 1} (b) {2, 3, …} (c) {1, 2} (d) {0, 1, 2} (e) none of these

15. Which of the following sets is a subset of {1, 2, 3, 4, 5}?
(a) $\{x \in \mathfrak{R}: x^2 = 4\}$ (b) $\{x \in \mathfrak{R}: x^2 \le 4\}$ (c) $\{x \in Z: x^2 = 4\}$
(d) $\{x \in Z: x^2 \le 4\}$ (e) none of these

16. Which of the following sets is a subset of {1, 2, 3, 4, 5, 6, 7, 8, 9}?
(a) $\{x \in Z: x < 10\}$ (b) $\{x \in Z: x^2 < 100\}$ (c) $\{x \in N: x^2 = 10\}$
(d) $\{x \in N: x^2 < 100\}$ (e) none of these

3-2 VENN DIAGRAMS & COUNTING

The word **count** usually means "one, two, three, four, ..." In mathematics, however, it has a wider usage. Perhaps a more appropriate word is "enumerate," meaning "to ascertain the number of." In mathematics the expression "count the number of ..." does not mean to count the objects individually, rather to determine how many objects there are altogether. Sometimes enumerating a large set can be a complicated task. Indeed, the root of the word "enumerate" is the Latin *numerus*, meaning "nimble."

As an example of counting in a mathematical sense, imagine a classroom that has 12 rows of chairs with 10 chairs in each row. If all but six of the chairs are occupied, how many people are sitting in the chairs? A counting technique for answering this question is to multiply 12 by 10, then subtract 6, to get 114. The chairs did not have to be counted individually to arrive at this answer.

This section introduces a basic counting principle, called the Inclusion-Exclusion Principle. It also explains and illustrates this Principle by Venn diagrams.

Number of Elements

Let $n(X)$ denote the number of distinct elements in a finite set X. The expression "count the number of elements in X" means to determine $n(X)$ by some method. Counting techniques are efficient methods for computing $n(X)$ directly.

If $A = \{1, 2, 3, 4, 5\}$ and $B = \{a, e, i, o, u, 3, 4\}$ then $n(A) = 5$ and $n(B) = 7$. Since $A \cap B = \{3, 4\}$ it follows that $n(A \cap B) = 2$. The Venn diagram in Figure 3–4 depicts the distribution of the elements in A and B. It also shows that $n(A \cup B) = 10$. Notice that the formula $n(A \cup B) = n(A) + n(B) - n(A \cap B)$ holds since $10 = 5 + 7 - 2$. This is a particular case of a basic counting technique called the **Inclusion-Exclusion Principle**. It holds in this case because when forming the union $A \cup B$ by first listing the elements in A and then listing the elements in B, the two elements in $A \cap B$ are listed twice, so they must be subtracted once in order to compensate.

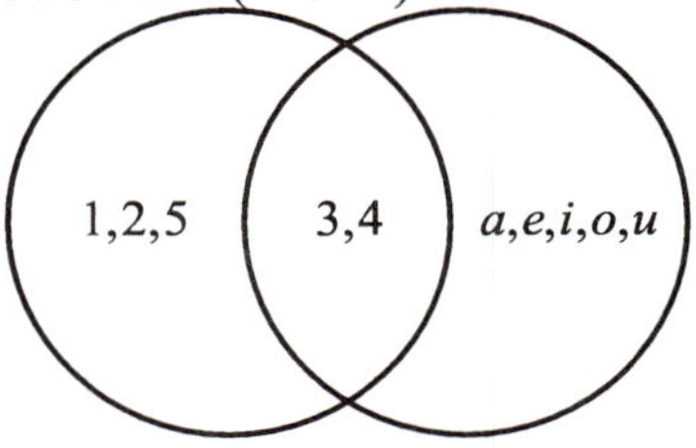

Figure 3–4

> **Inclusion-Exclusion Principle**
>
> If X and Y are any sets then $n(X \cup Y) = n(X) + n(Y) - n(X \cap Y)$.

EXAMPLE 1

A survey of 100 people asks:

Have you eaten at McDonald's this year?	Y	N
Have you eaten at Wendy's this year?	Y	N

The respondents circle "Y" or "N" for each question. If 68 respondents circled "Y" for McDonald's, 42 circled "Y" for Wendy's, and 27 circled "Y" for both questions, how many respondents circled "N" for both questions?

Solution. Let M denote the set of respondents who circled "Y" for McDonald's and let W denote the set of respondents who circled "Y" for Wendy's. Then $n(\mathrm{M}) = 68$ and $n(\mathrm{W}) = 42$. Since 27 circled both "Y", it follows that $n(\mathrm{M} \cap \mathrm{W}) = 27$. By the Inclusion-Exclusion Principle

$$n(\mathrm{M} \cup \mathrm{W}) = n(\mathrm{M}) + n(\mathrm{W}) - n(\mathrm{M} \cap \mathrm{W})$$
$$= 68 + 42 - 27$$
$$= 83$$

Therefore, 83 respondents circled "Y" for either McDonald's or Wendy's (or both). Since there were 100 respondents and since 100 − 83 = 17, seventeen respondents circled "N" for both questions.

The Venn diagram is shown in Figure 3–5. Note that the sum of the four numbers inside the diagram is 100, which is the total number of people surveyed.

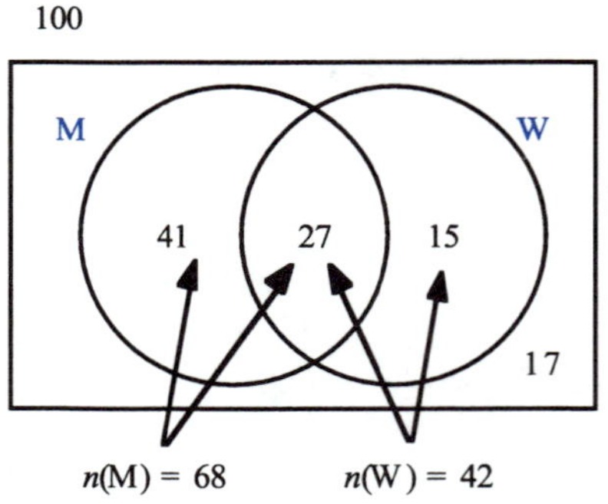

Figure 3–5

Recall that license plates in standard format S consist of three letters followed by three digits. License plates in *reverse format* R consist of three digits followed by three letters. Example 2 counts the number of license plates in either one of these formats. The numbers of license plates in each format will be derived in this chapter's Case Study.

EXAMPLE 2

Given that $n(\mathrm{S}) = 14{,}866{,}000$ and $n(\mathrm{R}) = 16{,}208{,}000$, how many license plates are either in the standard format or the reverse format?

Solution. The problem reduces to finding $n(\mathrm{S} \cup \mathrm{R})$ because $\mathrm{S} \cup \mathrm{R}$ represents all usable license plates in either the standard format or the reverse format. Each license plate in S begins with a letter and each license plate in R begins with a number, so there are no license plates in both formats. Therefore $n(\mathrm{S} \cap \mathrm{R}) = 0$. By the Inclusion-Exclusion Principle,

$$n(\mathrm{S} \cup \mathrm{R}) = n(\mathrm{S}) + n(\mathrm{R}) - n(\mathrm{S} \cap \mathrm{R})$$
$$= 14{,}866{,}000 + 16{,}208{,}000 - 0$$
$$= 31{,}080{,}000$$

This means that over 31 million license plates can be made using either the standard format or the reverse format. The sets S and R in Example 2 are disjoint, consequently, there is no overlap between the circles as shown in Figure 3–6.

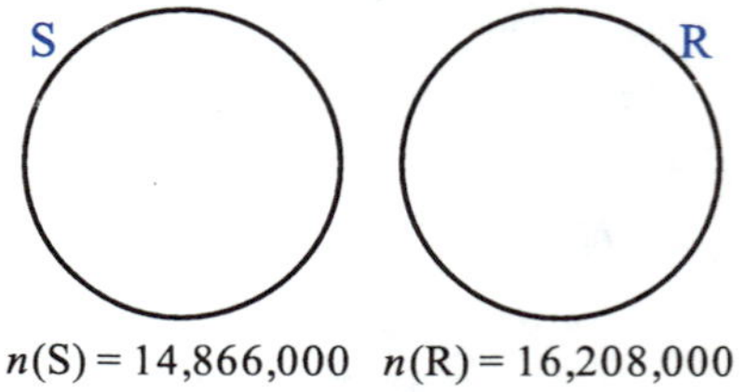

Figure 3–6

Because S and R are disjoint, the formula for $n(S \cup R)$ presents a particular case of the Inclusion-Exclusion Principle.

Inclusion/Exclusion Principle for Disjoint Sets

If X and Y are disjoint sets then $n(X \cup Y) = n(X) + n(Y)$.

Three Circles

One subset divides a universal set U into two regions. (See Figure 2 in Section 3-1 Two subsets that are not disjoint divide U into four regions. (See Figure 2 in this section.) Three subsets in which no two subsets are disjoint divide U into eight regions, as Example 3 will show.

Table 3–4 lists the 15 stocks among the 30 Dow Industrials with the highest dividend yield. Dividend yield is computed by dividing a company's annual dividend by its price. All data is as of June 6, 1997.

Company	Price	Annual Dividend	Dividend Yield
Philip Morris	42.63	1.60	3.75%
AT&T	35.75	1.32	3.69%
General Motors	56.63	2.00	3.53%
Chevron	72.38	2.32	3.21%
J.P. Morgan	111.50	3.52	3.16%
Exxon	61.25	1.64	2.68%
3M	92.50	2.12	2.29%
DuPont	110.75	2.52	2.28%
Kodak	85.00	1.76	2.07%
Int'l Paper	49.25	1.00	2.03%
Goodyear	59.63	1.12	1.88%
Sears	50.00	0.92	1.84%
Merck	92.25	1.68	1.82%
General Electric	62.25	1.04	1.67%
Union Carbide	45.88	0.75	1.63%

Table 3–4

EXAMPLE 3

Refer to Table 3–4 and define the following three events:

X: companies whose price is \$50.00 or less
Y: companies whose dividend is less than 1.50
Z: companies whose dividend yield is less than 2%

(a) Which companies are in all three sets?
(b) Which companies are in X and Y but not in Z?

Solution. Refer to Figure 3–7, which partitions the companies into the three sets. From Table 3–4, we use the first letter of each company's name to denote that company, except that we use B to denote General Motors, F to denote General Electric, and T to denote 3M. Then

X = {P, A, I, S, U}

Y = {A, I, G, S, F, U}

Z = {G, S, M, F, U}

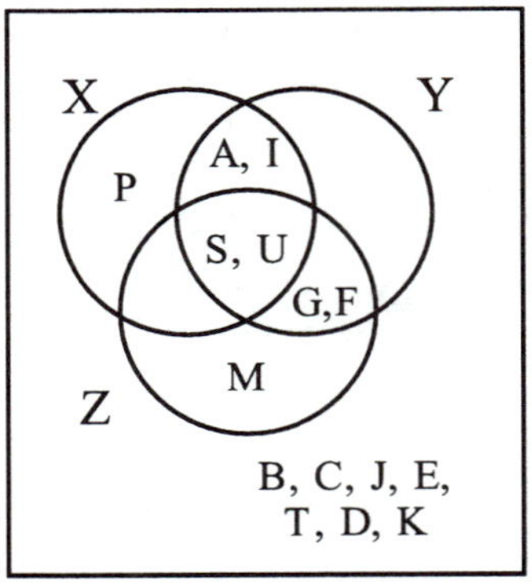

Figure 3–7

(a) The set X ∩ Y ∩ Z represents those companies in all three sets. The only companies in all three sets (the innermost region of the Venn diagram, are S and U. Hence X ∩ Y ∩ Z = {S, U}.

(b) The companies in X and Y are those in the intersection X ∩ Y = {A, I, S, U}. The companies from this intersection that are not in Z are A and I. Therefore, the set {A, I} is comprised of those companies in X and Y that are not in Z.

Example 4 is typical of problems involving the partition of a set by three of its subsets. It builds on the concepts introduced in Example 3.

EXAMPLE 4

A survey asked 200 people which sports they played regularly. It revealed that 32 played tennis, 23 played golf, 40 played softball, 10 played both tennis and golf, 20 played both tennis and softball, 15 played both golf and softball, and 8 played all three sports. Construct a Venn diagram for the three sets.

Solution. Let T = tennis, G = golf, and S = softball. Then, three circles are needed to represent these sets, and these three circles create a total of seven different regions. [See Figure 3–8(a).]

Always start by entering the number of objects that are in all three sets. In this example, 8 people play all three sports, and this number is entered in the innermost region of the Venn diagram, as in Figure 3–8(b).

Next, move to the numbers that refer to the overlap of two sets, for example, the 15 people that played both golf and softball. To use this information to continue filling in the Venn diagram, notice that there are two regions that are in both set G and set S. Therefore, the total of these two regions must be 15. Since 8 people have already been accounted for in the innermost region, 7 people are left over, and this number is then entered in the diagram as in Figure 3–8(c).

Figure 3–8(d) then completes all of the overlaps between two sets

Now, consider the information that concerns only one set at a time to fill in the remaining three regions. For example, 32 people played tennis. Since the three regions in the tennis circle that are filled in total 22, 10 more people are needed to complete the circle. Similarly, 6 more are needed for a total of 23 in the golf circle, and 13 more are needed for a total of 40 in the softball circle. [See Figure 3–8(e).]

Finally, since 200 people were surveyed, a box indicating the universal set is placed around the three circles, and the number 200 is written outside of the box. Adding up the numbers inside the circles gives 58, therefore, there are 142 people left over, and this number is placed inside the box, but outside of the circles. [See Figure 3–8(f).]

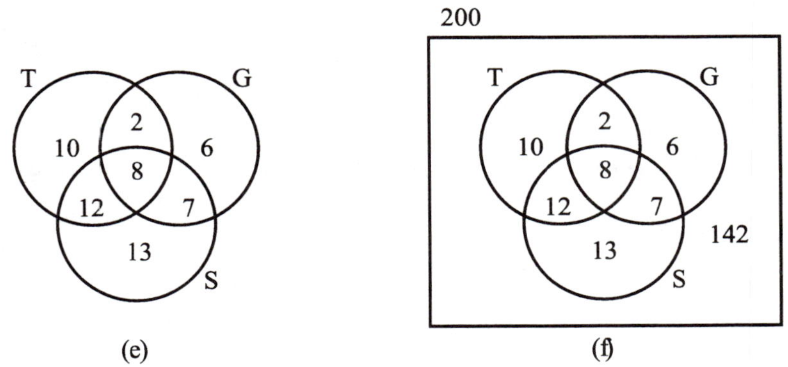

Figure 3–8

EXAMPLE 5

Use the information and the Venn diagram from Example 4 to answer each of the following questions.
(a) How many of those surveyed played both golf and softball but did not play tennis?
(b) How many of those surveyed played only golf?
(c) How many of those surveyed played none of these sports?

Solution.
(a) People who fall into this category must lie in the overlap between the golf and softball circles, but must be outside of the tennis circle. Therefore, 7 people played both golf and softball but did not play tennis.
(b) To answer this question, look at the region that is inside the golf circle, but outside of both the tennis and the softball circles. Therefore, 6 people played only golf.
(c) To find out how many people played none of these sports, refer to the number that is inside the box but outside of all three circles. Therefore, 142 people played none of these sports.

Example 5(c) is indicative of ways in which people can be misled by numbers. Altogether 200 people were surveyed. Of them, 30 played tennis, 25 played golf, and 40 played softball. The sum of these three numbers is 95, so many people conclude that 105 did not play any of these sports. Because of overlapping among the three sets, we are dealing with fewer than 95 individuals, so many more than 105 played none of these sports.

EXERCISE SET 3-2

In problems 1 to 6 let U = {a, b, c, ... , z}, V = {a, e, i, o, u}, and W = {v, o, w, e, l}. Find the number of elements in each of the following sets.

1. W' 2. V' 3. $V \cup W$
4. $V \cap W$ 5. $V' \cap W$ 6. $V \cup W'$

In problems 7 and 8 find $n(A \cup B)$.

7. $n(A) = 7$, $n(B) = 24$, and $n(A \cap B) = 2$
8. $n(A) = 6$, $n(B) = 12$, and $n(A \cap B) = 3$

In problems 9 and 10 find $n(A \cap B)$.

9. $n(A) = 12$, $n(B) = 17$, and $n(A \cup B) = 20$.
10. $n(A) = 6$, $n(B) = 12$, and $n(A \cup B) = 14$.

11. A lecture section of psychology consists of 145 students, of whom 80 are women. Fifteen of the women and 21 of the men are part-time students. How many of the men are not part-time students?

12. A third-grade class has 31 students, of whom 12 are girls. Three of the girls and 10 of the boys eat lunch at school while the others go home for lunch. How many of the boys go home for lunch?

13. The combined membership of the MAA and the AMS is approximately 45,000, of whom 27,000 belong to the MAA and 23,000 belong to the AMS. How many people belong to both organizations? (The MAA is the Mathematical Association of America, and the AMS is the American Mathematical Society.)

14. A credit union has 2,105 customers who have either a savings account or a checking account (or both a savings and a checking account). If 1,215 of the customers have a savings account and 1,760 have a checking account, how many of the customers have both a savings and a checking account?

15. A cable company has 7,350 subscribers. Of these subscribers, 2,560 subscribe to HBO, 950 subscribe to Cinemax, and 4,420 subscribe to neither HBO nor Cinemax. Draw a Venn diagram and determine the number of people who subscribe to both HBO and Cinemax.

16. A survey of 150 people found that 80 of them watch the Super Bowl, 45 watch the World Series, and 40 watch neither the Super Bowl nor the World Series. Draw a Venn diagram and determine the number of people surveyed who watch both the Super Bowl and the World Series?

In problems 17 - 20 draw a Venn diagram and use it to answer the questions.

17. A survey of 350 college freshmen at a particular university found that 194 of them held a part-time job, 47 participated in sports at the university, and 114 belonged to a fraternity or sorority. Of these students, 12 held a part-time job and participated in sports, 32 participated in sports and belonged to a fraternity or sorority, 53 held a part-time job and belonged to a fraternity or sorority, and 7 held a part-time job, participated in sports, and belonged to a fraternity or sorority.

    (a) How many held a part-time job and participated in sports, but did not belong to a fraternity or sorority?
    (b) How many held a part-time job but did not participate in either of the other two activities?
    (c) How many belonged to a fraternity or sorority, but did not participate in sports?
    (d) How many participated in none of the three activities?

18. A survey of 200 people revealed that 173 own a VCR, 42 own a computer, and 23 own a big-screen TV. Of these people, 30 own a VCR and a computer, 18 own a VCR and a big-screen TV, 15 own a computer and a big-screen TV, and 10 own all three.

    (a) How many own a VCR and a big-screen TV, but not a computer?
    (b) How many own only a VCR and neither of the other two items?
    (c) How many people own a big-screen TV but not a VCR?
    (d) How many people own none of these three items?

19. In a particular city, 2300 football season ticket holders were surveyed. 400 of them had hockey season tickets, 450 had baseball season tickets, and 200 had basketball season tickets. 250 of these people had both hockey and baseball season tickets, 75

had both hockey and basketball season tickets, 100 had both baseball and basketball season tickets, and 25 had season tickets for all three sports.

(a) How many had baseball season tickets, but did not have hockey or basketball season tickets?
(b) How many had baseball and basketball season tickets, but did not have hockey season tickets?
(c) How many people had hockey season tickets but did not have baseball season tickets?
(d) How many people had season tickets to none of these three sports?

20. 700 employees participate in the retirement plan at a particular company. Of these people, 360 invest in a stock fund, 330 invest in a bond fund, 220 invest in a money market fund, 180 invest in both stock and bond funds, 50 invest in both stock and money market funds, 90 invest in both bond and money market funds, and 30 invest in all three types of funds.

(a) How many employees invest in stock and bond funds but not in a money market fund?
(b) How many employees invest in a money market fund but in neither of the other two types of funds?
(c) How many employees invest in stock funds but not in bond funds?
(d) How many employees invest in none of these three funds?

In problems 21 and 22 create a Venn diagram to find $n(A \cup B \cup C)$.

21. $n(A) = 110$, $n(B) = 70$, $n(C) = 30$, $n(A \cap B) = 35$, $n(B \cap C) = 15$, $n(A \cap C) = 20$, and $n(A \cap B \cap C) = 5$.

22. $n(A) = 60$, $n(B) = 50$, $n(C) = 70$, $n(A \cap B) = 19$, $n(B \cap C) = 20$, $n(A \cap C) = 13$, and $n(A \cap B \cap C) = 11$.

In problems 23 and 24, assume that A and B are disjoint sets. Find $n(A)$ and $n(B)$.

23. $n(A \cup B) = 42$, and A contains six times as many elements as B.

24. $n(A \cup B) = 140$, and A contains 50 fewer elements than B.

In problems 25 and 26, assume that A and B are disjoint sets. Draw an appropriate Venn diagram and find $n(A \cup B \cup C)$.

25. $n(A) = 25$, $n(B) = 12$, $n(C) = 17$, $n(A \cap C) = 6$, and $n(B \cap C) = 4$.

26. $n(A) = 8$, $n(B) = 9$, $n(C) = 12$, $n(A \cap C) = 7$, and $n(B \cap C) = 1$.

In problems 27 and 28 draw a Venn diagram of the universal set U and subsets A and B. Use the given information to fill in the number of elements for each region.

27. $n(U) = 30$, $n(A) = 15$, $n(B') = 19$, and $n(A' \cap B) = 7$

28. $n(U) = 20$, $n(A) = 4$, $n(B') = 8$, and $n(A \cup B) = 13$

In problems 29 and 30 find $n(B)$.

29. $n(A \cap B) = 24$, $n(A \cap B') = 17$, $n(A \cup B) = 78$.

30. $n(A \cap B') = 13$, $n(A' \cap B) = 22$, $n(A \cup B) = 40$.

31. Is it possible for $n(A \cap B) = n(A \cup B)$?

32. Is it possible for $n(A \cap B) > n(A \cup B)$?

33. A business section article on the stock market revealed that of the 100 stocks studied one week, 55 stocks advanced, 24 stocks had a net change of more than 10%, 18 stocks split within the last 12 months, 18 of the stocks that advanced during the week also had a net change of more than 10%, 10 of the stocks that advanced during the week also split within the last 12 months, 8 stocks that had a net change of more than 10% also split within the last 12 months, 3 stocks that advanced had a net change of more than 10% and also split within the last 12 months.

    (a) How many stocks advanced but neither split nor had a net change of more than 10%?
    (b) How many stocks had a net change of more than 10% but neither advanced nor split?

34. Solve the preceding problem if none of the stocks that advanced had a net change of more than 10% and also split within the last 12 months.

35. An article on the banking industry revealed that of the 200 banks surveyed one month, 62 banks lowered their prime rate, 54 banks lowered their mortgage rate, 38 banks lowered their CD rate, 30 of the banks that lowered their prime rate also lowered their mortgage rate, 24 of the banks that lowered their prime rate also lowered their CD rate, 20 of the banks that lowered their mortgage rate also lowered their CD rate, and 18 banks lowered all three rates. How many of the banks lowered none of the rates?

36. Solve the preceding problem if the number of banks that lowered all three rates was 10 instead of 18.

37. In this section, we looked at the Inclusion/Exclusion Principle for only two sets. What is the analogous formula when three sets, A, B and C, are present? Use your answers to Problems 21 and 22 above to check your formula.

38. What does the Inclusion/Exclusion Principle look like when four sets are present? What is the general method for creating the Inclusion/Exclusion Principle formula?

39. If X is a set with $n(X) = 3$, how many subsets does X have?

40. If X is a set with $n(X) = k$, how many subsets does X have?

41. Is there a difference between $n(\emptyset)$ and $n(\{\emptyset\})$?

42. Is there a difference between $\emptyset$ and $\{\emptyset\}$?

Referenced Exercise Set 3-2

Problems 1 to 4 refer to a study in sociology on the economic mobility of the poor.[5] The accompanying table presents the status of the one-year poor, temporarily poor (1-2 years), and persistently poor (7-8 years) from that study.

	One-Year Poor	Temporarily Poor	Persistently Poor	Total
Females				
White	494	112	327	933
Black	562	307	122	991
Males				
White	472	40	1,102	1,614
Black	247	198	225	670
Total	1,775	657	1,776	4,208

Find $n(X)$, where X is the set of:

1. White people who are persistently poor.
2. Black people who are persistently poor.
3. Men who are one-year poor.
4. Women who are one-year poor.

In 1985 New Zealand's Prime Minister, David Lange, proposed that his country withdraw from the Anzus pact instead of permitting U.S. nuclear submarines to land in its ports. A national opinion poll[6] revealed that 70 percent favored remaining in the alliance while 60 percent approved Mr. Lange's course. Problems 5 and 6 refer to the poll.

5. What percent agreed with both positions?
6. What percent favored Mr. Lange's proposal and did not want to remain in the alliance?

7. In one National Hockey League season there were 234 games in which at least one team scored exactly two goals.[7] The non-losing team scored two goals in 38 games and the non-winning team scored two goals in 217 games. How many games ended in a tie?
8. Draw a Venn diagram with six sets.[8] How many regions are there?

---

[5]Greg J. Duncan, et al., *Years of Poverty, Years of Plenty*, Ann Arbor: Survey Research Center, Institute for Social Research, 1984, p. 49.

[6]Source: *New York Times*, October 6, 1985, p. 24.

[7]Benny Ercolani and Jeff Boyle, *NHL Data, 1984-85 Season*, Montreal: National Hockey League Communications Department, 1985.

[8]The Venn diagram is drawn in J. Chris Fisher, E. Koh, and Branko Grunbaum, "Diagrams Venn and How," *Mathematics Magazine*, February 1988, pp. 36-40.

9. The language problem that has dogged Canada since its founding reached a peak in 1990. The figures stated here are taken from an analysis of that situation by well-known political commentator John Burns.[9]

   (a) In Quebec, 6,035,316 people speak French, 2,595,821 speak English, and 2,226,756 speak both languages. How many people in Quebec speak either French or English?
   (b) In the rest of Canada, 18,187,255 people speak English, 1,978,585 speak French, and 1,829,415 speak both languages. How many Canadians outside Quebec speak either French or English?
   (c) How many Canadians altogether speak either French or English?

Cumulative Exercise Set 3-2

1. List the elements in the set of integers between 0 and 8 inclusive.
2. Determine whether A is a subset of B.

   (a) A = {1, 2, 3, 4, 5}, B = {2, 3, 4}
   (b) A = {a, b, c}, B = {1, 2, a, b, c, d}
3. List the elements in $A \cup B$ and $A \cap B'$.

   (a) U = {0, 1, 2, 3, 4, 5, 6, 7, 8}, A = {1, 2, 3, 4}, B = {2, 3, 4, 5, 6}
   (b) U = {a, b, c, d, e, f, g, h}, A = {a, b, c}, B = {b, c, d, g}
4. List the elements in $A \cap B'$, where A is the set of divisors of 12, B is the set of integers from 2 to 5, inclusive, and U = {0, 1, 2, 3, ... , 14}.
5. Find $n(A \cup B)$ when $n(A) = 20$, $n(B) = 15$, and $n(A \cap B) = 3$.
6. A club has 40 members, of whom 21 are female and 18 are business majors. If 5 of the males are business majors, how many females are not business majors?
7. A survey of 80 restaurants revealed that 36 served wine, 30 served pizza, 32 served hamburgers, 13 served wine and pizza, 19 served pizza and hamburgers, 8 served wine and hamburgers, and 6 served all three. How many served none of the foods?
8. Find $n(A \cup B)$ when $n(U) = 34$, $n(A') = 21$, $n(B) = 20$, and $n(A \cap B) = 3$.
9. An article in *Investment Dealer's Digest* on the trading of recent stock issues revealed that of the 30 stocks studied, 7 were initial price offerings (IPOs), 19 had a high price offer (HPOs) or more than \$10, 11 had a high percent change (HPCs) of greater than 10% from the previous month, 2 IPOs were also HPOs, 10 HPCs were also HPOs, and not one of the IPOs was an HPC. How many of the HPOs that were not IPOs did not have an HPC?
10. An article in *The New York Times* of December 30, 1990, on the performance of mutual funds listed 40 funds and categorized each fund according to whether it was a no-load (NL) fund, meaning it charged no sales tax, a growth (G) fund, or a high yield (HY) fund, meaning its return over the last five years was greater than 90%. There were 17 NL funds, 30 G funds, and 9 HY funds. There were 13 funds that

---

[9]John Burns, "Canada's crucial seven day countdown," *New York Times*, June 17, 1990, Sec. 4 (Week in Review), p. 3.

were NL and G, 6 funds that were G and HY, 2 funds that were NL and HY, and there were no funds with all three properties. How many funds had none of these properties?

Multiple-Choice Exercise Set 3-2

1. A classroom has five rows of desks with six desks in each row. If three desks are unoccupied, how many are occupied?
(a) 33 (b) 22 (c) 27 (d) 23 (e) none of these

2. If A denotes the sides of a coin then $n(A) =$
(a) 0 (b) 1 (c) 2 (d) 3 (e) none of these

3. If A denotes the faces on a die then $n(A) =$
(a) 3 (b) 4 (c) 5 (d) 6 (e) none of these

4. If A denotes the cards in an ordinary deck then $n(A) =$
(a) 13 (b) 4 (c) 12 (d) 52 (e) none of these

5. If A = {a, e, i, o, u} and B = {b, e, a, v, i, s} then $n(A \cap B) =$
(a) 3 (b) 8 (c) 9 (d) 11 (e) none of these

6. If A = {a, e, i, o, u} and B = {b, e, a, v, i, s} then $n(A \cup B) =$
(a) 3 (b) 8 (c) 9 (d) 11 (e) none of these

7. If the universal set is the months in a year and B those months that end in "ber" then $n(B') =$
(a) 8 (b) 4 (c) 3 (d) 9 (e) none of these

8. If D denotes the days in a week and S denotes those days spelled with exactly six letters then $n(D \cap S) =$
(a) 5 (b) 3 (c) 2 (d) 6 (e) none of these

In problems 9 to 12, A and B are the names of clubs in which 12 people belong to A, 20 belong to B, and 7 belong to both A and B.

9. $n(A \cap B) =$
(a) 5 (b) 12 (c) 20 (d) 27 (e) none of these

10. $n(A \cup B) =$
(a) 25 (b) 12 (c) 39 (d) 27 (e) indeterminable

11. The number of people who belong to either A or B is
(a) 25 (b) 12 (c) 39 (d) 27 (e) none of these

12. The number of people who belong to B but not A is
(a) 15 (b) 20 (c) 8 (d) 13 (e) indeterminable

In problems 13 and 14 let H denote heads and T denote tails.

13. A coin is flipped and it lands with $n(H) = 4$ and $n(T) = 2$. How many times was the coin flipped?
(a) 6 (b) 4 (c) 2 (d) 1 (e) indeterminable

14. Two coins are flipped and they land with $n(2H) = 4$ and $n(2T) = 2$. How many times were the coins flipped?
(a) 6 (b) 4 (c) 2 (d) 1 (e) indeterminable

Problems 15 and 16 refer to a card being drawn from an ordinary deck of 52 cards. Let C denote clubs and F denote cards that are fives

15. $n(C \cap F) =$
(a) 1 (b) 4 (c) 13 (d) 16 (e) none of these

16. $n(C \cup F) =$
(a) 1 (b) 4 (c) 13 (d) 16 (e) none of these

In problems 17 to 20 two dice are tossed. Let A denote those pairs of dice that add up to 6, and D denote those pairs of dice that land doubles.

17. $n(A) =$
(a) 1 (b) 5 (c) 6 (d) 10 (e) none of these

18. $n(D) =$
(a) 1 (b) 5 (c) 6 (d) 10 (e) none of these

19. $n(A \cap D) =$
(a) 1 (b) 5 (c) 6 (d) 10 (e) none of these

20. $n(A \cup D) =$
(a) 1 (b) 5 (c) 6 (d) 10 (e) none of these

Problems 21 to 24 refer to three sets A, B, and C with $n(A) = 12$, $n(B) = 13$, $n(C) = 14$, $n(A \cap B) = 4$, $n(A \cap C) = 3$, $n(B \cap C) = 5$, and $n(A \cap B \cap C) = 1$.

21. $n(A \cup B) =$
(a) 38 (b) 25 (c) 21 (d) 4 (e) none of these

22. $n(A \cap B') =$
(a) 8 (b) 15 (c) 21 (d) 4 (e) indeterminable

23. $n(A \cap B' \cap C) =$
(a) 38 (b) 25 (c) 3 (d) 2 (e) indeterminable

24. $n(A \cup B \cup C) =$
(a) 28 (b) 25 (c) 21 (d) 52 (e) none of these

3-3 TREE DIAGRAMS & THE MULTIPLICATION PRINCIPLE

Hundreds of thousands of people plunk down a dollar or so daily in hopes of "hitting the big one" in the state lottery. As their ticket chunks out of the computer, visions of expensive cars and wall-to-wall sound systems pop up. The odds against winning are monumental, but on they play. However, if you *knew* what the number was going to be by some foolproof method, would you only bet $1? Of course not! You would purchase tickets in a flurry. On the other hand, knowing the number is highly unlikely because officials have taken great precautions to prevent even the slightest chance of cheating.

Why then did a white Cadillac pull up in front of the Dew Drop Inn in South Philadelphia and its driver purchase 904 lottery tickets? That was just the beginning. By the end of the day the driver purchased over 10,000 tickets, all on numbers that consisted of 4's and 6's. Throughout the state, accomplices purchased another 10,000 tickets.

The Pennsylvania daily lottery is run by selecting a Ping-Pong ball from each of three urns with ten balls numbered 0 to 9. In 1980 Violet Lowery, a senior citizen, drew the number 666, making many of those 20,000 tickets big winners. The gang held tickets worth over $1.75 million.

The next day many of the sellers of the tickets, most of whom are local shop owners who also pay out the winnings for a fee from the state, complained that their payoffs were unusually high. The number 666 was very suspicious. The Bible calls it "the number of the beast." But how suspicious was it?

Here are three questions raised by the possible rigging of the "daily number." What are the chances of the number 666 being drawn? Is 666 less likely, the same, or more likely to be drawn than any other three-digit number? What facts would the state need in order to launch a full-scale investigation to determine if the fix was on? All three questions will be answered in the epilogue to this section.

Tree Diagrams

A **tree diagram** is a convenient way to visualize and to enumerate the number of outcomes in an event that occurs in two or more stages. It helps analyze a complex problem by breaking down the problem into smaller, more manageable, parts. A tree diagram begins with the possible outcomes of the first stage joined by lines to one common point. These lines form the branches of the tree. Each of these outcomes is joined to the outcomes of the second stage by another set of branches. The tree, which is lying on its side, grows from left to right with each successive stage.

For instance, consider the dinner menu below consisting of options: salad, entree, and dessert. If the restaurant wants its customers to order their meals by number, how many numbers are necessary?

Salad	Entree	Dessert
House (h)	Turkey (t)	Pie (p)
Spinach (s)	Veal (v)	Cake (c)
	Fish (f)	Ice Cream (i)
	Burger (b)	

The tree diagram in Figure 3–9 displays all the possibilities. At the first stage there are two possibilities, the house salad and a spinach salad. This stage has two branches emanating from the starting point. Each of the salads can be ordered with one of the four entrees, resulting in eight selections for these two courses. There are three desserts that can be ordered for each of these eight selections, so altogether 24 complete meals can be ordered. They are numbered in Figure 3–9. Note that each meal is defined by three branches so that, for instance, meal 18 consists of a spinach salad, veal, and ice cream.

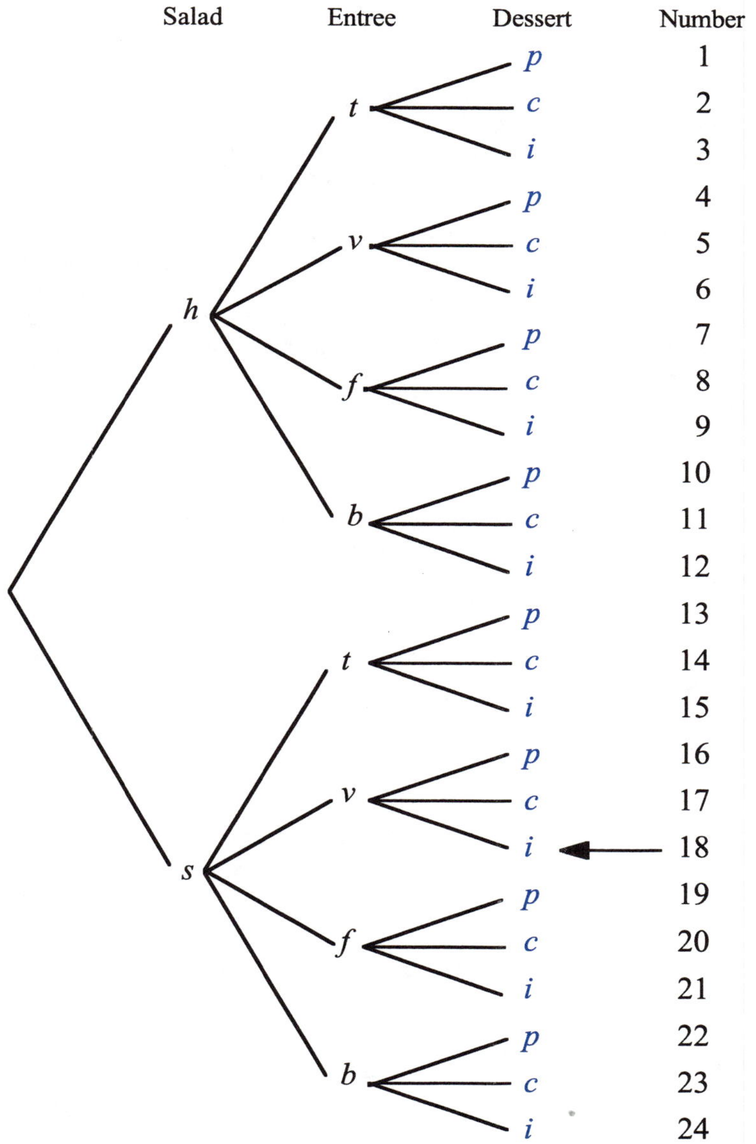

Figure 3–9

A similar counting problem revolves around creating macros on a computer. For example, Macintosh computer keyboards have four keys – the *shift* key , the *control* key, the *option* key, and the *command* key – that can change the functionality of any other key on the keyboard by simultaneously pressing one of these four keys and the other key. For example, pressing *shift* and the letter "o" types a capital "O" while pressing *command* and "o" opens a new file. In addition, multi-step tasks that are performed often can be done quickly and easily by assigning a combination of keys. For example, a person who owns a computer will have many occasions to type her name, and therefore, she might assign the combination of *control* and "o" as a shortcut for typing her name – Ophelia. Example 1 explores the number of ways in which these keys can be used.

EXAMPLE 1

Create a tree diagram to list all of the different combinations of keys that are possible if one of the four keys (*shift*, *control*, *option*, or *command*) is pressed at the same time as one of the letters *a*, *b*, or *c*. How many such key-combinations are possible?

Solution The tree diagram is shown in Figure 3–10. The first stage has four branches, one for each of the function keys, and each of these branches has three branches, one for each of the three letters. Therefore, the total number of key combinations that are possible is 12, which is found by counting the number of branches in the last stage. (Another way to compute this total is to use the fact that four groups of three objects gives a total of $4 \cdot 3 = 12$ objects.)

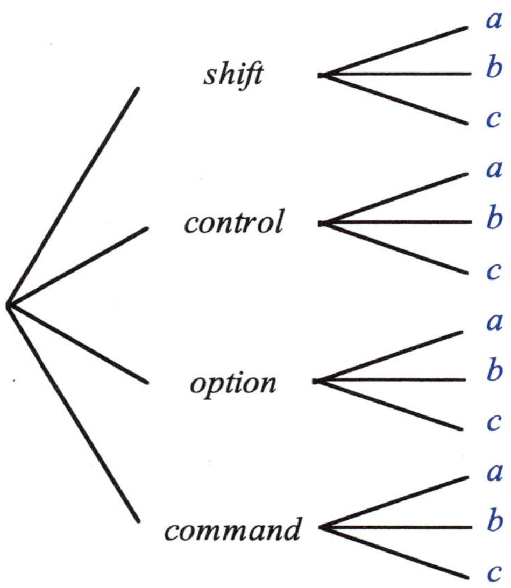

Figure 3–10

Trimmed Trees

The introduction to this section discussed the rigging of a state lottery. Here we begin to analyze the situation in mathematical terms.

Suppose a daily number is formed by drawing three balls from an urn containing ten balls numbered 0 to 9. The second ball is drawn from the urn without replacing the first ball, and the third ball is drawn from the urn without replacing the two that have already been drawn. This method of selection is called **selection without replacement** because the balls are not replaced after they are drawn. A tree diagram could present all the possibilities but it would be very unwieldy. We modify it by drawing only one line for each stage and by writing above that line the number of branches that would have connected that stage with the previous. The trimmed tree is:

$$\frac{10}{1^{\text{st}}\text{ draw}} \quad \frac{9}{2^{\text{nd}}\text{ draw}} \quad \frac{8}{3^{\text{rd}}\text{ draw}}$$

The number 10 represents the ten branches corresponding to the ten balls in the urn. Each of these branches would be connected to nine branches leading to the 2nd, and each of these would be connected to eight branches leading to the 3rd draw. Altogether there are $10 \cdot 9 \cdot 8 = 720$ ways of selecting the daily number without replacement.

Now suppose the daily number is formed differently by selecting three balls from the urn in such a way that the ball is selected, its number is recorded, and the ball is then returned to the urn for another selection. This method of selection is called **selection with replacement**. The trimmed tree can be written with only the numbers and not the branches:

$$\frac{10}{1^{\text{st}}\text{ draw}} \quad \frac{10}{2^{\text{nd}}\text{ draw}} \quad \frac{10}{3^{\text{rd}}\text{ draw}}$$

Consequently there are $10 \cdot 10 \cdot 10$ or 1000 such daily numbers altogether.

There is another, more direct, way to count the 1000 daily numbers. Since they are numbered from 000 to 999, there are 1000 such numbers altogether. We proceeded by trimmed trees, however, in order to provide background for a very important counting principle.

The Multiplication Principle

The **Multiplication Principle** provides a method for counting the number of ways that a particular task can be done when the task can be divided into distinct *stages*. In the restaurant example, there were three stages – picking a salad, then an entree and finally a dessert. In the computer problem, there were two stages – pick a function key, and then pick one of the letters *a*, *b*, or *c*. to picking Then in the daily number problem, there were three stages, one for each of the balls chosen. In order to determine the total number of ways to complete the task, the number of ways to complete each stage was first found, and then these numbers were multiplied together to give the final answer.

These examples then lead us to a statement of the Multiplication Principle.

Multiplication Principle

If a counting problem can be divided into two stages such that at stage 1 there are n_1 ways to make a selection and at stage 2 there are n_2 ways to make a selection, then the total number of ways to make the two selections is

$$n_1 \cdot n_2$$

Similarly, if the problem requires a third stage in which there are n_3 ways to make the third selection, then the total number of ways to make the three selections is

$$n_1 \cdot n_2 \cdot n_3$$

In general, if there are m stages, and if at the ith stage there are n_i ways to make a selection, then the total number of ways to make the m selections is

$$n_1 \cdot n_2 \cdot n_3 \cdot \ \ n_m$$

Example 2 applies the Multiplication Principle to license plates in standard format (three letters followed by three digits without regard to restrictions.)

EXAMPLE 2

(a) How many standard-format license plates are possible?
(b) How many standard-format license plates are possible if no letters and no digits can be repeated?

Solution. A license plate in standard format can be viewed as a tree with six stages: three letters (L) and three digits (D).

(a) There are 26 possible letters for each of the first three stages and ten possible digits for each of the last three stages. The trimmed tree is

26	26	26	10	10	10
L	L	L	D	D	D

Therefore, there are $26 \cdot 26 \cdot 26 \cdot 10 \cdot 10 \cdot 10 = 17{,}576{,}000$ license plates in standard format.

(b) If no letters can be repeated, then there are 26 possible letters for the first stage, 25 possible for the second stage, and 24 possible for the third stage. Similarly, there are there are 10 possible digits for the fourth stage, 9 possible digits for the fifth stage, and 8 possible digits for the sixth stage.

26	25	24	10	9	8
L	L	L	D	D	D

Therefore, there are $26 \cdot 25 \cdot 24 \cdot 10 \cdot 9 \cdot 8 = 11{,}232{,}000$ license plates in standard format if no repeated letters or digits are allowed.

In 1982 California had more than 18 million licensed vehicles. Example 2 shows that they could not all have license plates in standard format. So California added the reverse format (three digits followed by three letters.) By the Multiplication Principle the reverse format also covers 17,576,000 license plates, allowing over 35 million license plates in these two formats. Why, then, did California adopt yet a third format? The Case Study for this chapter answers the question.

The results of Example 3 will be needed repeatedly in the next two chapters.

EXAMPLE 3

(a) If a coin is flipped two times, in how many ways can it land?
(b) If a coin is flipped three times, in how many ways can it land?
(c) If a coin is flipped ten times, in how many ways can it land?

Solution.

(a) There are two choices for the first flip of the coin, heads (H) or tails (T). There are also two choices for the second flip. By the Multiplication Principle there are $2 \cdot 2 = 4$ ways that the coin can land. These four ways are listed in the following tree diagram:

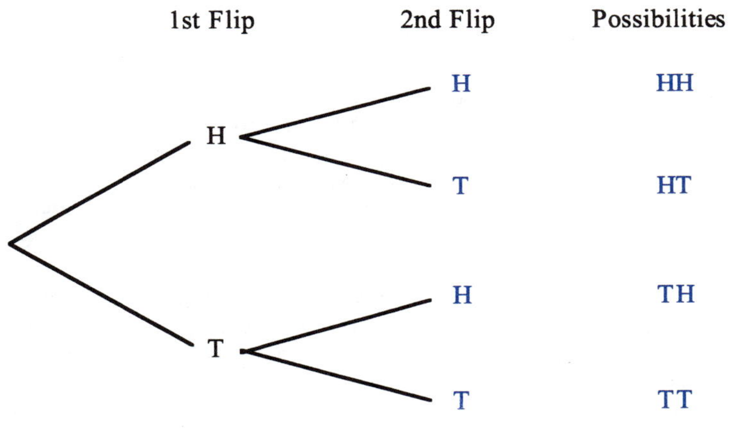

Figure 3–11

(b) By the Multiplication Principle there are $2 \cdot 2 \cdot 2 = 2^3$ ways the coin can land when it is flipped three times. Thus there are eight ways altogether.

(c) By the Multiplication Principle there are $2^{10} = 1024$ ways the coin can land when it is flipped ten times.

Notice in Example 3 that in each case, the exponent of 2 in the number of ways that the coin can land is equal to the number of times that the coin is flipped. Therefore if a coin is flipped n times it can land 2^n ways. We will use this observation repeatedly.

Sometimes the statement of a problem makes a situation seem very complex. The problem might state, "If five coins are flipped, in how many ways can they land?" The crucial ingredient in analyzing such a problem is the fact that:

"flipping n coins at once" is equivalent to "flipping one coin n different times"

Since flipping five coins is equivalent to flipping one coin five times, and a coin that is flipped five times can land in 2^5 ways, it follows that five coins can land in $2^5 = 32$ different ways when flipped. You will find that this kind of approach to a complex problem – breaking it down into a sequence of individual stages – can be very fruitful.

Epilogue: The Lottery Fix Revisited

We can now answer the questions raised in the introduction to this section. When 666 was drawn there should not have been any alarm that the lottery was fixed because 666 is simply one of the 1000 possibilities.

At first, state officials vehemently denied that foul play could have been associated with the drawing. There were large payoffs before, and there are certain numbers that people choose more often than others. The payoff for 666 was extremely large but it could be attributed to its unusual character.

Within a few days, however, several investigative reporters were on to the scent. They discovered that all numbers consisting entirely of 4's and 6's were bet heavily under very similar waging patterns. How many such numbers are there? By the Multiplication Principle there are three stages with two selections (4 or 6) at each stage. This is like flipping a coin three times, so there are $2 \cdot 2 \cdot 2 = 8$ ways of arranging the two numbers as a three-digit number.

The gamblers were confident that one of these eight numbers would be drawn. They placed heavy bets on all eight numbers. This betting provided the authorities with good evidence of a conspiracy but it was still not known how they pulled it off.

Pennsylvania lost \$1 million. Did state authorities recognize immediately that the fix was on? Are you kidding? Of course not! In fact, State Revenue Secretary Howard Cohen, whose department was responsible for running the lottery, stated emphatically for months that it was "impossible" to rig the lottery. Six months later, when the facts were uncovered, Cohen was forced to resign.

Can you guess how the lottery was rigged? In each urn, those balls numbered 4 and 6 were weighted to ensure that the tubes that pull one ball from each urn would pull a 4 or a 6. The non-weighted balls continued to whirl around, giving the appearance that the lottery was fair.

Two men were sentenced to prison terms of up to seven years for their roles in the lottery fix. Ironically, the day on which they went to prison Delaware's daily number came up 555. There was no unusually heavy betting on 555, however, so no investigation was begun.

EXERCISE SET 3-3

1. A delicatessen offers hoagies in 2 different sizes (short, long), in 3 different varieties (Italian, American, Tuna), and with or without hot peppers. Draw a tree diagram to list all of the possible hoagies. How many hoagies are possible?

2. A study classifies human beings according to gender (male, female), height (tall, medium, short), and weight (heavy, normal, light). classifications. How many classifications are there?

3. A particular model of car comes in one of three configurations (coupe, 2-door sedan, 4-door sedan), three colors (green, red, blue), and with two types of transmission (four-speed automatic, five-speed manual). Draw a tree diagram to list the possible choices. How many different ways are there to choose one of these cars?

4. A competitor in the triathlon must choose from one of 3 swim suits (Jantzen, Laguna, Catalina), 2 bicycles (Schwinn, Raleigh), and 4 running shoes (New Balance, Nike, Adidas, Puma). Draw a tree diagram to list the possible choices. How many choices are there?

5. You need to pick one more course to fill out your roster for the Spring Semester. To fulfill requirements, you can pick from a set of 4 different courses. Each of these courses has 3 different sections. Draw a tree diagram to list the possible choices. How many choices are there?

6. A coin is flipped and then a die is tossed. Draw a tree diagram to list the possible outcomes. How many outcomes are there?

7. A multiple-choice exam has 8 questions. Each question has 4 choices for the answer. How many different ways are there to write the answers to the questions?

8. A computer test bank for this chapter contains 12 problems on Section 3-1, 15 on Section 3-2, and 20 each on Sections 3-3 and 3-4. If a test consists of one problem from each section, how many different tests can be constructed?

9. A particular slot machine has three wheels. Each wheel of the slot machine contains a cherry, an orange, a plum, a lemon, and a blank space. How many different groups of three objects are possible?

10. A menu in a Chinese restaurant lists ten dinners in column A and seven dinners in column B. How many selections can be made with one dinner from column A and one dinner from column B?

11. A three-letter monogram consists of the three initials of a person's name. How many three-letter monograms are possible if no letter is to be repeated?

12. How many three-letter monograms are possible if repeated letters are allowed?

13. License plates in Hawaii consist of two letters followed by three digits. How many such license plates are possible?

14. License plates in Louisiana consist of three digits, followed by one letter, followed by three digits. How many such license plates are possible?

15. A combination lock has a dial with ten digits. Three numbers are chosen to form the combination that opens the lock. (For example, the combination 3-5-3 means to turn the dial to the right to 3, then to the left to 5, and then to the right to 3.) How

many sequences of three numbers (sometimes called combinations) can be formed if the digits are allowed to repeat?

16. If the dial of a lock has 30 numbers and repetition of the digits is allowed, how many sequences of 3 numbers can be formed if there are no restrictions on the numbers?

17. In many locks with combination consisting of a sequence of three digits, the first and second digits cannot be the same number, but the second and third can. Given this restriction, how many such sequences are possible if the dial has ten digits?

18. If the dial of a lock has 30 numbers, how many sequences of three numbers can be formed if the first and second numbers cannot be the same, but the second and third can?

19. In the first round of the NBA playoffs, two teams play each other until one of them wins three games (called a best-of-five series). Suppose that the Boston Celtics (C) play the Orlando Magic (M) in such a series. Construct a tree diagram and use it to list the possible series outcomes.

20. Five cards are placed into a box. Two of these cards are marked with an "X," and three of them are marked with an "O." You then pick one card at a time without replacement. You win if you pick the three O's before you pick the two X's. Draw a tree diagram and use it to list all of the possible winning plays.

21. If two dice are tossed, in how many ways can they land?

22. If three dice are tossed, in how many ways can they land?

23. If eight dice are tossed, in how many ways can they land?

24. If n dice are tossed, in how many ways can they land?

In problems 25 to 28 at each intersection the driver of the car can either make a left-hand turn, make a right-hand turn, or proceed straight ahead.

25. How many different routes are possible if two intersections are encountered?

26. How many different routes are possible if three intersections are encountered?

27. How many different routes are possible if ten intersections are encountered?

28. How many different routes are possible if n intersections are encountered?

29. License plates in several states consist of six alphanumeric characters. (An *alphanumeric character* can be one of the 26 letters or one of the 10 digits.) How many such license plates are possible?

30. How many seven-digit telephone numbers can be formed if the first digit can be neither 0 nor 1?

31. License plates in Washington, D.C., consist of six digits. How many such license plates are possible if the first digit cannot be zero and if all six digits cannot be the same?

32. How many seven-digit telephone numbers can be formed if the first three digits cannot be 800? (For example, the number 800-1234 is not allowed, but 812-1234, 806-1234, and 400-1234 are all allowed.)

33. In how many ways can the letters of the name LISA be arranged?

34. In how many ways can the letters of the name FRANK be arranged?

35. Make a list of the different ways in which the letters of the name PETE can be arranged. How many ways are there?

36. Make a list of the different ways in which the letters of the name BOBBY can be arranged. How many ways are there?

37. How many different arrangements of four characters can be formed if the first two characters are letters and the last two characters are digits?

38. How many different arrangements of four characters can be formed if the first two characters are different letters and the last two characters are different digits?

39. A bag contains numerous \$1 and \$5 bills. You reach in and grab one bill at a time. You stop whenever you grab at least \$6 or whenever three bills have been grabbed. Draw a tree diagram to list the possible grabs.

40. A bag contains numerous \$1 and \$5 bills. You reach in and grab one bill at a time until you have grabbed at least \$5. Draw a tree diagram to list the possible grabs.

41. There are 20 students in a class and each student must receive a grade of A, B, C, D, or F. How many different ways are there for the grades to be assigned?

42. A developer is building a development with three models of homes: Cape-Cod, split-level, and colonial. All models come with a one-car or two-car garage. The Cape-Cods come with one or two bedrooms, the split-level with two, three, or four bedrooms, and the colonial with three, four, or five bedrooms. If the builder wants to construct a sample house of each possible type, how many sample houses must be constructed?

43. Suppose that a ballot lists the names of four candidates for governor, three for lieutenant governor, and six for state treasurer. How many different ways can the offices be filled if one person is chosen for each position?

44. How many license plates consist of two letters followed by two digits if the first letter cannot be I or O and the second letter cannot be I, O, Q, or Z?

45. How many license plates consist of three letters followed by two digits if the third letter cannot be I, O, Q, or Z and the first digit cannot be 1 or 0?

46. A man has five shirts, four pairs of pants, and three pairs of shoes. How many different outfits can he wear if an outfit consists of a shirt, a pair of pants, and a pair of shoes?

47. How many arrangements of four letters begin with one consonant, end with a different consonant, and have two different vowels in between? (The vowels are a, e, i, o, u. All other letters are consonants.)

48. How many different sets of three initials have exactly two initials the same?

49. How many different sets of three initials have exactly two initials the same or all three initials the same?

Referenced Exercise Set 3-3

1. Over 1000 bombings take place in the United States each year.[10] Federal authorities trace the explosives used in the bombs by identifying microscopic chips of layered paints called "taggants." A typical identification taggant consists of ten different colored layers. How many different taggants can be made from ten different colors?

2. The IBM disk operating system for the PC-AT microcomputer[11] used *directories* to organize computer programs (called *files*) into groupings. Draw a tree diagram which has one root directory that leads to four subdirectories (WP, DATA, SPREAD, and UTILITIES), where WP has files for WordPerfect, Multimate, Microsoft Word, and Word Star, DATA has files for dBase, DataErase, and Oracle, SPREAD has files for Lotus, Excel, and Quattro Pro, and UTILITIES has files for Norton and PC Tools.

3. Maalox Plus is an antacid that is packaged with six tablets on a cardboard card.[12] Each tablet is white on one side and yellow on the other. Two possible arrangements on a card are

W	Y	W		Y	Y	W
			and			
W	Y	Y		W	Y	Y

   How many different arrangements are possible on a card?

4. Max Wertheimer, the founder of Gestalt psychology, proposed the following brainteaser to his friend, Albert Einstein.[13] Answer the question using the Multiplication Principle.

   An amoeba propagates by simple division; it takes three minutes for each split. I put such an amoeba into a glass container with a nutrient fluid; the rate of multiplication rises, of course; it takes one hour until the vessel is full of amoebas. Question: How long would it take to fill the vessel likewise, if I start not with one amoeba, but two?

---

[10] Richard D. Lyons, "Tagging Bombs, Trapping Bombers," *New York Times*, June, 24, 1979.

[11] IBM, *Reference Manual: Disk Operating System Version 3.20*, Boca Raton, FL, 1986.

[12] Bonnie Litwiller and David Duncan, "Maalox Lottery: A Novel Probability Problem," *The Mathematics Teacher*, September 1987, pp. 455-456.

[13] Edith H. and Abraham S. Luchins, "The Einstein-Wertheimer Correspondence on Geometric Proofs and Mathematical Puzzles," *The Mathematical Intelligencer* **12**(2), 1990, pp. 35-43. See the English translation of the postscript to the letter on page 40.

Cumulative Exercise Set 3-3

1. List the elements in the set $S = \{x \in N : x \text{ is even and } x < 10\}$.

2. Find X′ for each choice of U.
   (a) U is the set of days in a week and X consists of those days that begin with the letter T.
   (b) U = Z (the integers) and $X = \{x \in Z : x \leq 0\}$

3. Let the universal set U consist of the months of the year. Let A be the set of months having of summer and B be the set of months having exactly 30 days.
   (a) List the elements in U.
   (b) List the elements in A.
   (c) List the elements in B′.
   (d) Express the set B′ in words.
   (e) List the elements in $A \cap B$.
   (f) List the elements in $A \cup B$.

4. Find $n(A \cup B)$ where
   (a) $n(A) = 3$, $n(B) = 4$, and A and B are disjoint sets.
   (b) $n(A) = 3$, $n(B) = 4$, and $n(A \cap B) = 2$.
   (c) $n(A' \cap B) = 3$, $n(A \cap B') = 4$, and $n(A \cap B) = 2$.

5. If $n(A) = 8$, $n(B) = 4$, and $n(A \cap B) = 4$, how are the sets A and B related?

6. A club has 40 members, of whom 22 are married, 5 are married women, and 8 are unmarried men. How many women are unmarried?

7. A ballot has four candidates for president and 2 candidates for secretary. In how many different ways can the two offices be filled?

8. In how many ways can the grades of A, B, C, D, or F be assigned to a class of six students?

9. Of the 110 students in class, 60 are freshmen, 7 are veterans, 65 are enrolled in business, 2 are who are veterans, 5 are veterans who are enrolled in business, 32 are freshmen who are enrolled in business, and 2 are freshmen who are veterans and enrolled in business.
   (a) How many are freshmen and veterans but are not enrolled in business?
   (b) How many are freshmen but neither veterans nor enrolled in business?
   (c) How many are neither freshmen nor veterans nor enrolled in business?

10. Draw a tree diagram to list the possible outcomes when a die is tossed and then two coins are flipped.

11. Define a set of initials to consist of exactly three letters.
    (a) How many sets of initials are there?
    (b) Is the outcome JFK different from the outcome FJK?
    (c) How many sets of initials have all three letters different?

12. If a penny, nickel, dime, and quarter are flipped, in how many ways can they land so that
    (a) three of the coins land with heads facing up?
    (b) at least one coin lands with heads facing up?

13. A builder can build homes with two, three, or four bedrooms, with four possible exteriors, with three different floor plans, and with or without a basement. If the builder would like to display a sample house of each possible type, how many sample houses must be constructed?

Multiple-Choice Exercise Set 3-3

1. A tree diagram describes events that occur in two or more
(a) stages (b) outcomes (c) ways (d) permutations

2. The outcomes of an event are joined in a tree diagram by
(a) subsets (b) branches (c) arrows (d) combinations

3. A meal-day consists of one of four selections for breakfast, one of three selections for lunch, and one of 12 selections for dinner. How many different meal-days are possible?
(a) 19 (b) 144 (c) 96 (d) 24 (e) none of these

4. How many Daily Numbers can be formed by selecting one ball from three containers, each with ping-pong balls numbers 1 through 5?
(a) 10·10·10 (b) 10·9·8 (c) 5·5·5 (d) 5·4·3 (e) none of these

5. How many Daily Numbers can be formed by selecting one ball from two containers, each with ping-pong balls numbers 1 through 10?
(a) 10·10·10 (b) 10·10 (c) 9·9·9 (d) 9·9 (e) none of these

6. If an event occurs in five stages, and there are two possible selections at each stage, how many outcomes are possible?
(a) 10 (b) 4 (c) 25 (d) 7 (e) none of these

7. If a coin is flipped five times, in how many ways can it land?
(a) 10 (b) 4 (c) 25 (d) 7 (e) none of these

8. Morse code consists of sequences of characters, where each character is either a dot or a dash. How many codes have sequences of five characters?
(a) 25 (b) 10 (c) 32 (d) 7 (e) none of these

9. If a coin is flipped 99 times, in how many ways can it land?
(a) 299 (b) 198 (c) 101 (d) 298 (e) none of these

10. How many different answer sheets are possible for a six-question true- false test?
(a) 12 (b) 64 (c) 36 (d) 32 (e) none of these

11. How many different answer sheets are possible for a six-question multiple-choice test in which each question has five possible answers?
(a) 65 (b) 56 (c) 36 (d) 30 (e) none of these

12. How many different answer sheets are possible for a 20-question multiple-choice in which each question has five possible answers?
(a) 205 (b) 520 (c) 400 (d) 100 (e) none of these

13. If an event occurs in two stages, and there are six possible selections at each stage, is the total number of ways to make the two selections?
(a) 12 (b) 24 (c) 8 (d) 10 (e) none of these

14. A pair of dice is tossed. How many different sums are possible?
(a) 36 (b) 12 (c) 11 (d) 6 (e) none of these

15. How many codes consist of a letter followed by a digit?
(a) 36 (b) 52 (c) 234 (d) 260 (e) none of these

16. How many codes consist of a digit followed by a letter?
(a) 36 (b) 52 (c) 234 (d) 260 (e) none of these

17. How many codes consist of two letters followed by a digit?
(a) $26 \cdot 260$ (b) $26 \cdot 26 \cdot 10$ (c) $26 \cdot 25 \cdot 10$ (d) $26 \cdot 25 \cdot 9$ (e) none of these

18. How many signals consist of two digits followed by a letter?
(a) $26 \cdot 260$ (b) $26 \cdot 26 \cdot 10$ (c) $26 \cdot 25 \cdot 10$ (d) $26 \cdot 25 \cdot 9$ (e) none of these

19. Chris and Pat know they will pass Math 55 with a grade of A, B, or C. How many different pairs of grades are possible for the two students?
(a) 9 (b) 3 (c) 6 (d) 2 (e) none of these

20. If a nickel, dime, and quarter are flipped, in how many ways can they land so that two of the coins land with tails facing up?
(a) 1 (b) 2 (c) 3 (d) 4 (e) none of these

21. A nickel, dime, and quarter are flipped. In how many ways can they land so that one coin lands with heads facing up?
(a) 1 (b) 2 (c) 3 (d) 4 (e) none of these

22. How many three-digit numbers can be formed if each digit is either a 4 or a 6?
(a) 8 (b) 24 (c) 72 (d) 96 (e) none of these

3-4 PERMUTATIONS & COMBINATIONS

Many useful counting techniques involve permutations and combinations. This section explains the meanings of these terms and develops rules for computing them. Chapter 4 will apply these rules to probability problems.

Order

Suppose that a scholarship committee is going to award a total of $1200 to two students and that the committee has narrowed its list to four candidates – Alex, Brenda, Chris, and Darnel. Suppose also that the committee is undecided whether to grant two $600 awards or to give one award of $800 and another award of $400. Examples 1 and 2 address these two cases.

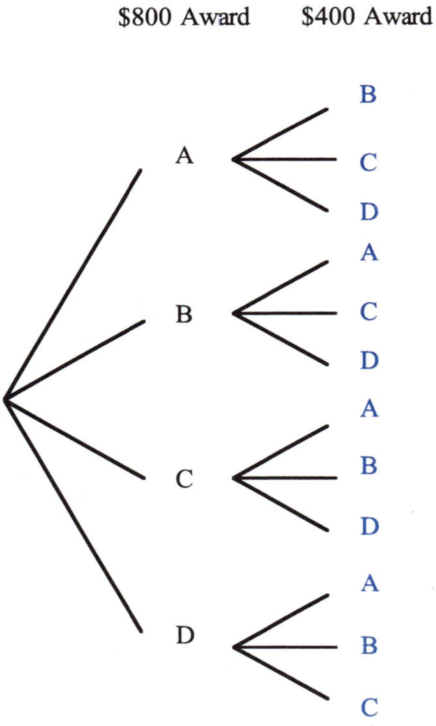

Figure 3–12

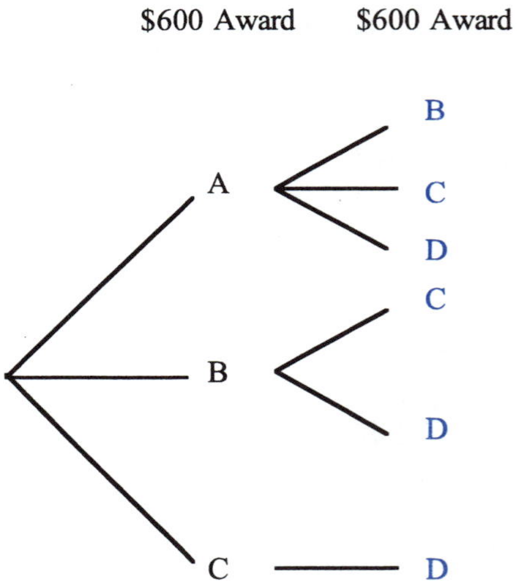

Figure 3–13

EXAMPLE 1

In how many ways can the scholarship committee select two of the four students so that one student is awarded \$800 and the other student \$400?

Solution. Use the Multiplication Principle with two stages, where each stage represents a scholarship. The first stage has four choices for the \$800 scholarship while the second stage has three choices for the \$400 scholarship. No student can receive both scholarships. Figure 3–12 shows the $4 \cdot 3 = 12$ ways that two of the students can receive the scholarships. Note the difference between Brenda getting \$800 and Chris getting \$400 versus Chris getting \$800 and Brenda getting \$400.

EXAMPLE 2

In how many ways can the scholarship committee select two of the four students so that each student wins \$600?

Solution. This problem cannot be solved by means of the Multiplication Principle alone. Certain pairs of students in Figure 3–12 reduce to one possibility here. Initially the pairs B-C and C-B seem to refer to separate awards, but here the order of the students is not relevant because each award is \$600, so the two pairs reduce to the one set {B, C}. Figure 3–12 displays 20 pairs, with each pair of students listed twice. However, when each pair is listed once, there are only ten possibilities, as shown by the tree diagram in Figure 3–13.

The concept of order distinguishes Example 1 from Example 2. In Example 1 it mattered which student came first because one award was \$800 and the other \$400. In Example 2 the order of the names was immaterial because both awards were for \$600. The following definition distinguishes these two concepts for objects that are selected without replacement.

DEFINITIONS

A list of objects in a specific order is called a **permutation**. A list of objects in which the order is not considered is called a **combination**.

The 20 lists in Example 1 are permutations, while the 10 lists in Example 2 are combinations.

Factorials

The expression $n!$ is read "n **factorial**." For a positive integer n it is defined by the formula

$$n! = n(n-1)(n-2)\cdots 3 \cdot 2 \cdot 1$$

Thus 5! = 120 because

$$5! = 5 \cdot 4 \cdot 3 \cdot 2 \cdot 1$$

The factorial symbol is also defined when $n = 0$, however, 0! does not fit the definition above. 0! is defined to be equal to 1.

Many calculators have a key for computing factorials, so that pressing 9 followed by the $n!$ key yields 362,880. Factorials play a central role in permutations and combinations, so it is important to be able to evaluate them in various forms. For instance, $(7 - 2)! = 5! = 120$. Example 3 shows how to evaluate two other forms.

EXAMPLE 3

Evaluate (a) $\dfrac{8!}{(8-5)!}$ (b) $\dfrac{8!}{(8-5)!\cdot 5!}$

Solution. Expand each factorial and cancel the common factors.

(a) $$\frac{8!}{(8-5)!} = \frac{8!}{3!} = \frac{8\cdot7\cdot6\cdot5\cdot4\cdot3\cdot2\cdot1}{3\cdot2\cdot1} = 8\cdot7\cdot6\cdot5\cdot4 = 6720$$

(b) $$\frac{8!}{(8-5)!\cdot5!} = \frac{8!}{3!\cdot5!} = \frac{8\cdot7\cdot6\cdot5\cdot4\cdot3\cdot2\cdot1}{3\cdot2\cdot1\cdot5\cdot4\cdot3\cdot2\cdot1} = \frac{8\cdot7\cdot6}{3\cdot2\cdot1} = 56$$

Formulas for Permutations and Combinations

The notation $P(n, r)$ denotes the number of permutations of n objects taken r at a time. Thus P(10, 4) stands for the number of permutations of ten objects taken four at a time. By the Multiplication Principle, $P(10, 4) = 10\cdot9\cdot8\cdot7$. This number can be written in terms of factorials.

$$P(10,4) = 10\cdot9\cdot8\cdot7 = \frac{10!}{6!} = \frac{10!}{(10-4)!}$$

The final form is a specific case of the general formula for evaluating permutations. The derivation of the general form is similar.

$$P(n,r) = n(n-1)(n-2)\cdots(n-r+1) = \frac{n!}{(n-r)!}$$

Formula for Permutations

The number of ways of permuting n distinct objects taken r at a time is

$$P(n,r) = \frac{n!}{(n-r)!}$$

EXAMPLE 4

How many permutations of three letters can be selected from the alphabet?

Solution. Since the alphabet has 26 letters, $n = 26$. Since three letters are to be chosen from the alphabet, $r = 3$. The number of permutations is therefore,

$$P(26,3) = \frac{26!}{(26-3)!} = \frac{26!}{23!} = 26 \cdot 25 \cdot 24 = 15{,}600$$

An important case of the formula for $P(n, r)$ occurs when $n = r$. Then

$$P(n,n) = \frac{n!}{(n-n)!} = \frac{n!}{0!} = \frac{n!}{1} = n!$$

Therefore, the number of permutations of n objects is $n!$. Example 5 uses this fact. It also provides the reasoning for the way in which the formula for combinations is derived from the formula for permutations.

EXAMPLE 5

Use Permutations to answer each of the following questions.

(a) How many governing boards can be formed from a law firm with eight partners, where a governing board consists of a president, vice-president, and secretary.

(b) How many three-member investigative committees can be formed from a law firm with eight partners?

Solution.

(a) The order in which partners are selected to a governing board is important because the three positions are distinguishable. Therefore the problem deals with permutations. There are eight partners so $n = 8$. There are three officers so $r = 3$. The number of governing boards is

$$P(8,3) = \frac{8!}{(8-3)!} = 8 \cdot 7 \cdot 6 = 336$$

Therefore 336 different governing boards can be formed.

(b) When forming three-member committees the order of selection is not important because the positions are not distinguishable. The problem reduces to finding the number of combinations of eight partners taken three at a time. We derive this number from the formula for $P(8, 3)$.

Consider a particular committee consisting of Adams (A), Buchanan (B), and Carter (C). The committee can be regarded as the set {A, B, C}. By the Multiplication Principle there are six different arrangements (permutations) of the partners in the set {A, B, C}:

A-B-C A-C-B B-A-C B-C-A C-A-B C-B-A

In a case as in part (a) these are different arrangements; for instance, the arrangement A-B-C (Adams is president, Buchanan vice-president, and Carter secretary) is different from A-C-B (Adams is president, Carter vice-president, and Buchanan secretary). But as a three-member committee in which each member has equal standing, these six reduce to the one set, {A, B, C}, because for instance, with sets, {A, B, C} = {A, C, B}. It follows that for every six permutations there is one combination, so the number of combinations is one-sixth the number of permutations.

$$\frac{P(8,3)}{6} = \frac{336}{6} = 56$$

Altogether 56 committees, each with three members, can be formed.

The notation C(n, r) denotes the number of combinations of n objects taken r at a time. Part (b) of Example 5 showed that C(8, 3) is derived as follows:

$$C(8,3) = \frac{P(8,3)}{6} = \frac{P(8,3)}{3!} = \frac{8!}{(8-3)!\cdot 3!}$$

To evaluate C(n, r) for any n and r (where $r \le n$) means to count the number of subsets containing r elements of a set with n elements. There are P(n, r) arrangements which consist of r elements. However, because the order in which the elements in a subset are written does not matter, each subset of r elements corresponds to r! arrangements. Therefore the number of subsets is P(n, r) divided by r!. This can be expressed in a more useful form as follows:

$$C(n,r) = \frac{P(n,r)}{r!} = \frac{P(n,r)}{1}\cdot\frac{1}{r!} = \frac{n!}{(n-r)!}\cdot\frac{1}{r!}$$

Formula for Combinations

The number of combinations of n distinct objects taken r at a time is

$$C(n,r) = \frac{n!}{(n-r)!\cdot r!}$$

It is important to be able to distinguish between permutations and combinations because in many problems the essential ingredient is the proper identification of the concept being used. Keep in mind that a permutation is an "arrangement" that implies order, while a combination is a "collection" without regard to order. In both cases, however, the selection must come from one set of objects and repeats must not be allowed.

Example 6 applies the formulas for permutations and combinations to wagering. In horse racing, a "trifecta" refers to three horses that finish first, second, and third in a prescribed order. A "boxed trifecta" refers to three horses that finish first, second, and third in any order.

EXAMPLE 6

In a horse race that involves ten horses,
(a) how many trifecta tickets are possible?
(b) how many boxed trifecta tickets are possible?

Solution.

(a) Since there are ten horses and three winners, $n = 10$ and $r = 3$. Trifecta tickets are permutations because the order is important. Therefore, the number of trifecta tickets is given by P(10, 3) = 720.

(b) Boxed trifecta tickets are combinations because only the names of the winners must be listed, not their order of finish. Therefore there are C(10, 3) = 120 boxed trifecta tickets.

Example 7 combines the three counting techniques that were studied in the last two sections: the Multiplication Principle, permutations, and combinations. Since many problems deal with decks of ordinary cards, it might be helpful to refer to Table 3–5, which indicates three ways to partition the 52 cards.

- The top row of Table 3–5 shows the partition of the deck into two colors : red and black.
- The second row shows the partition into four suits : hearts, diamonds, spades, and clubs.
- The first column shows the partition of the deck into 13 values : ace, king, queen, jack, 10, 9, 8, 7, 6, 5, 4, 3, 2.

(Note: The values "King," "Queen," and "Jack" are referred to as "face cards." Therefore, there are 12 face cards in a deck: four kings, four queens, and four jacks.)

	Red cards		Black cards	
Value	Hearts	Diamonds	Spades	Clubs
Ace	A♥	A♦	A♠	A♣
King	K♥	K♦	K♠	K♣
Queen	Q♥	Q♦	Q♠	Q♣
Jack	J♥	J♦	J♠	J♣
10	10♥	10♦	10♠	10♣
9	9♥	9♦	9♠	9♣
8	8♥	8♦	8♠	8♣
7	7♥	7♦	7♠	7♣
6	6♥	6♦	6♠	6♣
5	5♥	5♦	5♠	5♣
4	4♥	4♦	4♠	4♣
3	3♥	3♦	3♠	3♣
2	2♥	2♦	2♠	2♣

EXAMPLE 7

A poker hand consists of 5 cards drawn from an ordinary deck of 52 cards where the order in which the cards are drawn is unimportant. How many poker hands consist of three aces and two kings?

Solution. The procedure will be to find the number of ways of selecting three aces, then to find the number of ways of selecting two kings. Each desired hand consists of three aces *and* two kings. By the Multiplication Principle the total number of such hands is obtained by multiplying these two numbers.

There are four aces, so there are C(4, 3) = 4 ways to select three aces from them. Similarly, ther C(4, 2) = 6 ways to select two kings from the four kings. By the Multiplication Principle there are 24 hands that consist of three aces and two kings.

Since the most difficult part of a counting problem is to decide whether to use a permuta combination, or the Multiplication Principle, we end this section with a table outlining when ea appropriate.

Permutations	Combinations	Multiplication Principle
1. Selection must come from one set of objects 2. No repeats allowed 3. Order counts	1. Selection must come from one set of objects 2. No repeats allowed 3. Order does not count	Used when neither permutations nor combinations can be used

To illustrate how to use the table, consider the following problem. An employer has a workg consisting of six people. The employer has two bonus awards to present to some of the people i workgroup. In how many ways can the bonus awards be presented if:

(a) The workgroup is split into two subgroups, one award will be given to one of four peop one subgroup and another award will be given to one of the remaining two people i second subgroup?
(b) the awards are equal and any person can win two, one, or no award?
(c) the awards are equal but no person can win more than one award?
(d) one award is twice as large as the other and no person can win more than one award?

The first question to ask is whether the selection must come from one set of objects. In part (a selection will be from two different sets so neither permutations nor combinations can be used to the problem. The correct choice is the multiplication principle. The answer is 4 • 2 = 8 ways. In (b), (c) and (d) the answer to this question is "yes" since the selection comes from the set of six pe

The next question to ask is whether repeats are allowed. In part (b) the answer is "yes" because an person can win more than one award, and in parts (c) and (d) the answer is "no." This means th part (b) the solution involves neither permutations nor combinations because repeats are not allo Thus for part (b) use the multiplication principle. The answer to part (b) is 6 • 6 = 36.

In parts (c) and (d) repeats are not allowed and the selection is from the same set, so permutations or combinations are used to obtain the solution. Next ask the question whether or important. In part (c) the awards are equal so the order in which the two individuals are select immaterial since they receive the same award. Use combinations. In part (d) think of the p receiving the larger award being selected first and the person receiving the smaller award selected second. The order is important because the two awardees receive different amounts. permutations. The answer to part (c) is C(6, 2) = 15 and the answer to part (d) is P(6, 2) = 30.

ERCISE SET 3-4

In problems 1 to 8 evaluate the given expression.

1. (a) P(5, 2) (b) C(5, 2)
2. (a) P(7, 3) (b) C(7, 3)
3. (a) P(6, 2) (b) C(10, 2)
4. (a) P(12, 3) (b) C(8, 3)
5. (a) P(5, 0) (b) C(5, 0)
6. (a) P(7, 0) (b) C(7, 7)
7. (a) 0! (b) (8 – 3)!
8. (a) 1! (b) 52!/50!

In all of the following problems, determine whether the solution involves, permutations, combinations, or the Multiplication Principle, then use the appropriate formula to answer the question.

9. In how many ways can a club with ten members elect a president and a vice-president?
10. In how many ways can four people line up in a bakery?
11. If the call letters of a radio station must begin with, the letter W, how many different stations could be designated by using four letters, such as WIND, without using repetitions?
12. The Coast Guard uses flags raised on a flagpole to signal ships concerning such matters as weather conditions. If three different flags area available, how many signals can be sent using all three flags if the order of the arrangement is important?
13. The number of signals available in Problem 12 can be increased by using signals of two or three flags. How many such signals are possible if the order of the flags is important?
14. In how many ways can ten CDs be arranged in a horizontal holder?
15. In how many ways can the gold, silver, and bronze medals be won if ten women compete in figure skating?
16. In how many ways can four students be given assigned seats in a classroom with ten seats?
17. An agricultural expert has designed an experiment to test the effects of two types of insecticide and four types of fertilizer on six different types of crops. Each plot of land uses one insecticide, one fertilizer, and one crop. How many plots are needed?
18. A veterinarian wishes to test three different levels of dosage of a vaccine on four types of animals at two different age levels of the animals. How many classes of vaccinations are possible?

19. There are 20 candidates for an all-star basketball game and 5 will be selected for the first team. How many different first-teams can be chosen?

20. Twelve people work as programmers in a computer center, with teams of three programmers serving as consultants. In how many ways can teams be formed?

21. Seven astronauts have reached the final training level for a particular space flight.
    (a) How many two-person crews can be chosen from the seven?
    (b) How many two-person crews will have one person designated captain and the other designated copilot?

22. How many two-card hands can be dealt from an ordinary deck of 52 cards?

23. How many poker hands consist of three queens and two jacks?

24. How many poker hands consist of four queens and one jack?

25. How many poker hands consist of two queens, two jacks, and one king?

26. How many poker hands consist of one queen, three jacks, and one king?

27. Evaluate the expression:

$$C(6, 6) + C(6, 5) + C(6, 4) + C(6, 3) + C(6, 2) + C(6, 1) + C(6, 0).$$

28. Evaluate the expression:

$$C(4, 4) + C(4, 3) + C(4, 2) + C(4, 1) + C(4, 0).$$

29. A professional basketball team has 5 starters who are selected from the 12 players.
    (a) How many starting teams are possible?
    (b) How many starting teams are possible if two of the players' contracts specify that they must start every game?

30. (a) How many poker hands can be drawn from an ordinary deck of 52 cards?
    (b) How many poker hands contain all four aces?

31. In how many ways can three initials be chosen so that two initials are the same but not all three?

32. If a penny, nickel, and dime are tossed in a fountain, in how many ways can they land so that two of them are heads?

33. How many poker hands consist of three red cards and two black cards?

34. How many poker hands consist of three clubs and two spades?

35. How many three-digit sequences can be formed if repetition is allowed? (For example, the sequence 989 is allowed. So is 007.)

36. How many three-digit sequences can be formed if no repetitions are allowed?

37. How many four-digit sequences can be formed if repetition is allowed? (For example, the sequences 9892 and 0007 are allowed.)

38. How many four-digit sequences can be formed if no repetitions are allowed?

39. How many boxed three-digit sequences can be formed?

40. How many three-digit sequences have exactly two of the digits the same?

41. In how may ways can four walls be painted if four colors are available and:
   (a) repetition is allowed?
   (b) repetition is not allowed?

42. In how many ways can a sorority with 20 members elect:
   (a) a president and a vice-president?
   (b) a three-member committee?

43. There are six finalists for two awards. In how many ways can the awards be given if:
   (a) the awards are equal and any finalist can win two, one, or no award?
   (b) the awards are equal but no finalist can win more than one award?
   (c) one award is twice as large as the other and no finalist can win more than one award?

44. A pharmaceutical salesperson wants to hand out $1000 worth of samples to ten doctors. In how may ways can the samples be given if:
   (a) all ten get an equal amount worth $100?
   (b) five doctors are chosen at random and are given equal amounts worth $200?
   (c) two doctors are chosen at random, one is given $600 worth of samples and the other is given $400 worth?

45. In wagering, an *exacta* refers to two objects that finish first and second in a specified order. If a dog race contains nine greyhounds, how many different exacta tickets can be purchased?

46. Many harness races consist of exactly nine horses.
   (a) How many different trifecta tickets can be purchased?
   (b) How many different boxed trifecta tickets can be purchased?

47. (a) Explain why $C(14, 5) = C(14, 9)$.
   (b) Explain why $C(n, r) = C(n, n - r)$.

48. In Figure 3–14, a person wants to drive from Point A to Point B by following a route along the streets as shown. At each intersection, the car proceeds either north or east only. How many routes from Point A to Point B are possible if the car must pass through Point G to get gas? Explain how the Multiplication Principle can be used to help answer the question.

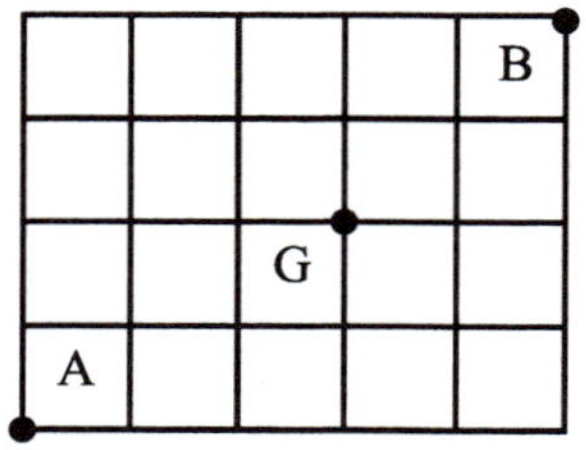

Figure 3–14

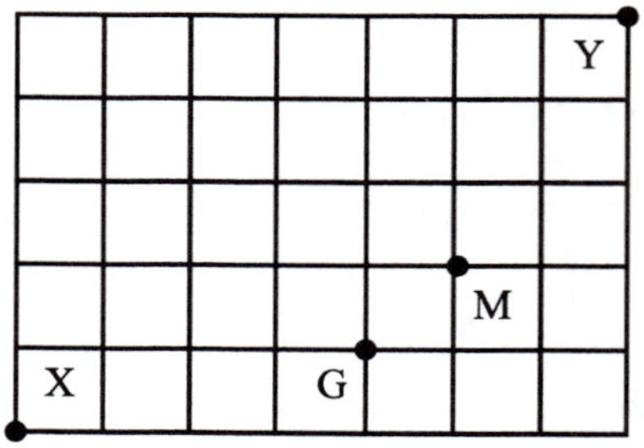

Figure 3–15

49. In Figure 3–15, a person wants to drive from Point X to Point Y by following a route along the streets as shown. At each intersection, the car proceeds either north or east only. How many routes from Point X to Point Y are possible if the car must pass through Point G to get gas and Point M to drop off mail? Explain how the Multiplication Principle can be used to help answer the question.

Referenced Exercise Set 3-4

1. Campanology is the art of ringing bells.[14] Since modern bells often weigh more than a ton, it takes about two seconds to advance or retard each bell. Consequently a "ring" of bells is an orderly sequence of permutations on the bells instead of a melody. An *extent* is the maximum number of rings, so that the extent of six bells (called a Minor) is $6! = 720$. What is the extent of a Major, which consists of eight bells? (Plain Bob Major, which was played in 1963 and 1977, required over 18 hours to ring!)

2. In music theory, chords and lines are basic terms.[15] A chord refers to a subset of the white keys on a piano, while a line is a sequence of distinct white keys.
(a) Which of these two concepts refers to a permutation?
(b) Which of these two concepts refers to a combination?

Problems 3 and 4 require a knowledge of the programming language BASIC. They are based on an article that describes how computer programs can be used to introduce permutations and combinations.[16]

---

[14]Arthur T. White, "Ringing the Cosets," *American Mathematical Monthly*, October 1987, pp. 721-746.

[15]John Clough and Gerald Myerson, "Musical Scales and the Generalized Circle of Fifths," *American Mathematical Monthly*, November 1986, pp. 695-701.

[16]Clark Kimberling, "Permutations and Combinations," *The Mathematics Teacher*, May 1987, pp. 403-407.

3. What is the output of this program, and how is it related to permutations or combinations?

```
10  FOR I = 1 TO 4
20      FOR J = 1 TO 4
30          IF I ≥ J THEN 50
40      PRINT I; J
50      NEXT J
60  NEXT I
```

4. What is the output of the program in Problem 3 if the inequality sign in line 30 is replaced by "=", and how is it related to permutations or combinations?

Multiple-Choice Exercise Set 3-4

1. A <u>permutation</u> is an arrangement of objects
   (a) without repetition and with regard to order
   (b) without repetition and without regard to order
   (c) with repetition and with regard to order
   (d) with repetition and without regard to order

2. A <u>combination</u> is an arrangement of objects
   (a) without repetition and with regard to order
   (b) without repetition and without regard to order
   (c) with repetition and with regard to order
   (d) with repetition and without regard to order

3. In which list of objects is order important?
   (a) the selection of five cards from a deck
   (b) the sum of two dice
   (c) the selection of school board members
   (d) the assigning of grades to three students
   (e) none of these

4. Which list of objects describes a permutation?
   (a) five cards selected from a deck
   (b) the sum of two dice
   (c) school board members
   (d) grades that three students receive
   (e) none of these

5. In which list of objects is order <u>not</u> important?
   (a) five cards selected from a deck
   (b) awards for $25 and for $10
   (c) finishing first, second, and third in a race
   (d) grades that three students receive
   (e) none of these

6. Which list of objects describes a combination?
   (a) five cards selected from a deck
   (b) awards for \$25 and for \$10
   (c) finishing first, second, and third in a race
   (d) grades that three students receive
   (e) none of these

7. $4! =$
   (a) 4 (b) 12 (c) 0.25 (d) 120 (e) none of these

8. $(9 - 3)! =$
   (a) $9! - 3!$ (b) 24 (c) $6!$ (d) 120 (e) none of these

9. $(12 - 7)! =$
   (a) $12! - 7!$ (b) 720 (c) $6!$ (d) 120 (e) none of these

10. $\dfrac{5!}{(5-2)!} =$
    (a) 1/2 (b) 10 (c) 12 (d) 20 (e) none of these

11. $\dfrac{5!}{(5-2)! \cdot 2!} =$
    (a) 1/2 (b) 10 (c) 12 (d) 20 (e) none of these

12. $\dfrac{5!}{(5-2)!}$ describes
    (a) a permutation of 5 objects taken 2 at a time
    (b) a set of 3 elements
    (c) a combination of 5 objects taken 2 at a time
    (d) subsets with 3 elements in a set with 5 elements
    (e) none of these

13. $\dfrac{5!}{(5-2)! \cdot 2!}$ describes
    (a) a permutation of 5 objects taken 2 at a time
    (b) all subsets of a set with 5 elements
    (c) a combination of 5 objects taken 2 at a time
    (d) all subsets of a set with 3 elements
    (e) none of these

14. How many permutations of the ten digits are there?
    (a) 10 (b) $10!/2!$ (c) $10!/(2!\,8!)$ (d) 3,628,800 (e) none of these

15. How many people can have different sets of three initials?
    (a) $26 \cdot 26 \cdot 26$ (b) $26 \cdot 25 \cdot 24$ (c) P(26, 23) (d) C(26, 23) (e) none of these

16. How many poker hands have four aces?
    (a) 5 (b) C(52, 4) (c) C(13, 4) (d) P(13, 4) (e) none of these

17. At *Hurts Car Rental* you can choose from 6 different sizes, 12 different makes, and 2 options. Altogether how many different arrangements of these three choices are possible?
    (a) 3 (b) 144 (c) 20 (d) 124 (e) none of these

18. A red die and a green die are tossed. In how many ways can they land so that their sum is 4?
(a) 1 (b) 2 (c) 3 (d) 4 (e) none of these

19. How many 7-digit phone numbers can be dialed if the first digit cannot be 0, 1, 8, or 9?
(a) 6,000,000 (b) 4,000,000 (c) 5,999,999 (d) 28 (e) none of these

20. How many Morse codes have five characters, where each character is either a dot or a dash?
(a) 5 (b) 10 (c) 25 (d) 32 (e) none of these

CASE STUDY LICENSE PLATES AND AREA CODES[17]

In the nineteen eighties most of the fifty states of the US adopted the *standard format* of auto licenses, consists of three letters followed by three digits. Some states soon found themselves running out of usable license plates so they also adopted the *reverse format* as well, which consists of three digits followed by three letters.

In 1984 there were about twenty million licensed vehicles in California, more than in any other state. The Golden State decided not only to use the standard format and the reverse format for its license plates but also to use a third format. The new format consists of a single digit followed by three letters and three digits; we will call it the *7-Cal format*.

How many usable license plates do these three formats allow? We will answer this question by defining the term "usable license plate" and by applying the counting techniques we have developed in this chapter.

Usable License Plates

A convenient way to view the standard format is LLLDDD, where L stands for one of the 26 letters and D for one of the 10 digits. By the Multiplication Principle there are 17,576,000 possible license plates in standard format.

26	26	26	10	10	10
L	L	L	D	D	D

This is the *theoretical number* of license plates because not all license plates are usable. For instance, JKI123 is not usable because the third letter can easily be confused with the first digit. This is an example of a license plate being restricted for reasons of "clarity." Other license plates are reserved for disabled persons and disabled veterans. Two examples are DPW123 and VET789.

The set of *usable license plates* refers to those license plates which are available to the general public. This set excludes license plates which are restricted for clarity or reserved for disabled persons and disabled veterans. Example 1 counts the number of usable license plates in standard format.

EXAMPLE 1

The number of theoretical license plates in standard format LLLDDD is 17,576,000. The number of usable license plates is

$$17{,}576{,}000 - n(\mathrm{C} \cup \mathrm{D})$$

where

---

[17]Adapted from the article: Edwin G. Landauer, "Counting Using License Plates and Phone Numbers: A Familiar Experience," *The Mathematics Teacher*, March 1984, pp. 183-187, 234.

C denotes the set of license plates restricted for clarity

D denotes the set of license plates reserved for the disabled.

The clarity rule is as follows:

1. The letters I, O, Q, and Z are not allowed in the third-letter position. Since four letters cannot be used in the third position

$$n(C) = 26 \cdot 26 \cdot 4 \cdot 10 \cdot 10 \cdot 10 = 2{,}704{,}000$$

There are two rules for reserved license plates:

2. The combinations DPW, DPX, DPY, and DPZ are designated for vehicles of disabled persons.
3. The combinations VET, VTN, and VTR are designated for vehicles of disabled veterans.

By the Multiplication Principle there are 1000 license plates of the form DPW _ _ _ because each blank can have ten possible digits. Similarly, each combination DPX, DPY, and DPZ eliminates 1000 license plates, making 4000 license plates that are reserved for disabled persons by rule 2. Three such combinations are also reserved for disabled veterans by rule 3, so $n(D) = 7000$.

It follows from rules 1 and 2 that a license plate lies in both C and D only if the letters are DPZ. There are 1000 license plates with the letters DPZ, so

$$n(C \cap D) = 1000$$

By the Inclusion-Exclusion Principle

$$n(C \cup D) = 2{,}704{,}000 + 7000 - 1000 = 2{,}710{,}000$$

Therefore the number of usable license plates in standard format is

$$17{,}576{,}000 - 2{,}710{,}000 = 14{,}866{,}000$$

The theoretical number of license plates in reverse format is the same as the theoretical number in standard format. But the number of restrictions is different. Exercises 1 to 3 count the number of usable license plates in reverse format.

The 7-Cal Format

The 7-Cal format consists of one digit from 1 to 9 followed by the standard format. It can be written ELLLDDD, where E represents one of the digits from 1 to 9. The theoretical number of license plates in 7-Cal format is

$$9 \cdot 26 \cdot 26 \cdot 26 \cdot 10 \cdot 10 \cdot 10 = 158{,}184{,}000$$

This number is over half the population of the entire United States! It stands to reason that the 7-Cal format can accommodate all of the licensed passenger vehicles in the state of California. However, the 7-Cal format has four restrictions.

1. The letters I and O are not allowed in the first-letter position.
2. The letters I, O, Q, and Z are not allowed in the third-letter position.
3. The three-character combination 1DP is reserved for vehicles of disabled persons.

4. The three-character combination 1VT and the four-character combination 1VET are reserved for vehicles of disabled veterans.

Example 2 counts the number of usable license plates in the 7-Cal format by an approach that is quite different from the approach in Example 1.

EXAMPLE 2

By the Multiplication Principle the number of usable license plates in the 7-Cal format ELLLDDD is

$$9 \cdot 24 \cdot 26 \cdot 22 \cdot 10 \cdot 10 \cdot 10 = 123{,}552{,}000$$

The number 24 comes from rule 1 and the number 22 from rule 2. This total includes the restrictions for clarity but not the reservations for disabled persons.

By rule 3 the characters 1DP are followed by one letter and three digits. The letter cannot be I, O, Q, or Z because of rule 2, so the number of license plates reserved for disabled persons is

$$22 \cdot 10 \cdot 10 \cdot 10 = 22{,}000$$

Next consider rule 4. There are 22,000 license plates beginning with 1VT. Also, 1000 license plates begin with 1VET. Thus 23,000 license plates are reserved for disabled veterans. The sets of license plates that are reserved for disabled persons and reserved for disabled veterans are disjoint, so the number of reserved license plates is given by 22,000 + 23,000 = 45,000.

Altogether there are 123,552,000 B 45,000 = 123,507,000 usable license plates in the 7-Cal format.

The following table summarizes the number of usable license plates in California.

Format	Number of Usable License Plates
Standard	14,866,000
Reverse	16,208,000
7-Cal	123,507,000
Total	154,581,000

Future Needs

More than 31 million usable license plates are available in either the standard or the reverse format. Yet in 1980 California, with less than 18 million licensed vehicles, began issuing license plates in the 7-Cal format. Why did the Golden State adopt a third format?

For two reasons. First, California does not reissue regular license plates, except for vanity plates. This reason alone creates the need for a new format, but it is indigenous to California. Second, all states have lists of unacceptable combinations of letters and numbers. Most of them are obscene, so we will not list them here (for reasons of taste as well as litigation).

What format does your state use? How many usable license plates are available. The exercises examine Louisiana, Montana, and California.

Area Codes

The necessity for new area codes in the US has a parallel history to the need for additional license plates. For decades the *standard format* for area codes served the public's need. An area code in standard format is a three digit number that has the following properties:

1. The first digit is neither 0 nor 1.
2. The second digit is 0 or 1.
3. The third digit is not 0.

To place a long distance telephone call you first send a signal to the telephone system that the number being dialed is a long distance call. this is done by dialing either 0 or 1, depending upon whether it is to be a regular call (if "1" is dialed) or it is a call requiring operator assistance (if "0" is dialed). When 0 or 1 is dialed it is a signal to the telephone system that the number being dialed is a long distance call, so the next three digits will be the area code. The first rule in the standard format is designed to make sure that area codes are not confused with this signal to the system. The second and third rules provide further evidence to the system that the call is indeed a long distance call.

As phones became more common in the nineteen eighties there was a need to increase the number of area codes in use because the number of telephone numbers issued was increasing at an accelerated rate, especially in issued in large cities. The *modified format* was instituted. It requires only rules 1 and 2. This resulted in only a relatively small increase in the number of codes available, which was acceptable then because only large metropolitan areas were being squeezed out of available telephone numbers. Then in the nineteen nineties, with the advent of fax machines, cellular phones and pager numbers, a large increase in area codes was needed. Even smaller cities were running out of available telephone numbers. The *modern format* was instituted. It consists of applying only rule 1 to a three-digit number for it to be an allowable area code. Example 3 shows how these changes in the rules have dramatically increased the number of area codes available.

EXAMPLE 3

By the Multiplication Principle the number of area codes available in the standard format is

$$8 \cdot 10 \cdot 9 = 144$$

The number of area codes available in the modified format is

$$8 \cdot 2 \cdot 10 = 160$$

This change in the nineteen eighties increased the number of area codes by 16, from 144 to 160. By the nineteen nineties, however, a dramatic increase was needed. The number of area codes in the modern format is

$$8 \cdot 10 \cdot 10 = 800$$

This is an increase of 640 numbers, from 160 to 800.

It doesn't look like we will run out of available area codes very soon, but who know what new inventions lie on the horizon that will require new types of uses of telephones that would require new numbers? If we need more area codes, what can be done? Some answers are given in the exercises.

CASE STUDY EXERCISE SET

1. For reasons of clarity, the letters I and O are not allowed in the first-letter position of license plates in the reverse format DDDLLL. How many license plates does this restriction eliminate?

2. Two sets of three-letter sequences are reserved for license plates in the reverse format DDDLLL. Those designated for vehicles of disabled persons are DPW, DPX, DPY, DPZ, RDP, SDP, TDP, UDP, VDP, WDP, XDP, YDP, and ZDP. Those designated for vehicles of disabled veterans are VET, VTN, and VTR. How many license plates do these reservations eliminate?

3. How many usable license plates are in the reverse format DDDLLL? (You must make use of the answers to Problems 1 and 2.)

4. How many usable license plates consist of a letter followed by three digits and three letters, subject to the restrictions stated in Problems 1 and 2?

5. License plates in Louisiana consist of three digits, then a letter, and three digits. The letter designates the state police troop area that issued the plate (A, B, C, D, E, F, G, H, I, K, L). The letter X means that the plate was issued by mail. How many usable license plates does this format allow?

6. Montana's license plate format consists of one or two digits, followed by a dot, and then up to five more digits. The first group of digits runs from 1 to 56; each group designates the county in which the owner resides. The second group runs from 1 to the number of cars licensed in that county. How many usable license plates does this format allow?

7. One way to increase the number of area codes is to delete the restriction imposed by the first rule of the standard format. How many additional codes are gained by this action?

8. Another way to increase area codes is to adopt a four digit area code system instead of the current three-digit system. One way to accomplish this is to add the fourth digit 0 to all existing codes and then new codes would have any four digit number except those with a fourth digit 0. How many additional codes would be added with this action?

CHAPTER REVIEW

Key Terms

3-1 Sets

Set
Standard Format
Element
Natural Number
Integer
Rational Number
Subset
Venn Diagram
Equal Sets
Complement
Universal Set
Intersection
Union
Irrational Number
Real Number
Empty (Null) Set
Disjoint Sets

3-2 Venn Diagrams and Counting

"Count"

3-3 Tree Diagrams and the Multiplication Principle

Tree Diagram
Selection Without Replacement
Selection With Replacement
The Multiplication Principle

3-4 Permutations and Combinations

Permutations
Factorial
Combinations

REVIEW EXERCISE SET

1. (a) Is the description "nice numbers" a set or a collection?
   (b) Is the description "U.S. cities whose population in the 1980 census exceeded 100,000" a set or a collection?

In problems 2 and 3 let U = {a, b, c, ... , z},V = {a, e, i, o, u}, and A = {a, b, c, d, e}.

2. List the elements in each set.
   (a) $V \cup A$
   (b) $V \cap A$.
3. List the elements in each set.
   (a) A'
   (b) $V' \cap A'$
   (c) $(V \cup A)'$.
4. (a) Is the set $A = \{x \in Z: x^2 = -1\}$ a subset of the set of rational numbers Q?
   (b) Is the set $B = \{x \in Z: x^2 > 9\}$ a subset of the set of rational numbers Q?
5. Find $n(A \cup B)$ where $n(A) = 9$, $n(B) = 5$, and $n(A \cap B) = 4$.
6. Find $n(A \cup B)$ where $n(A \cap B') = 9$, $n(A' \cap B) = 5$, and $n(A \cap B) = 4$.

7. Let the universal set consist of all real numbers and define X to be the set of rational numbers. Describe the set X′.

8. The 30,000 members of the Mathematical Association of America can subscribe to three journals: *Monthly* (M), *Math Magazine* (G), and *College Math Journal* (C). In 1988, the numbers of subscribers (in thousands) were:

   $n(M \cap G \cap C) = 1,$
   $n(M \cap G) = 5,$
   $n(M \cap C) = 7,$
   $n(G \cap C) = 1,$
   $n(M) = 18,$
   $n(G) = 7$
   $n(C) = 11$

   How many members subscribed to
   (a) both *Math Magazine* and the *College Math Journal*?
   (b) the *College Math Journal* only?
   (c) none of the three journals?

9. Refer to Figure 3–16 to determine the number of elements in each set.
   (a) $A \cap B$
   (b) $A \cup B$
   (c) A′
   (d) U

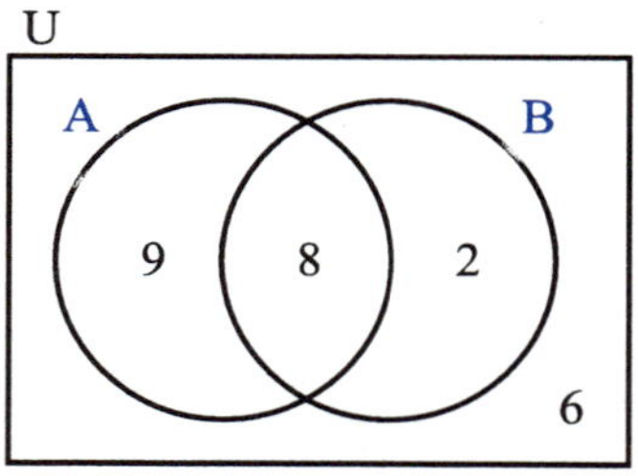

Figure 3–16

10. In how many different ways can four people be seated in four seats in a row?

11. In the game of "Clue" a suggestion consists of a suspect, a weapon, and a room. There are six suspects, six weapons, and nine rooms. How many suggestions are possible?

12. How many sequences of four letters begin with one consonant, end with a different consonant, and have two different vowels in between?

13. Four people enter the elevator in a building on the ground floor. The elevator can stop at floors 1 through 12.
    (a) In how many different ways can the four people exit?
    (b) In how many different ways can the four people exit if no two exit on the same floor?

14. Each ticket in the game of "Lotto" has six different numbers punched from the numbers 1 to 40. How many different tickets are possible?

15. In how many different ways can the letters of the word MAJORITY be arranged?

In problems 16 to 18 a tournament that starts with 64 teams awards four prizes. A team is eliminated after it loses, while a winning team continues to play.

16. How many orderings of the first through the fourth place teams are possible?

17. How many combinations of teams can make the "final four?"

18. The Final Four teams in the 1991 NCAA basketball tournament were Duke, UNLV, North Carolina, and Kansas. In how many different ways could these four teams have finished first through fourth?

19. Twenty dancers audition for roles in *A Chorus Line*.
    (a) How many different selections of three of them can be made for the first, second, and third leading roles?
    (b) How many different selections of three of the remaining 17 dancers can be made for three indistinguishable understudy roles?

20. Two dice (one red and the other green) are rolled and their sum is recorded. In how many different ways can the dice land so that the sum is equal to 10?

4 PROBABILITY BASICS

Chapter Overview

In many aspects of mathematics a formula or process is used to predict the future. Sometimes the prediction is exact but often events take place randomly, and it is not possible to predict exactly what will happen. A model must be constructed that takes this random nature of outcomes into effect. For example, a meteorologist cannot tell precisely when, or even if, it will rain but, instead, assigns probabilities to the possible types of weather to expect. The first part of the chapter shows how to construct probability models and illustrates their fundamental properties. The later sections cover more specific topics that govern when we are able to add, subtract, multiply and divide probabilities. We will also calculate probabilities using the counting techniques of Chapter 3.

Throughout the chapter games of chance are used as examples because they are easy to explain and understand. We also mention several significant applications of probability theory in the real world. In fact, the first example is from a study of automobile accidents on a major Boston thoroughfare that used probability theory in its analysis.

Case Study Preview

Suppose you face a trial and your lawyer says it is a toss-up whether to select a trial by jury or a trial by judge alone. Which would you choose? Two political scientists researched this question, and their results, based on probabilistic reasoning, are given in the case study in this chapter.

4-1 SAMPLE SPACES & EVENTS

Many events in the world that seem to happen by chance actually occur according to certain underlying patterns. We can often uncover an order to their occurrence by studying their behavior. The theory of probability is the branch of mathematics that helps find these patterns of these seemingly unpredictable events.

Introduction: An Example

In the early 1970's, the Massachusetts Department of Public Works conducted a study of speed limits in an effort to improve motor vehicle safety. Highway accident data was analyzed on a suburban stretch of Route 2, a major road entering Boston, MA from the west. In an article[1] about the study, William B. Fairley raised the critical question: what season of the year had the most accidents? The year was divided into four quarters, the first being January to March, the second April to June, and so forth. Data was gathered for each quarter from 1970 to 1972. The second column in Table 4–1 gives the quarterly count, called the **frequency**. For example, during the three years of the study, there was a total of 302 accidents in the first quarter: 138 in 1970, 74 in 1971 and 90 in 1972.

The third column of the table gives the fraction of all accidents that occurred in each quarter. These fractions, written in decimal form and called **relative frequencies**, are computed by dividing each quarterly count by the total number of accidents in all quarters, 1269. Thus the fraction of accidents of all accidents that occurred in the first quarter is 302/1269 = 0.24.

If we select an accident at random from the total of 1269, we say that the **probability** that it occurred in Quarter 1 is 0.24. To say that the probability of this event is 0.24 means that, since it has occurred 24% of the time in the past, it is reasonable to predict that it will happen approximately 24% of the time in the future. In this way probability is used to predict future trends. That is, under the same conditions, 24% of the accidents on Route 2 will occur in Quarter 1.

Quarter	Quarterly Count	Quarterly Ratio
1	302	0.24
2	244	0.19
3	283	0.22
4	440	0.35
	1269	1.0

Table 4–1 Accidents on Route 2, 1970-1972

The phrase "under the same conditions" spells the difference between studies of real world phenomena and games of chance, such as tossing coins, rolling dice, or playing cards or roulette. The conditions in one of these games are repeated exactly every time the game is played, but it is much more difficult to repeat real world experiments under the same conditions. Therefore games are used to illustrate the concepts of probability

[1]William B. Fairley, "Accidents on Route 2: Two-way Structures for Data," *Statistics and Public Policy*, Addison-Wesley, (Boston, MA) 1977, pp. 23-50.

because they are familiar, are easy to demonstrate, and are repeated under identical conditions on each play.

Sample Spaces

The basic idea in probability is to study results of probability experiments. By a **probability experiment** we mean an action whose possible results can be determined and recorded. For example, four simple probability experiments are:

Experiment A: Toss a coin and record "heads" or "tails."
Experiment B: Roll a die and record the number on the top face.
Experiment C: Take a 10-question multiple-choice quiz and record the number of wrong answers.
Experiment D: Choose an accident along Route 2 and record the quarter of the year in which it occurs.

Each time a probability experiment is conducted, a result is observed and recorded. An experiment can be viewed as consisting of two steps: an action followed by the recording of an observation about the action. Each possible observation is called an **outcome**. The set of all outcomes of a probability experiment is called its **sample space**. For example, Experiment A has two outcomes, "heads" and "tails." If we let "H" represent observing "heads" and T represent observing "tails," then the sample space of Experiment A is given by:

$$S = \{H, T\}$$

DEFINITION

The **sample space** of an experiment is the set of all possible **outcomes** of the experiment. Each time the experiment is conducted, exactly one outcome is recorded.

Example 1 shows how to determine the sample spaces for Experiments B, C and D.

EXAMPLE 1

Find the sample space of Experiments B, C and D.

Solution. Experiment B has six possible outcomes: 1, 2, 3, 4, 5, 6. Hence the sample space for Experiment B is

$$S = \{1, 2, 3, 4, 5, 6\}$$

In Experiment C a person can get anywhere from zero to 10 wrong answers. Therefore, the sample space for Experiment C is

$$S = \{0, 1, 2, 3, 4, 5, 6, 7, 8, 9, 10\}$$

Experiment D has 4 outcomes. Its sample space is

$$S = \{\text{quarter 1, quarter 2, quarter 3, quarter 4}\}$$

Events

Often we are interested in looking at the possibility of more than one outcome from the sample space being observed. For example, consider Experiment D. The sample space is

S = {quarter 1, quarter 2, quarter 3, quarter 4}

Suppose we wanted to determine how many accidents occurred in the first six months of the year. The corresponding outcomes would be quarter 1 and quarter 2. In mathematical language, this is the set

E = {quarter 1, quarter 2}

The set E is a subset of the sample space, and in general, a subset of the sample space is called an **event**.

DEFINITION

An **event** of an experiment is a subset of the sample space. The event, E, **occurs** if the outcome of the experiment is an element of the subset E.

EXAMPLE 2

Referring to Experiment B, describe in terms of sets the two events E and F, where E and F are defined as

E: An odd number occurs.
F: A number less than 5 occurs.

Solution. Event E consists of the outcomes, 1, 3 and 5. Thus

E = {1, 3, 5}

Event F consists of the outcomes 1, 2, 3 and 4. Thus

F = {1, 2, 3, 4}

When experiments become more complex, the sample space can quite often be given in more than one way. For instance, Example 3 illustrates two different sample spaces that could result from tossing *two* coins.

EXAMPLE 3

Toss two coins. Record (a) the number of heads, and (b) the sequence of heads and tails. For each experiment, list the elements in the sample space and in the event, E, that corresponds to "at least one head occurs."

Solution. (a) Since two coins are tossed, there are three outcomes: 0, 1 and 2, where the outcome "0" means that 0 heads occur (both coins come up tails); the outcome "1" means that 1 head occurs, (one coin comes up heads, and the other one tails); and the outcome "2" means that 2 heads occur (both coins come up heads). Hence the sample space is

$$S = \{0, 1, 2\}$$

"At least one head occurs" means that either 1 or 2 heads occur. Therefore,

$$E = \{1, 2\}$$

(b) To determine the possible sequences of heads, refer to the tree diagram from Example 3 in Section 3-3.

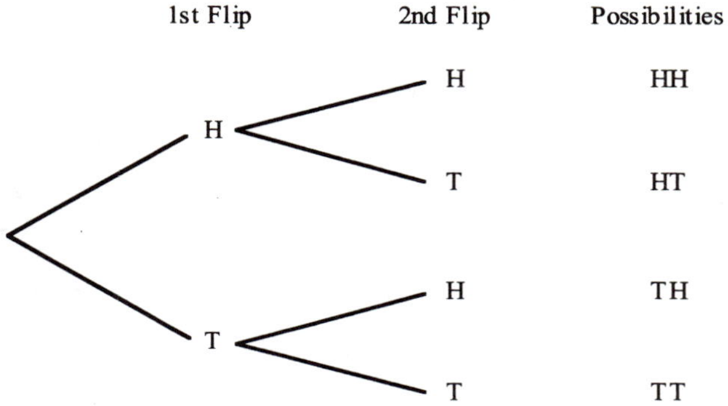

Figure 4–1

Since tossing two coins is equivalent to tossing one coin twice, the possible sequences of heads and tails are given in Figure 4–1. Therefore, there are 4 outcomes for the experiment: HH, HT, TH and TT, and the sample space is

$$S = \{HH, HT, TH, TT\}$$

The event E then contains all of these outcomes that contain 1 or more heads.

$$E = \{HH, HT, TH\}$$

One way to understand the necessity for differentiating between "HT" and "TH" is to imagine that the two coins are of different denominations, such as a penny and a dime. Then "HT" would represent "heads on the penny and tails on the dime," while "TH" would represent "tails on the penny and heads on the dime." These two situations are then clearly distinct.

Example 4 examines one situation where order is unimportant and another where the order is important.

EXAMPLE 4

Suppose that three people – Arnold, Barbara, and Christine – are running for student government, and that two of them will be elected.

(a) List the elements in the sample space if the two positions are identical in their responsibilities. Also list the elements in the event, E, defined by "Barbara is elected."

(b) List the elements in the sample space if the two positions are different in their responsibilities, for example if the most-popular candidate becomes president and the second most-popular becomes vice-president. Also list the elements in the event, F, defined by "a female is elected president."

Solution. In writing the outcomes, let each person be represented by his or her first initial.

(a) Since the positions are identical, there are three outcomes – Arnold and Barbara are elected, Barbara and Christine are elected, or Arnold and Christine are elected. Therefore, the sample space can be written as

S = {AB, BC, AC}

Event E includes all elements from the sample space that include Barbara. Therefore,

E = {AB, BC}

(b) Since the positions are different in their responsibilities, the outcome AB is different from the outcome BA. The former means that Arnold is elected president while Barbara is elected vice-president, while the latter means that Barbara is elected president while Arnold is elected vice-president. Therefore, the sample space is given by

S = {AB, AC, BA, BC, CA, CB}

The event F then excludes all outcomes where Arnold is elected president.

F = {BA, BC, CA, CB}

When you play games like Monopoly™ or Sorry™ you are interested in the sum of the numbers on the top faces. But when you play games like Backgammon, the numbers themselves on the top faces are important. For example, in Monopoly, you move eight squares if you roll a two and a six or if you roll a three and a five. But in Backgammon these two rolls produce different moves because you can move one piece two spaces and another six spaces with the first roll, but on the second can you move one piece three spaces and another five spaces.

The two sample spaces are generated by recording either the sum of the numbers on the top faces or by recording the actual numbers on the dice. The latter experiment consists of rolling two dice and recording the numbers on the top two faces as an ordered pair. So if a "1" appears on each die we record (1, 1). We find these two sample spaces in the next example.

EXAMPLE 5

Roll two dice. Record (a) the sum of the numbers on the top faces, and (b) the numbers on the top faces as an ordered pair. Find the sample space for each experiment.

Solution.

(a) There are 11 outcomes for the experiment where 2 is the smallest sum that can occur and 12 is the largest possible sum. Thus the sample space is

S = {2, 3, 4, 5, 6, 7, 8, 9, 10, 11, 12}

(b) As done in the previous two examples with coins, in this experiment distinguish between the two dice. Let the first die be white and the second be black. Thus "5 on the white die and 6 on the black die" is represented by the ordered pair (5, 6). This ordered pair is a different outcome than (6, 5) since (6, 5) represents "6 on the white die and 5 on the black die." The sample space consists of the 36 ordered pairs. Thus

$$
\begin{aligned}
S = \{ & (1, 1), (1, 2), (1, 3), (1, 4), (1, 5), (1, 6), \\
& (2, 1), (2, 2), (2, 3), (2, 4), (2, 5), (2, 6), \\
& (3, 1), (3, 2), (3, 3), (3, 4), (3, 5), (3, 6), \\
& (4, 1), (4, 2), (4, 3), (4, 4), (4, 5), (4, 6), \\
& (5, 1), (5, 2), (5, 3), (5, 4), (5, 5), (5, 6), \\
& (6, 1), (6, 2), (6, 3), (6, 4), (6, 5), (6, 6) \}
\end{aligned}
$$

Union, Intersection and Complement

Since events are defined in terms of sets, we can use set theory to define new events from other events using the notions of union, intersection and complement.

Unions and intersections are employed in the exact same manner as in Chapter 3, however, an event is often defined by a description rather than by listing its elements. In this case the word "or" generally represents the union of two events while the word "and" refers to the intersection.

The concept of complements is virtually the same as in set notation also, however, we use the sample space of an experiment in place of the universal set. Therefore, the complement of an event is the set of all outcomes in the sample space that are not in E.

The next example illustrates the uses of these three concepts in terms of events.

EXAMPLE 6

Two dice are rolled, and the sum of the faces is recorded. Events E and F are defined as follows

E: An odd sum occurs
F: A sum of at least 9 occurs

List the elements in each of the following events and describe each in terms of unions, intersections, or complements when appropriate.

(a) Event E
(b) Event F
(c) The event "an odd sum or a sum of at least 9 occurs"
(d) The event "an odd sum and a sum of at least 9 occurs"
(e) The event "a sum less than 9 occurs"

Solution. The sample space is $S = \{2, 3, 4, 5, 6, 7, 8, 9, 10, 11, 12\}$.
(a) $E = \{3, 5, 7, 9, 11\}$.
(b) $F = \{9, 10, 11, 12\}$.
(c) The word "or" indicates "union," so this event is given by $E \cup F$.

$$E \cup F = \{3, 5, 7, 9, 10, 11, 12\}$$

(d) The word "and" indicates "intersection," so this event is given by $E \cap F$.

$$E \cap F = \{9, 11\}$$

(e) The event "a sum less than 9 occurs" is the complement of event F and is therefore given by

$$F' = \{2, 3, 4, 5, 6, 7, 8\}$$

Two events that cannot both occur at the same time are called **mutually exclusive** events. For example, when two dice are rolled, the events "an even sum" and "an odd sum" cannot both occur, so they are mutually exclusive. In terms of sets, two events E and F are mutually exclusive if $E \cap F = \emptyset$. Example 7 looks at

EXAMPLE 7

Two dice are rolled, and the numbers on the faces are recorded as ordered pairs. Events E and F are defined as follows

E: A sum of 7 or 11 occurs
F: "Doubles" occur

List the ordered pairs and the number of elements in each of the following events.

(a) E
(b) F
(c) $E \cap F$

Solution. The sample space consists of the 36 ordered pairs listed in Example 5.

(a) Event E consists of all ordered pairs that add to either 7 or 11. Therefore,

$$E = \{(1, 6), (2, 5), (3, 4), (4, 3), (5, 2), (6, 1), (5, 6), (6, 5)\}$$

Consequently, $n(E) = 8$. In other words, there are eight ways out of 36 to get either a sum of 7 or a sum of 11.

(b) Event F consists of all ordered pairs where one number is repeated.

$$F = \{(1, 1), (2, 2), (3, 3), (4, 4), (5, 5), (6, 6)\}$$

Therefore, $n(F) = 6$, and there are six ways out of 36 to roll "doubles."

(c) Since both "7" and "11" are odd sums, and since all rolls the land in "doubles" produce even sums, events E and F have no ordered pairs in common. Therefore,

$$E \cap F = \emptyset$$

and E and F are *mutually exclusive*. In addition $n(E \cap F) = 0$ since there are no elements in the empty set.

Notice the different uses of the symbol "Ø" and the number "0" in Example 7. The empty set symbol, "Ø," is used when we write sets or events. The number "0" is used when we are counting the number of objects in a set or event and find that there are none.

EXERCISE SET 4-1

In problems 1 to 16 find the sample space of the experiment.

1. Roll a 4-sided die and record the number on the top face.
2. Roll a 10-sided die and record the number on the top face.
3. Select a letter from the word "letter."
4. Select a letter from the word "mathematics."
5. An ordinary deck of cards is shuffled and five cards are selected at random, and the number of aces drawn is recorded.
6. An ordinary deck of cards is shuffled and five cards are selected at random, and the number of hearts drawn is recorded.
7. Two 4-sided dice are thrown, and the sum of the faces is recorded.
8. Two 4-sided dice are thrown, and the numbers on the faces are recorded as an ordered pair.
9. Three coins are thrown, and the number of heads is recorded.
10. Four coins are thrown, and the number of heads is recorded.
11. Three coins are thrown, and the sequence of heads and tails is recorded.
12. Four coins are thrown, and the sequence of heads and tails is recorded.
13. 10 Green, 1 Red, and 2 Black marbles are placed in a box, three are selected without replacement, and the colors chosen (without regard to order) are recorded.
14. 10 Green, 1 Red, and 2 Black marbles are placed in a box, three are selected with replacement, and the colors chosen (without regard to order) are recorded.
15. 10 Green, 1 Red, and 2 Black marbles are placed in a box, three are selected without replacement, and the sequence of colors (with regard to order) is recorded.
16. 10 Green, 1 Red, and 2 Black marbles are placed in a box, three are selected with replacement, and the sequence of colors (with regard to order) is recorded.

In Problems 17 to 20 list the elements in event E if two six-sided dice are rolled and the sum of the faces is recorded.

17. E is the event "a sum of at most 6 occurs."
18. E is the event "a sum greater than 4 occurs."
19. E is the event "a sum of at least 7 occurs."
20. E is the event "a sum of less than 4 occurs."

In Problems 21 to 24 list the elements in event F if two six-sided dice are rolled and the numbers on the faces are recorded in an ordered pair.

21. F is the event "a sum of 2, 3 or 12 occurs."
22. F is the event "a sum of 6 or 8 occurs."

23. F is the event "a sum of at least 9 occurs."

24. F is the event "a sum no more than 4 occurs."

In Problems 25 to 30 assume that two six-sided dice are rolled and that the numbers on the faces are recorded in an ordered pair. For the given events E and F, find (a) $E \cup F$ (b) $E \cap F$, and (c) $E' \cap F$.

25. E is the event "a sum of at most 4 occurs."
F is the event "a sum of 2, 3 or 12 occurs."

26. E is the event "a sum of at least 9 occurs."
F is the event "a sum of 2, 3 or 12 occurs."

27. E is the event "a sum of at least 9 occurs."
F is the event "doubles occur."

28. E is the event "a sum of at least 8 occurs."
F is the event "the first die is greater than the second die."

29. E is the event "a sum of 2 or 12 occurs."
F is the event "the first die is a four."

30. E is the event "both dice are even."
F is the event "the second die is a six."

In Problems 31 to 34 assume that a four-sided die and a six-sided die are rolled.

31. Assuming that the sum of the faces of the dice is recorded, list the elements in the sample space.

32. Assuming that the numbers on the faces of the dice are recorded as an ordered pair with the number on the four-sided die listed first, how many elements are in the sample space?

33. Referring to the sample space from Problem 32, list the elements in the event E defined by "a sum of 7 occurs."

34. Referring to the sample space from Problem 32, list the elements in the event E defined by "an even number occurs on each die."

In Problems 35 to 44 consider to the experiment where a committee of two individuals is chosen from a list of six people whose last initials are A, B, C, R, S, T. and where the events E, F, and G are defined as follows.

E: A, B or C is on the committee.
F: Neither A nor B is on the committee.
G: Either C or S is on the committee.

35. List the elements of the events E, F and G.

36. List the elements in event E'

37. List the elements in event $F \cap G'$

38. List the elements in event $E' \cap G'$

39. List the elements in event $E' \cup G$

40. List the elements in event $F \cup G'$

41. List the elements in event $(E \cup G)'$
42. List the elements in event $(E \cup F)'$
43. List the elements in event $E \cap F \cap G$
44. List the elements in event $E \cup F \cup G$

For Problems 45 to 48, refer to the following data from the Fairley study that lists the number of accidents by month on the second row.

Jan	Feb	Mar	Apr	May	Jun	Jul	Aug	Sep	Oct	Nov	Dec
120	85	97	81	88	75	95	84	104	149	143	148

The experiment is to select an accident at random and to record its month. Define the following events.

E: the accident occurred in the third quarter
F: the accident occurred in June, July, or August
G: the accident occurred in a month in which at least 100 accidents occurred

List the elements in each of the following events.

45. $E \cap F$
46. $E \cap G$
47. $F \cap G$
48. $E \cap F'$
49. $E' \cap G$
50. $F' \cap G'$

In problems 51 to 54 the sample space S of an experiment is given along with two events E and F. Determine whether E and F are mutually exclusive events.

51. $S = \{a, b, c, d\}$, $E = \{a, c\}$, $F = \{b, d\}$.
52. $S = \{a, b, c, d\ e, f\}$, $E = \{a, b, c\}$, $F = \{b, d, f\}$.
53. $S = \{2, 3, 4, 5, 6, 7, 8, 9, 10, 11, 12\}$, $E = \{2, 3, 4\}$, $F = \{4, 5, 6, 7\}$.
54. $S = \{-3, -2, -1, 0, 1, 2, 3, 4\}$, $E = \{-3, -2, 0, 1\}$, $F = \{2, 3, 4\}$.
55. For any event E, is it true that E and E' are mutually exclusive? Why?
56. Let E and F be mutually exclusive events. Must E' and F' be mutually exclusive?
57. Let E and F be mutually exclusive events. Must $(E \cup F)'$ and $E \cap F$ be mutually exclusive?

Referenced Exercise Set 4-1

1. In a study[2] designed to explore the possibility of a link between air pollution and ill health the investigators discussed the effects of their research on mortality rates characterized by age, race, and sex. They defined the following events.

   I: Infant (under one year old)
   C: Child (from 1 to 14 years old)
   A: Adult (over 14 years old)
   N: Nonwhite
   W: White
   F: Female
   M: Male

   Define each of the following events in words.
   (a) $I \cup A$, (b) $A \cup F$, (c) $C \cap M$, (d) $N \cap F$, (e) F', (f) $A' \cap W'$, (g) $I \cap N \cap M$

2. In a medical experiment[3] designed to study the effect of Reye's syndrome on the inner ear, patients were asked whether they had contracted influenza infection, whether they had taken aspirin during their illness, and whether they had an arsenide-deficient diet. Define the events I, A and D as follows.

   I: the patient had influenza
   A: the patient took aspirin
   D: the patient had an arsenide-deficient diet

   Define the following events in words.
   (a) $I \cup A$, (b) $A \cup D$, (c) $I \cap D$, (d) $A \cap D$,(e) I', (f) $I' \cap A'$, (g) $I \cap A \cap D$

Multiple-Choice Exercise Set 4-1

1. If an accident is selected at random among the 1269 accidents studied in the Fairley Study the probability that it occurred in the second quarter is approximately
   (a) 0.24 (b) 0.19 (c) 0.22 (d) 0.35 (e) none of these

2. If an accident is selected at random among the 1269 accidents studied in the Fairley Study the probability that it occurred in the third quarter is exactly
   (a) 1/4 (b) 244/1269 (c) 283/1269 (d) 4/365 (e) none of these

3. The sample space for flipping a coin and recording whether it lands heads or tails is
   (a) {H, T} (b) {H} (c) {T} (d) {HH, HT, TH, TT}
   (e) none of these

4. The sample space for flipping two coins and counting the number of heads is
   (a) {H, T} (b) {H} (c) {T} (d) {HH, HT, TH, TT}
   (e) none of these

---

[2]Pratt, John W. (ed.), *Statistical and Mathematical Aspects of Pollution Problems*, Marcek Dekker, Inc., New York, pp. 223-47, 1974.

[3]Rarey, K.E., et. al., "Effects of Influenza, Aspirin, and an Arsenite-deficient Diet on the Inner Ear in Reye's Syndrome," *Annals of Otology, Rhinology and Laryngology*, Vol. 93, pp. 551-557, Nov-Dec, 1993.

5. The sample space for selecting a card from a deck and recording its suit is
   (a) {red, black} (b) {king, queen, jack}
   (c) {two, three, four, … , ace} (d) {heart, club, spade, diamond}
   (e) none of these

6. The sample space for the sexes (B = boy, G = girl) of the children in a family with two children is
   (a) {B, G} (b) {0, 1, 2}
   (c) {BB, BG, GG} (d) {BB, GB, BG, GG} (e) none of these

7. A marble is selected from a bag with 3 red (R) marbles, 2 blue (B) marbles, and 1 green (G) marble. If its color is recorded, what is the sample space?
   (a) {3, 2, 1} (b) {R, B, G}
   (c) {RR, BB, GG} (d) 6 (e) none of these

8. Two marbles are selected at once from a bag with 7 red (R) marbles, 2 blue (B) marbles, and 1 green (G) marble. If their color is recorded, what is the sample space?
   (a) {7, 2, 1} (b) {R, B, G}
   (c) {RR, BB, GG} (d) 10 (e) none of these

9. A subset of a sample space is called a(n)
   (a) element (b) outcome (c) event (d) subspace (e) none of these

10. If a day of the week is selected, the event for a day on the weekend is
    (a) {Mon, Tue, Wed, Thur, Fri, Sat, Sun}
    (b) {Mon, Tue, Wed, Thur, Fri}
    (c) 2
    (d) 2/7
    (e) none of these

11. If E = {1, 3} then the event E′ in the sample space S = {1, 3, 5, 7, 9} is
    (a) $\{x \in Z : x < 10\}$ (b) $\{x \in Z : 0 \le x \le 10\}$ (c) {5, 7, 9}
    (d) $\{x \in Z : 1 \le x \le 9\}$ (e) none of these

12. $E \cup F$ occurs whenever
    (a) both E and F occur (b) E or F or both occur
    (c) neither E nor F occurs (d) E or F occurs but not both (e)none of these

13. An example of mutually exclusive events E and F when a card is selected from a deck is
    (a) E = heart, F = king (b) E = face card, F = diamond
    (c) E = black card, F = heart (d) E = ace, F = red card (e) none of these

14. If E and F are events with $E \cap F = \emptyset$ then E and F are called
    (a) mutually exclusive (b) empty (c) subsets
    (d) independent (e) none of these

In problems 15 to 17 the sample space for an experiment is S = {a, b, c, d, e, f}. Events E, and F are defined by E = {a, b, d, e} and F = {b, d, e, f}. Describe the given event.

15. $E \cup F$ =
    (a) {a, b, c, d, e, f} (b) {a, b, d, e} (c) {a, b, d, e, f}
    (d) {b, d, e} (e) none of these

16. $E \cap F$ =
(a) {b, d, e, f} (b) {a, b, d, e} (c) {b, d} (d) {b, d, e} (e) none of these

17. $E \cap F'$ =
(a) {c} (b) {a, b, d, e} (c) {b, d} (d) {a} (e) none of these

In problems 18 to 20 the experiment is to roll two dice and record the sum of the numbers on the top faces. Find the number of ordered pairs corresponding to each of the following events.

18. {2, 3}
(a) 2 (b) 5 (c) 6 (d) 3 (e) 36

19. {9, 10, 11}
(a) 3 (b) 30 (c) 36 (d) 8 (e) 9

20. {6, 7, 8, 9, 10}
(a) 5 (b) 30 (c) 22 (d) 23 (e) 24

4-2 ASSIGNING PROBABILITIES

In the Fairley study in the last section, we said that the relative frequencies with which accidents occurred could be used to predict the likelihood of accidents in the future. In this section, we will look at some of the other examples from the previous section, such as coin tossing and dice rolling, and discuss the probability of their outcomes. One of the main concepts will concern whether or not the individual outcomes in a sample space have the same chance of occurring.

Theoretical and Empirical Probabilities

In this section we will study how to predict the likelihood of an outcome occurring in a probability experiment. There are two basic ways to determine the likelihood of events occurring. When we consider a study using statistics, like the Fairley study of accidents on Route 2, we look at the percentage of the time each outcome occurs. But when we consider a probability experiment, such as tossing two coins or rolling two dice, we usually discuss how often an outcome is likely to occur using a discussion of whether the outcomes are equally likely.

In this section we assign **probabilities** to the outcomes in the sample space. The probability of an event is a measure of how likely the event is to occur. If we assign probabilities to events by designing a sample space so that the outcomes are equally likely, we call them **theoretical probabilities**. On the other hand, if the sample space is derived by observing the frequencies of events in the real world, such as observing how many accidents occur on a certain stretch of highway during the twelve months of the year, then the probabilities are the relative frequencies and are called **empirical probabilities.**

Equally Likely Events

When we toss one coin, the two usual outcomes, "heads" and "tails," are assumed to be equally likely and such a coin is called a "fair" or "balanced" coin. To express this simple idea mathematically, let us define the sample space with H representing "heads" and T representing "tails," so the experiment has the sample space S = {H, T}. Then we say that the **probability of the event H is 1/2**, which we express as **Pr(H) = 1/2**. In the same way, **Pr(T) = 1/2**.

The fact that the probability for heads is one-half does not mean that heads will occur *exactly* half of the time. Rather, this probability means that if a coin is tossed many times, say 1000 times, we would expect heads to occur about half of the time, or *about* 500 times. On the other hand, if a coin is tossed only a few times, say 10 times, then the chances that heads occurs approximately half of the time, say 4, 5, or 6 times, is not extremely high.

A **probability distribution** is a table listing all of the outcomes of a sample space together with the probability of each outcome. Since Pr(H) = 1/2 and Pr(T) = 1/2, and since these two are the only outcomes, the probability distribution for the experiment of tossing one coin is given by **Table 4–2**.

TOSS ONE COIN

Outcome	Probability
H	1/2
T	1/2

Table 4–2

Another simple probability experiment is that of rolling one die and recording the number on the top face. As in tossing a coin, we assume the six outcomes, 1 through 6, are equally likely to occur, assuming that the die is fair. The outcomes being equally likely means that none of the outcomes is more likely to occur than any of the others. Therefore, we should expect each outcome to occur one in every six rolls on average. In general, if an experiment is conducted many times and each of the outcomes occurs with the same frequency over the long run, then the events are said to be **equally likely**. For example if we rolled a die many times, let's say 6000 times, each face would come up approximately 1000 times.

ROLL ONE DIE

Outcome	Probability
1	1/6
2	1/6
3	1/6
4	1/6
5	1/6
6	1/6

Table 4–3

Once we have determined that the outcomes of a probability experiment are equally likely, the next step is to express this fact in a probability distribution. To find the probability distribution for the experiment of rolling one die and recording the number on the top face, we assign the probability of 1/6 to each outcome. The probability distribution is then given by Table 4–3.

This leads to the following general rule for assigning probabilities to outcomes in a sample space if the outcomes are equally likely.

Probability of an Event

Suppose a sample space of an experiment has N outcomes that are equally likely. Then the **probability of each outcome is 1/*N***. If an event E in such a sample space consists of m distinct outcomes, then the **probability of E**, written Pr(E), is

$$\Pr(E) = \frac{m}{N}$$

The next example shows how to compute the probability of events that consist of more than one outcome.

EXAMPLE 1

Find the probability of each of the following events for the probability experiment of rolling one die and recording the number on the top face.

(a) E = {3, 4}
(b) F = {1, 3, 4, 5, 6}
(c) G is the event that an even number occurs
(d) D = F′

Solution. Recall that the sample space is given by S = {1, 2, 3, 4, 5, 6} and that the outcomes are equally likely.

(a) There are two elements in E. Since the sample space has six elements we have $m = 2$ and $N = 6$ in the formula and so Pr(E) = 2/6 = 1/3.
(b) There are five elements in F. In the formula let $m = 5$ and $N = 6$ to get Pr(F) = 5/6.
(c) G = {2, 4, 6} so there are three elements in G. In the formula let $m = 3$ and $N = 6$ to get Pr(G) = 3/6 = 1/2.
(d) D = F′ = { 2 } so there is one element in D. In the formula let $m = 1$ and $N = 6$ to get Pr(D) = 1/6.

Outcomes That Are Not Equally Likely

In many cases, the elements of a sample space are not equally likely. In these cases, we will have to list the elements of the sample space in a different way so that we can construct a sample space that does have equally likely outcomes and is related to our original sample space.

For example, when a fair coin is tossed more than once, the first sample space that comes to mind, namely the number of heads that occurs, is not very useful because the outcomes are not equally likely.

Consider the experiment of tossing two coins and recording the number of heads observed. The sample space is S = {0, 1, 2}. You might guess that the outcomes are equally likely, but a simple test will cause you to begin to doubt this guess. If you were to toss two coins several times, say 100 times. You would probably see that "1 Head" occurs much more often than either of the other two outcomes. Such a result should bring into question the guess that all three outcomes are equally likely.

To determine the correct probabilities we need another sample space in which the outcomes are equally likely. To create this sample space, use the tree diagram for tossing two coins that results in the sample space

S = {HH, HT, TH, TT}

In fact, these four outcomes are equally likely. This assertion is not too hard to believe. To start, just consider the first coin by itself. We know that for a fair coin heads and tails are equally likely on one toss, so the first coin should be heads as often as it is tails. The regardless of how the first coin lands, the second coin should be heads as often as tails. Therefore, there is no reason for any one of the four outcomes to occur more often than any of the others. Since there are four outcomes that are equally likely, they should each occur one in every four tosses, meaning that the probability of each is 1/4, as shown in the probability distribution given in Table 4–4.

TOSS TWO COINS

Outcome	Probability
HH	1/4
HT	1/4
TH	1/4
TT	1/4

Table 4–4

EXAMPLE 2

Find the probability of each of the following events for the probability experiment of tossing two coins and recording the sequence of H's and T's.

(a) Let A be the event "exactly one head occurs."
(b) Let B be the event "at least one head occurs."
(c) Let C be the event "both coins are the same."

Solution. The sample space, S = {HH, HT, TH, TT} has four equally likely outcomes. Therefore, $N = 4$.
(a) Event A is given by the set {HT, TH}. Since there are two distinct outcomes of S in A, we have $m = 2$ and $N = 4$. Therefore, Pr(A) = 2/4 = 1/2.
(b) Event B is given by the set {HH, HT, TH}. Since there are three distinct outcomes of S in B, we have $m = 3$ and $N = 4$. Therefore, Pr(A) = 3/4.
(c) Event C is given by the set {HH, TT}. Since there are two distinct outcomes of S in C, we have $m = 2$ and $N = 4$. Therefore, Pr(A) = 2/4 = 1/2.

Now we can assign probabilities to the experiment of tossing two coins and recording the number of heads by using the probabilities in the previous distribution. In the sample space S = {HH, HT, TH, TT}, the outcome "TT" corresponds to "0 heads" occurring, the outcomes "HT" and "TH" correspond to "1 head" occurring, and the outcome "HH" corresponds to "2 heads occurring." Now, if the number of heads is recorded, the sample space is S = {0, 1, 2}. Since there is one way in four equally likely ways to get 1 head, Pr(0) = 1/4. Since there are two ways in four to get one head, Pr(1) = 2/4 = 1/2. Finally, since there is one way in four to get two heads, Pr(2) = 1/4. These results are summarized in the probability distribution given in Table 4–5.

TOSS TWO COINS,
Record # of Heads

Outcome	Probability
0	1/4
1	1/2
2	1/4

Table 4–5

The model of two coins can be used for any experiment with two equally likely outcomes that is performed twice. For example, you can use this idea to determine the probability of getting two True/False questions correct if you guess at the answer, or you could use the idea of two coins to determine the probability of having one boy and one girl if you are planning on having two children.

Let us look at another example of how to assign probabilities when the outcomes are not equally likely.

EXAMPLE 3

A box has three differently colored marbles - red, white, and blue. There are 9 red marbles, 6 white marbles, and 3 blue marbles. With the exception of their colors, the marbles are identical. Find the sample space and create the probability distribution for picking one marble from the box.

Solution. There are three possible outcomes - red, white and blue. So the sample space is {red, white, blue}. Obviously, these outcomes are not equally likely since, for example, there are more red marbles than white marbles. Even though the colors are not equally likely, each of the 18 marbles is equally likely. Therefore, we should think of this experiment as having 18 equally likely outcomes. Since there are 9 red marbles, Pr(red) = 9/18 = 1/2. Since there are 6 white marbles, Pr(white) = 6/18 = 1/3. Finally, since there are 3 blue marbles, Pr(blue) = 3/18 = 1/6. These results are summarized in the probability distribution given in Table 4–6.

Outcome	Probability
Red	1/2
White	1/3
Blue	1/6

Table 4–6

Problems that can be addressed in a manner similar to Example 3 would include any where a set of objects can be broken own into disjoint subsets where the subsets are not necessarily the same size. For example, if 55 men and 80 women attend a class reunion and a door prize is given out, the probability that the door prize will be won by a woman is 80/135. (Note that 135 = 80 + 55.)

Roll Two Dice and Record the Sum

The experiment of rolling two dice and recording the sum of the top faces is another example of a sample space whose outcomes are not equally likely. The sample space for this experiment is S = {2, 3, 4, 5, 6, 7, 8, 9, 10, 11, 12}. Though this sample space consists of the outcomes that are usually desired in games involving dice, it is not very useful for finding probabilities because the 11 outcomes are not equally likely.

As seen in the previous section, there is another sample space that is natural when two dice are rolled. If instead of recording the sums, the numbers on the faces are recorded as ordered pairs, the sample space consists of the following 36 elements.

S ={ (1, 1), (1, 2), (1, 3), (1, 4), (1, 5), (1, 6),
(2, 1), (2, 2), (2, 3), (2, 4), (2, 5), (2, 6),
(3, 1), (3, 2), (3, 3), (3, 4), (3, 5), (3, 6),
(4, 1), (4, 2), (4, 3), (4, 4), (4, 5), (4, 6),
(5, 1), (5, 2), (5, 3), (5, 4), (5, 5), (5, 6),
(6, 1), (6, 2), (6, 3), (6, 4), (6, 5), (6, 6) }

The important fact in this case is that these 36 outcomes *are equally likely*. The reason why they are equally likely is not too different from the reason that the sample space {HH, HT, TH, TT} consists of four equally likely outcomes. Essentially, *a die is a six-sided coin*. Instead of having two equally likely outcomes like a coin, a die has six equally likely outcomes. So looking at the two dice one-at-a-time, there is no reason to get, for example, a "1" any more or less often than any of the other five outcomes. Similarly, once the first die is rolled, there is no reason to get, for example, a "4" any more or less often than any of the other outcomes. Therefore none of the ordered pairs, such as (1, 4) should occur any more or less frequently than any of the other 35 ordered pairs. Therefore, the probability of each ordered pair is 1/36.

Finally, to determine each of the probabilities of the sums given in the sample space consisting of 2 through 12, we need only see how many of the ordered pairs correspond to each sum. For example, the ordered pairs (1, 4), (2, 3), (3, 2), and (4, 1) correspond to "a sum of 5." Since "a sum of 5" corresponds to 4 equally likely outcomes in a sample space of 36 equally likely outcomes, the probability of rolling a sum of 5 is 4/36, or 1/9. Since the number of ordered pairs that correspond to each sum plays such a crucial role, it is a good idea to create a table of these numbers as is done in Table 4–7. Note that the number of ways of obtaining a sum increases from 1 to 6 and then decreases back to 1. A sum of 7 will occur most frequently, while a sum of 2 and a sum of 12 are tied for the least frequent occurrence.

Sum of Two Dice	Number of Ordered Pairs
2	1
3	2
4	3
5	4
6	5
7	6
8	5
9	4
10	3
11	2
12	1

Table 4–7

EXAMPLE 4

Two dice are rolled, and the sum of the faces is recorded. Find the probability of each of the following events.

(a) E is the event that a sum of 7 or 11 occurs
(b) F is the event that a sum of 2, 3, or 12 occurs
(c) G is the event that a sum of at least 9 occurs
(d) H is the event that doubles occur

Solution. Refer to Table 4–7.

(a) From Table 4–7, there are six ordered pairs that correspond to "a sum of 7", and two other ordered pairs that correspond to "a sum of 11." Therefore, there are 6 ordered pairs corresponding to event E, and Pr(E) = 6/36 = 1/6.
(b) From Table 4–7, there is one ordered pair that corresponds to "a sum of 2," two other ordered pairs that correspond to "a sum of 3," and one more ordered pair that corresponds to "a sum of 12." Therefore, there are 4 ordered pairs corresponding to event F, and Pr(F) = 4/36 = 1/9.
(c) The event G is given by the set {9, 10, 11, 12}. From Table 4–7, there are four ordered pairs that correspond to "a sum of 9," three other ordered pairs that correspond to "a sum of 10," two more ordered pairs that correspond to "a sum of 11," and one more ordered pair that corresponds to "a sum of 12." Therefore, there are 10 ordered pairs corresponding to event G, and Pr(G) = 10/36 = 5/18.
(d) For event H, Table 4–7 can no longer be used since it only gives the sums of the dice, and here we are asking for more information. However, the probability of event H can still be found by listing the ordered pairs that satisfy event H. The ordered pairs producing doubles are given by the set {(1, 1), (2, 2), (3, 3), (4, 4), (5, 5), (6, 6)}. Since this event contains 6 of the 36 equally likely outcomes, Pr(H) = 6/36 = 1/6.

Finally, dividing every one of the outcomes in Table 4–7 by 36 will produce a probability distribution for the sum of two dice as shown in Table 4–8.

RECORDING THE SUM
OF TWO DICE

Outcome	Probability
2	1/36
3	2/36
4	3/36
5	4/36
6	5/36
7	6/36
8	5/36
9	4/36
10	3/36
11	2/36
12	1/36

Table 4–8

Odds

When probabilities are assigned in empirical studies, they are usually based on past experience. Theoretically, if the probability of 1/4 is assigned to the event E, it means that if the exact experiment were performed many times, E would occur about 1/4 of the time. But often the experiment will be conducted only once, like when a weather forecaster predicts the likelihood that a hurricane will hit land, or an economist predicts the likelihood that the Dow-Jones average will increase by at least 20%. Frequently this type of probability is expressed in terms of "odds." Consider the statements:

The odds against a hurricane reaching land are 3 to 1
The odds in favor of a 20% or more increase in the Dow-Jones average are 5 to 3

The first statement means that, under exactly the same weather conditions, for every three times the hurricane does not reach land, it will reach land one time. Let E be the event that the hurricane will not reach land. Then E will occur 3 out of 4 times, so Pr(E) = 3/4. Notice that the first statement mentions the "odds against" the event occurring while the second statement mentions the "odds in favor". One is the complement of the other, so when computing odds, we must be careful which odds are desired. If the odds against an event are 3 to 1, then the odds in favor of the event are 1 to 3. We generalize these remarks as follows.

DEFINITION

- If the odds in favor of an event E are m to n, then

$$\Pr(E) = \frac{m}{m+n}$$

- If $\Pr(E) = m/N$ then the odds in favor of E are m to $N - m$.
- If the odds in favor of E are m to n then the odds against E are n to m.

EXAMPLE 5

The odds in favor of a general strike are 6 to 5.
(a) What is the probability that a strike will take place?
(b) What are the odds against a strike?

Solution.
(a) Let E be the event "a strike will take place." Then $m = 6$ and $n = 5$ in the formula, so

$$\Pr(E) = \frac{6}{6+5} = \frac{6}{11}$$

(b) The probability that a strike will not occur is $1 - 6/11 = 5/11$ so the odds against a strike are 5 to 6.

An alternate formula for odds in favor of event E is

$$\text{Odds in favor of E} = \frac{\Pr(E)}{\Pr(E')} = \frac{\Pr(E)}{1-\Pr(E)}$$

This definition is used if Pr(E) is given and the odds are to be computed. The original definition is used when the odds are given and Pr(E) is to be computed. The next example illustrates that the two definitions are equivalent.

EXAMPLE 6

The odds in favor of an increase tomorrow in the Dow-Jones average are 5 to 3. Let E represent the event that an increase will take place tomorrow.
(a) Use the original formula to compute Pr(E).
(b) Use Pr(E) computed in (a) in the alternate formula to compute the odds in favor of E.

Solution.
(a) From the original formula, $m = 5$ and $n = 3$, so

$$\Pr(E) = \frac{5}{5+3} = \frac{5}{8}$$

(b) Since the odds are given as 5 to 3, the answer should be 5/3 if the definitions are equivalent. Substituting $\Pr(E) = 5/8$ and $\Pr(E') = 1 - \Pr(E)) = 1 - 5/8 = 3/8$ into the alternate formula yields

EXERCISE SET 4-2

In problems 1 to 14 the experiment is to roll a die and record the number on the top face. Find the probability of each event.

1. E = {5, 6}
2. E = {1, 4}
3. E = {1, 3, 5}
4. E = {1, 2, 3, 4, 5}
5. E is the event "a number less than 4 occurs"
6. E is the event "a number greater than 4 occurs"
7. E is the event "at least a 3 occurs"
8. E is the event "at least a 5 occurs"
9. E is the event "at most a 2 occurs"
10. E is the event "at most a 4 occurs"
11. E is the set of numbers not in event E of problem 9
12. E is the set of numbers not in event E of problem 10
13. E is the event "a 3 does not occur"
14. E is the event "a 1 does not occur"

In problems 15 to 20 find the probability of the following events for the probability experiment of tossing two coins and recording the sequence of heads and tails.

15. E = {HH, TH, TT}
16. E = {TH}
17. E is the event "an even number of heads occurs"
18. E is the event "an odd number of tails occurs"
19. E is the event "at most one head occurs"
20. E is the event "at least one head occurs"

In problems 21 to 24 a fair die has different colored sides. The experiment consists of tossing the die and recording the color of the top face. Find the sample space and assign probabilities to the outcomes in a probability distribution.

21. One side is red and five sides are blue.
22. One side is red, one side is green and four sides are blue.
23. One side is red, one side is green, two sides are white and two sides are blue.
24. One side is red, one side is green, one side is white and three sides are blue.

In problems 25 to 34 the experiment is to roll two dice and record the sum of the numbers on the top faces. Find the probability of each event.

25. E = {2, 12}

26. E = {3, 4}

27. E = {2, 3, 4, 7, 12}

28. E = {3, 4, 6, 8, 11}

29. E is the set of numbers less than 7.

30. E is the set of numbers greater than or equal to 9.

31. E is the set of multiples of 4.

32. E is the set of multiples of 3.

33. E is the set of numbers not in event E of problem 27.

34. E is the set of numbers not in event E of problem 28.

In problems 35 to 40 the experiment is to shuffle an ordinary deck of cards, a card is selected at random and its suit is recorded. Assuming that the outcomes are equally likely, find the probability of each event.

35. E is the event that the card is a red card.

36. E is the event that the card is a spade.

37. E is the event that the card is a ten.

38. E is the event that the card is a nine.

39. E is the event that the card is a face card.

40. E is the event that the card is not a face card.

In problems 41 to 44 find Pr(E).

41. The odds in favor of E are 4 to 7.

42. The odds in favor of E are 8 to 3.

43. The odds against E are 5 to 2.

44. The odds against E are 3 to 7.

In problems 45 to 48 find the odds in favor of E and the odds against F.

45. Pr(E) = 2/9, Pr(F) = 3/7

46. Pr(E) = 3/4, Pr(F) = 1/8

47. Pr(E) = .26, Pr(F) = .38

48. Pr(E) = .42, Pr(F) = .05

In problems 49 to 54 the experiment is to toss three coins and record the number of heads observed. Find the probability of each event.

49. E is the event "all three coins show heads."

50. E is the event "all three coins show heads."
51. E is the event "at least 2 heads occur."
52. E is the event "at least 1 head occurs."
53. E is the event "an even number of heads occurs."
54. E is the event "more tails than heads occur."

In problems 55 to 60 assume that the couple does in fact have the number of children that they plan.

55. A couple is planning on having two children, what is the probability that they have one boy and one girl?
56. A couple is planning on having two children, what is the probability that they have two girls?
57. A couple is planning on having three children, what is the probability that they have all boys?
58. A couple is planning on having three children, what is the probability that they have one boy and two girls?
59. A couple is planning on having three children, what is the probability that they have at least one girl?
60. A couple is planning on having three children, what is the probability that they have at most one girl?

For problems 61 to 70 assume that you are taking a True/False test, and that you run out of time before you get a chance to answer all of the questions. You then quickly write down either "True" or "False" for each question without reading the question.

61. If there are two questions remaining when you write down your responses, what is the probability that you get them both right?
62. If there are two questions remaining when you write down your responses, what is the probability that you get them both wrong?
63. If there are two questions remaining when you write down your responses, what is the probability that you get one right and one wrong?
64. If there are two questions remaining when you write down your responses, what is the probability that you get at least one of them right?
65. If there are three questions remaining when you write down your responses, what is the probability that you get all three right?
66. If there are three questions remaining when you write down your responses, what is the probability that you get two right and one wrong?
67. If there are three questions remaining when you write down your responses, what is the probability that you get at least one right?
68. If there are three questions remaining when you write down your responses, what is the probability that you get at least one wrong?
69. If there are four questions remaining when you write down your responses, what is the probability that you get all four right?

70. If there are four questions remaining when you write down your responses, what is the probability that you get all four wrong?

71. Suppose Pr(E) = .35 and Pr(F) = .65.
Find (a) the odds in favor of E and (b) the odds against F.

72. The odds in favor of the first proposal passing are 7 to 6, the odds of the second proposal passing are 9 to 4 and the odds that both will pass are 5 to 8. What are the odds that at least one will pass?

Cumulative Exercise Set 4-2

In problems 1 to 4 find the sample space of the experiment.

1. Two dice are rolled and the sum of the top faces is recorded.
2. Three dice are rolled and the sum of the top faces is recorded.
3. Four coins are tossed and the number of heads is recorded.
4. Seven coins are tossed and the number of heads is recorded.

In problems 5 and 6 the experiment has the sample space S = {1, 2, 3, 4, 5, 6, 7, 8, 9, 10}. Describe the following events in terms of sets.

5. The number is even or less than 6.
6. The number is a prime number and less than 8.

In problems 7 and 8 the sample space S for an experiment consists of the twelve months of the year. Events E, F and G are defined. Describe the following events: (a) $E \cup F$, (b) $E \cap F$, (c) G', (d) $E \cap G'$.

7. E consists of the first four months
F consists of the third through the sixth months
G consists of January, March and October

8. E consists of the last four months
F consists of the third through the ninth months
G consists of January, March and October

In problems 9 to 12 the experiment is to roll two dice and record the sum of the top faces. Find the probability of each event.

9. E = {1, 2, 3, 4, 5}
10. E is the set of numbers less than 5.
11. E is the set of numbers greater than 3.
12. E is the set of numbers greater than 5 and less than 9.

Multiple-Choice Exercise Set 4-2

In problems 1 to 10 the experiment is to roll two dice and record the sum of the top faces.

1. Pr(6) =
   (a) 5/36 (b) 3/36 (c) 2/36 (d) 4/36 (e) none of these

2. Pr(7) =
   (a) 1/36 (b) 4/36 (c) 5 (d) 5/36 (e) none of these

3. Pr(8 or 9) =
   (a) 7/36 (b) 8/36 (c) 9/36 (d) 5/36 (e) none of these

4. Pr(10 or 11) =
   (a) 3/36 (b) 4/36 (c) 7/36 (d) 8/36 (e) none of these

5. Pr(6, 7, or 10) =
   (a) 6/36 (b) 21/36 (c) 17/36 (d) 14/36 (e) none of these

In problems 6 to 10 let the events A, B, and C be defined as follows:

A: the sum is 2, 3 or 8
B: the sum is 3, 4 or 8
C: the sum is even

6. Pr(A) =
   (a) 9/36 (b) 2/36 (c) 7/36 (d) 8/36 (e) none of these

7. $Pr(A \cap B)$ =
   (a) 9/36 (b) 2/36 (c) 7/36 (d) 8/36 (e) none of these

8. $Pr(B \cap C)$ =
   (a) 9/36 (b) 2/36 (c) 7/36 (d) 8/36 (e) none of these

9. $Pr(A \cup B)$ =
   (a) 9/36 (b) 2/36 (c) 7/36 (d) 8/36 (e) none of these

10. $Pr(B \cup C)$ =
    (a) 9/36 (b) 20/36 (c) 7/36 (d) 8/36 (e) none of these

In problems 11 to 15 the experiment is to toss three coins and record the number of heads.

11. Pr(3) =
    (a) 1/8 (b) 2/8 (c) 3/8 (d) 4/8 (e) none of these

12. Pr(2) =
    (a) 1/8 (b) 2/8 (c) 3/8 (d) 4/8 (e) none of these

13. Pr(1) =
    (a) 1/8 (b) 2/8 (c) 3/8 (d) 4/8 (e) none of these

14. Pr(1 or 2) =
    (a) 1/8 (b) 2/8 (c) 3/8 (d) 4/8 (e) none of these

15. Pr(0, 1 or 2) =
    (a) 1/8 (b) 2/8 (c) 3/8 (d) 4/8 (e) none of these

4-3 FUNDAMENTAL PROBABILITY PROPERTIES

In the previous section we were able to compute the probability of an event by counting the number of outcomes contained in the event. This required us to use a sample space whose outcomes were equally likely, as in the experiment of tossing two coins and recording the sequence of H's and T's. If we started with a sample space whose outcomes were not equally likely, as in rolling two dice and recording the sum of the top faces, we had to express the events in another sample space whose outcomes are equally likely. In this section we will study rules that will allow us to compute probabilities of events in additional ways.

The first rule tells us when it is allowable to add probabilities. To explain its meaning we consider the Fairley study of accidents along Route 2 first introduced in Section 1. The number of accidents in a three year span was gathered per quarter. Recall that the quarterly average is m/N where m is the number of accidents that occurred in the quarter and N is the total number of accidents, 1269. For instance, in quarter 1, $m = 302$, and so $m/N = 302/1269 = 0.24$, rounded to two decimal places. Here again is the probability distribution.

Quarter	Quarterly Count	Quarterly Ratio
1	302	0.24
2	244	0.19
3	283	0.22
4	440	0.35
	1269	1.00

Table 4–9

This study can be expressed in terms of probabilities by regarding it as an experiment whose sample space is

{quarter 1, quarter 2, quarter 3, quarter 4}

The number of accidents, or "quarterly count" is not a probability, but the "quarterly ratio," which is the relative frequency of each number, is a probability. It is computed by dividing each quarterly count by the total number of accidents, 1269. We call the quarterly ratios **empirical probabilities** since they are derived from actual data. The probabilities of the four outcomes are, respectively, 0.24, 0.19, 0.22 and 0.35. That is, if we select an accident at random, then the probability that it occurred in quarter 1 is Pr(quarter 1) = 0.24, whereas the probability that it occurred in quarter 2 is Pr(quarter 2) = 0.19. Since these two probabilities are different, the four outcomes are not equally likely.

To picture the experiment with equally likely events, think of it in this way. Let the sample space consist of the 1269 individual accidents. If we select one accident at random, each is as likely to be chosen as any other, so each has probability 1/1269. With this sample space, the event "quarter 1" has 302 outcomes, each with probability 1/1269. To get the probability of "quarter 1" we add them to get 302/1269. We are allowed to add the probabilities because they are mutually exclusive.

Compare the two sample spaces by taking a closer look at the event "quarter 1" in each sample space. The first sample space has 4 outcomes that are not equally likely. The second sample space has 1269 outcomes that are equally likely. In the first sample space, "quarter 1" is a single outcome. In the second sample space "quarter 1" consists of 302 distinct outcomes. In each sample space its probability is the same, 302/1269.

When does it make sense to add two of these numbers? Suppose we wanted to find the probability that an accident chosen at random occurred in quarter 1 or quarter 2. Since there were 302 accidents in quarter 1 and 244 accidents in quarter 2 there were 546 (= 302 + 244) accidents total in quarters 1 and 2. Hence the probability that an accident chosen at random occurred in quarter 1 or 2 is 546/1269 which is 0.43 when rounded to two decimal places. This answer is the same one we would get if we added the two probabilities, 0.24 + 0.19 = 0.43. This example leads us to the following addition rule for outcomes.

Addition Rule for Outcomes

If two events E and F are comprised of one outcome each, then provided that these outcomes are distinct,

$$\Pr(E \text{ or } F) = \Pr(E \cup F) = \Pr(E) + \Pr(F)$$

This rule can be extended to include three or more outcomes. The next example illustrates how to apply this rule.

EXAMPLE 1

Table 4–10 lists the number of homeruns each National League baseball team had by May 30 during the 1997 season.. Compute the probabilities of the following events if one of the homeruns is selected at random.

(a) Find the probability that the homerun was hit by a team from California.

(b) Find the probability that the homerun was hit by a team that had more than 45 homeruns.

Team	Homeruns	Homerun Ratio
Colorado	76	.123
Montreal	57	.093
San Diego	49	.080
Atlanta	47	.076
New York	45	.073
Los Angeles	43	.070
Chicago	41	.067
Houston	39	.063
St. Louis	39	.063
Florida	38	.062
Pittsburgh	37	.060
San Francisco	36	.058
Cincinnati	35	.057
Philadelphia	34	.055
Total	616	1.000

Table 4–10

Solution. Each of the numbers in the "Homerun Ratio" column was obtained by diving the number of homeruns hit by that team by the total number of homeruns 616.

(a) Define event C to be "the homerun was hit by a team from California." Then

$$\Pr(C) = \Pr(\text{San Diego}) + \Pr(\text{Los Angeles}) + \Pr(\text{San Francisco})$$
$$= .080 + .070 + .058 = .208$$

(b) Let event F be "the homerun was hit by a team with more than 45 homeruns." Then

$$\Pr(F) = \Pr(\text{Colorado}) + \Pr(\text{Montreal}) + \Pr(\text{San Diego}) + \Pr(\text{Atlanta})$$
$$= .123 + .093 + .080 + .076 = .372$$

The key idea in the rule for the addition of the probabilities of outcomes is that two distinct outcomes cannot happen at the same time. Events are comprised of outcomes and should therefore follow a similar pattern. Recall that that two events E and F are "mutually exclusive" if $E \cap F = \emptyset$, in other words they cannot happen at the same time. Example 1 suggests that we can add probabilities of any two events as long as they are mutually exclusive. This suggestion allows us to extend the addition rule to events that are mutually exclusive, meaning that if two events are mutually exclusive, the probability of their union is the sum of their probabilities. This idea is expressed in the next principle.

Addition Principle for Mutually Exclusive Events

If two events E and F are mutually exclusive, that is, $E \cap F = \emptyset$, then

$$\Pr(E \text{ or } F) = \Pr(E \cup F) = \Pr(E) + \Pr(F)$$

The next two examples illustrate this principle. In Example 2 recall that the face cards are the kings, queens and jacks.

EXAMPLE 2

The experiment is to draw a card from an ordinary deck. The sample space consists of the 52 cards, each assigned the probability 1/52. Define the events

E: a heart is drawn
F: a black face card is drawn

Find $\Pr(E \cup F)$.

Solution. The events are mutually exclusive since they cannot occur at the same time. Since there are 13 hearts, $\Pr(E) = 13/52$, and since there are 6 black face cards, $\Pr(F) = 6/52$. Therefore,

$$\Pr(E \cup F) = \Pr(E) + \Pr(F) = \frac{13}{52} + \frac{6}{52} = \frac{19}{52}$$

In fact, the Addition Principle for Mutually Exclusive Events can be extended to any number of events as long as every possible pair of events is mutually exclusive.

EXAMPLE 3

Suppose that you purchase a \$1 scratch-off lottery ticket. One in every 10 of these tickets wins \$2; one in every 50 wins \$10, one in every 100 wins \$25, one in every 1000 wins \$100. Each ticket can win only one prize, and none of the other tickets are winners. What is the probability that you buy a winning ticket?

Solution. Here, we have four outcomes with which we are concerned. Since one ticket cannot win more than one prize, every pair of these events is mutually exclusive.

Therefore to calculate the probability of buying a winning ticket, add the probabilities of the four prizes.

$$\Pr(\text{winning}) = \frac{1}{10} + \frac{1}{50} + \frac{1}{100} + \frac{1}{1000} = .131$$

Basic Properties of Distributions

Probability distributions have several important basic properties. When we assign a probability to an outcome, it must be a number between 0 and 1, inclusively. An event that is certain to happen, called a **certain event**, has probability 1 while an **impossible event** has probability 0. For example, if a die is rolled a certain event would be "observe a number less than or equal to 6" and an impossible event is "observe a number greater than 6."

Notice that in Table 4–9 and Table 4–10 the columns "Quarterly Ratio" and "Homerun Ration" each add up to 1. This answer of "1" refers to 100% of the data being accounted for. Since probabilities are closely related to relative frequencies, it makes sense then that the sum of the probabilities of all outcomes in the sample space also add to 1. It is important to note that there should be no overlap between any two outcomes in the sample space for these rules to apply.

Basic Properties of Probability Distributions

Let S be the sample space for an experiment such that S consists of n outcomes. Let the probabilities of these outcomes in S be $p_1, p_2, \ldots, p_n$, respectively. Then

1. For all $i = 1, \ldots, n$, $0 \le p_i \le 1$.
2. $p_1 + p_2 + \ldots + p_n = 1$.
3. If E is the certain event, then Pr(E) = 1.
4. If E is the impossible event, then Pr(E) = 0.

The following two examples illustrate some of these properties.

EXAMPLE 4

A sample space for an experiment is comprised of three outcomes with S = {a, b, c}.
(a) If Pr(a) = .35 and Pr(b) = .47, determine Pr(c).
(b) If Pr(a) = .65 and Pr(b) = .39, determine Pr(c).

Solution. (a) According to Property 2, the sum of the probabilities of all three outcomes must be 1. Therefore,

$$\Pr(a) + \Pr(b) + \Pr(c) = 1$$

Since we know the probability of outcomes a and b, we can compute the probability for outcome c by subtracting the other two probabilities from 1.

$$Pr(c) = 1 - Pr(a) - Pr(b) = 1 - .35 - .47 = .18$$

(b) Similarly, we can find Pr(c) by calculating

$$Pr(c) = 1 - Pr(a) - Pr(b).$$

However, when we do so, we find that Pr(c) = 1 – .65 – .39 = –.04. Since Property 1 states that all probabilities must be between 0 and 1, we find that this probability distribution cannot occur.

EXAMPLE 5

A sample space for an experiment is comprised of three outcomes with S = {a, b, c}. If it is known that outcome a is twice as likely as outcome b, and that outcome b is five times as likely as outcome c, determine the probability distribution.

Solution. (a) According to Property 2, the sum of the probabilities of all three outcomes must be 1. Therefore,

$$Pr(a) + Pr(b) + Pr(c) = 1$$

The facts that "a is twice as likely as b" and "b is five times as likely as c" can be written in equation form as

$$Pr(a) = 2Pr(b)$$
$$Pr(b) = 5Pr(c)$$

Now in the equation Pr(a) + Pr(b) + Pr(c) = 1 substitute for Pr(a) and then for Pr(b) to get

$$\begin{aligned} Pr(a) + Pr(b) + Pr(c) &= 1 \\ 2\,Pr(b) + Pr(b) + Pr(c) &= 1 \\ 10\,Pr(c) + 5\,Pr(c) + Pr(c) &= 1 \\ 16\,Pr(c) &= 1 \\ Pr(c) &= \frac{1}{16} \end{aligned}$$

Once we know Pr(c) = 1/16, we can calculate Pr(a) and Pr(b).

$$Pr(b) = 5Pr(c) = 5(1/16) = 5/16$$
$$Pr(a) = 2Pr(b) = 2(5/16) = 10/16$$

Therefore the probability distribution is given by

Outcome	Probability
a	10/16
b	5/16
c	1/16

Complements

Suppose an urn contains 52 marbles, 4 of which are red and the rest are various other colors. If a marble is drawn at random from the urn, what is the probability that it is not red? There are 4 red marbles so there are 52 - 4 = 48 marbles that are not red, implying that the probability is 48/52 = 12/13. This idea can be expressed in terms of complements. Let E represent "the marble is red". Then the desired event, "the marble is not red" is the complement, E′. In terms of probabilities, the last computation can be expressed as

$$\Pr(E') = \frac{52-4}{52} = \frac{48}{52}$$

Still another way to write it is

$$\Pr(E') = \frac{52}{52} - \frac{4}{52} = 1 - \frac{4}{52} = 1 - \Pr(E)$$

This generalizes into the rule for complements.

Rule for Complements

If E is any event and E′ is its complement, then

$$\Pr(E') = 1 - \Pr(E)$$

Since E is the complement of E′ this rule can also be written in alternate form as

$$\Pr(E) = 1 - \Pr(E')$$

Verification of the Rule for Complements

Since $E \cap E' = \emptyset$, the addition rule for mutually exclusive events yields

$$\Pr(E \cup E') = \Pr(E) + \Pr(E')$$

By definition of E′, $E \cup E' = S$, the sample space, and, since S is the certain event, we have

$$\Pr(E \cup E') = \Pr(S) = 1$$

Hence, putting these two equations together, we get

$$\Pr(E) + \Pr(E') = 1$$

Therefore, $\Pr(E') = 1 - \Pr(E)$.

The next example shows how to apply the complement rule as well as the addition principle for mutually exclusive events.

EXAMPLE 6

The experiment is to roll two dice and record the sum of the numbers on the top faces. The sample space is

$$S = \{2, 3, 4, 5, 6, 7, 8, 9, 10, 11, 12\}$$

Define the following events.

A = A sum of at most 3 occurs
B = A sum of at least 10 occurs
C = A sum of at least 4 occurs

Find the probability of each of the following events.

(a) Pr(A′)
(b) Pr(B′)
(c) Pr(C)
(d) Pr(C′ ∪ B)

RECORDING THE SUM OF TWO DICE

Outcome	Probability
2	1/36
3	2/36
4	3/36
5	4/36
6	5/36
7	6/36
8	5/36
9	4/36
10	3/36
11	2/36
12	1/36

Solution. First list the outcomes in each of the events.

A = {2, 3}
B = {10, 11, 12}
C = {4, 5, 6, 7, 8, 9, 10, 11, 12}

(a) Since event A consists of the outcomes "2" and "3," Using the Addition Principle for Outcomes, we can add their probabilities to find Pr(A).

$$\Pr(A) = \Pr(2 \text{ or } 3) = \Pr(2) + \Pr(3) = \frac{1}{36} + \frac{2}{36} = \frac{3}{36} = \frac{1}{12}$$

Hence, using the rule for complements we get

$$\Pr(A') = 1 - \Pr(A) = 1 - \frac{1}{12} = \frac{11}{12}$$

(b) Pr(B) = Pr(10 or 11 or 12) = Pr(10) + Pr(11) + Pr(12) = 3/36 + 2/36 + 1/36 = 6/36 = 1/6. Hence

$$\Pr(B') = 1 - \Pr(B) = 1 - \frac{1}{6} = \frac{5}{6}$$

(c) It is easier to find C′ and then use the alternate form of the complement rule to find Pr(C). Since C′ = {2, 3}, we have Pr(C′) = Pr(2) + Pr(3) = 1/36 + 2/36 = 3/36 = 1/12. Therefore

$$\Pr(C) = 1 - \Pr(C') = 1 - \frac{1}{12} = \frac{11}{12}$$

(d) Note that we do not yet have a rule that allows us to add the probabilities of events C and B because they are not mutually exclusive. (In the next section we will discover that rule.) However, C′ and B are mutually exclusive, so we can add their probabilities. In part (c), we saw that Pr(C′) = 3/36, and in part (b), we saw that Pr(B) = 6/36. Then

$$\Pr(C' \cup B) = \Pr(C') + \Pr(B) = \frac{3}{36} + \frac{6}{36} = \frac{9}{36} = \frac{1}{4}$$

EXERCISE SET 4-3

In problems 1 to 4 a homerun is picked at random from those given in Table 4–10. Find the probability of each of the following events.

1. The homerun was hit by a team from Pennsylvania.
2. The homerun was hit by a team with an even number of homeruns.
3. The homerun was hit by a team in the National League East. (The National League East is comprised of Atlanta, Florida, Montreal, New York, and Philadelphia.)
4. The homerun was hit by a team in the National League West. (The National League West is comprised of Colorado, Los Angeles, San Diego and San Francisco.)
5. The homerun was hit by a team with fewer than 40 homeruns.
6. The homerun was hit by a team with between 40 and 50 homeruns.

In problems 7 to 12 a card is drawn from an ordinary deck. Determine whether the given pair of events is mutually exclusive. If they are mutually exclusive, then find $Pr(E \cup F)$.

7. E is the event "a red card is drawn"
   F is the event "a spade is drawn"
8. E is the event "a diamond is drawn"
   F is the event "a black card is drawn"
9. E is the event "the ace of spades is drawn"
   F is the event "a black two is drawn"
10. E is the event "a face card is drawn"
    F is the event "a ten is drawn"
11. E is the event "a face card is drawn"
    F is the event "a club is drawn"
12. E is the event "a red three is drawn"
    F is the event "a heart is drawn"

In problems 13 to 18 an experiment has a sample space S = {a, b, c, d, e, f}. Is the given table a possible probability distribution for the experiment? If not, state why.

13.

Outcome	a	b	c	d	e	f
Probability	.05	.1	.25	.3	.25	.05

14.

Outcome	a	b	c	d	e	f
Probability	.1	.2	.3	.3	.05	.05

15.

Outcome	a	b	c	d	e	f
Probability	.1	.2	.3	.3	.2	.1

16.

Outcome	a	b	c	d	e	f
Probability	.1	.1	.1	.2	.2	.2

17.

Outcome	a	b	c	d	e	f
Probability	.4	.3	.1	.1	.2	−.1

18.

Outcome	a	b	c	d	e	f
Probability	.3	.1	.2	−.1	.25	.25

In problems 19 to 22 use complements and the data in Table 4–10 to answer the question.

19. What is the probability that a randomly selected homerun was hit by a team from the United States?

20. What is the probability that a randomly selected homerun was not hit by Colorado?

21. What is the probability that a randomly selected homerun was hit by a team with an odd number of homeruns?

22. What is the probability that a randomly selected homerun was hit by a team with fewer than 50 homeruns?

23. If a day of the year is chosen at random, what is the probability that the day is not in January? (Ignore the possibility of a leap year.)

24. If a day of the year is chosen at random, what is the probability that the day is not in June or July? (Ignore the possibility of a leap year.)

In problems 25 to 30 an experiment has the sample space S = {a, b, c, d} with the following probability distribution:

Outcome	a	b	c	d
Probability	.18	.43	.22	p

Define the events E = {a} and F = {c, d}.

25. Find p.

26. Find Pr(E′).

27. Find Pr(F).

28. Find Pr(F′).

29. Find Pr(E ∪ F).

30. Find Pr(E ∪ F)′.

In problems 31 to 36 an experiment has the sample space S = {a, b, c, d, e, f} with the following probability distribution:

Outcome	a	b	c	d	e	f
Probability	.05	35	.1	.25	.2	p

Define the events E = {a, b, c}and F = {e, f}

31. Find p.

32. Find Pr(E) and Pr(F).

33. Find Pr(E ∪ F).

34. Find Pr(E′)

35. Find Pr(E ∪ F)′.

36. Find Pr(F ∪ F′).

In problems 37 and 238 let E and F be events such that Pr(E) = .3, Pr(F) = .4 and Pr(E ∩ F) = 0.

37. Find Pr(E ∪ F)

38. Find Pr[(E ∪ F)′]

39. In an introductory business course students were asked if they were accounting majors or not. The survey also recorded the sex of the student. The results are given in the table.

	Accounting Major	Other Major
Men	7	15
Women	9	10

Define the events.

E: The student was a female accounting major
F: The student was a male accounting major

Find (a) Pr(E), (b) Pr(F), (c) Pr(E ∪ F).

40. In an introductory political science course students were asked if they were registered Democratic or Republican and whether they were in favor of a business tax increase or not. The results are given in the table.

	Favor Increase	Oppose Increase
Democrat	6	11
Republican	2	12

Define the events.

E: The student was a Democrat who opposed the increase.
F: The student was a Republican who favored the increase.

Find (a) Pr(E), (b) Pr(F), (c) Pr(E ∪ F)

41. Find the probability distribution if an experiment has sample space S = {a, b, c} where outcome a is three times as likely as outcome b, and outcome b is twice as likely as outcome c.

42. Find the probability distribution if an experiment has sample space S = {a, b, c} where outcome a is four times as likely as outcome b, and outcome b is ten times as likely as outcome c.

43. Find the probability distribution if an experiment has sample space S = {a, b, c, d} where outcomes a and b are equally likely, outcome c is twice as likely as outcome a, and outcome d is three times as likely as outcome a.

44. Find the probability distribution if an experiment has sample space S = {a, b, c, d} where outcomes a, b and c are equally likely, and outcome d is five times as likely as outcome a.

In problems 45 and 46 consider an American roulette wheel which has 38 slots. Two of the slots are green and are labeled 0 and 00. The other slots are numbered 1 to 36 and are evenly divided between red and black.

45. If a ball lands in a slot at random, what is the probability that it lands in a green slot or a black slot?

46. If a ball lands in a slot at random, what is the probability that it lands in a slot numbered 0 or 00 or in a slot whose number is between 25 and 36?

In problems 47 to 50 the experiment is to pick a five-letter password. Determine whether the two events E and F in each pair are i) complements, ii) mutually exclusive but not complements or iii) not mutually exclusive.

47. E: the letters in the password are all different
F: at least two of the letters in the password are different

48. E: there are no repeated letters in the password
F: exactly two of the letters in the password are the same

49. E: there are no repeated letters in the password
F: the letters in the password are not all the same

50. E: the letters in the password are all different
F: there are no repeated letters in the password

Referenced Exercise Set 4-3

1. (Graduate admissions) An important issue in our society is determining whether discrimination because of sex or ethnic identity is being practiced. In an article by Peter Bickel, et al[4], the following data were gathered while studying the graduate admissions at the University of California, Berkeley, for the fall quarter, 1973.

	Outcome	
Applicants	Admit	Deny
Men	3738	4704
Women	1494	2827

The question studied was whether the decision to admit or deny was influenced by the sex of the applicant. Define the events.

E: The applicant was male and was admitted
F: The applicant was male and was denied admittance
G: The applicant was female and was admitted
H: The applicant was female and was denied admittance

Find (a) Pr(E ∪ F), (b) Pr(G ∪ H), (c) Pr(E ∪ G), (d) Pr(F ∪ H).

---

[4]Bickel, Peter, et al, "Sex Bias in Graduate Admissions: Data From Berkeley," *Science*, vol. 187(February 7, 1975), pp. 398-404.

2. In an article on the possible underestimating of catastrophic accident probabilities[5], William B. Fairley reviewed data supplied to the Federal Power Commission by importers of liquefied natural gas. A large spill in a port could cause a catastrophic fire, resulting in the loss of lives and property similar to a major accident at a nuclear power plant. Fairley defined the following probability distribution based on natural gas shipments to Staten Island and Providence in a ten-year period.

Outcome	Probability
Accident occurred based on officially-reported data	8/10,000
Accident occurred based on unreported accounts	32/10,000
No accident occurred	p

(a) Determine Pr(No accident occurred)
(b) Determine Pr(An accident occurred)
(c) What is the factor by which Fairley multiplied the historical probability of an accident in order to obtain the probability of an unreported accident? That is, how much larger is the latter than the former?

Cumulative Exercise Set 4-3

1. Find the sample space of the experiment of tossing a die, recording the number on the top face, and then choosing either the color red or blue.

2. An experiment consists of choosing a committee of two individuals from a list of four people whose last initials are A, B, C, D. Determine the sample space of the experiment and list the elements of the events E, F and G.

   E: A and B are on the committee
   F: Neither A nor D is on the committee
   G: Either B, C or D is on the committee

3. The experiment consists of tossing a coin three times and recording the number of heads. Find the sample space and determine the elements in the two events:

   E: More than no heads
   F: An odd number of heads

   Find (a) $E \cup F$, (b) $E \cap F$, (c) F'

4. Find the probability of the events in problem 3 parts (a), (b) and (c).

5. An experiment has the sample space S = {a, b, c, d, e, f} with the following probability distribution:

Outcome	a	b	c	d	e	f
Probability	.15	.25	.10	.05	.20	.25

---

[5]Fairley, William, "Evaluating the 'Small' Probability of a Catastrophic Accident from the Marine Transportation of Liquefied Natural Gas", *Statistics and Public Policy*, by Fairley and Mosteller, ed., Addison-Wesley, Reading, MA, 1977, pp. 331-353.

Define the events

E = {a, b}, F = {c, d, f} and G = {a, f}.

Find (a) Pr(E ∪ F), (b) Pr(E ∪ G), (c) Pr(F′), and (d) Pr[(E ∪ F)′].

6. In an introductory philosophy course students were asked if they were science majors or not. The survey also recorded the sex of the student. The results are given in the table.

	Science major	Not science major
Men	22	10
Women	18	5

E: The student was a science major
F: The student was a female accounting major

Find (a) Pr(E), and (b) Pr(F)

7. Two dice are tossed and the sum of the top faces is recorded. Find Pr(E ∪ F) for the given events E and F.

E: an even sum
F: a sum less than 8

8. A marble is selected at random from an urn with 2 red, 3 white, 5 blue and 4 green marbles. Find Pr(E ∪ F) for the given events E and F.

E: a red or white marble
F: a blue or green marble

9. An experiment is conducted with the following probability distribution.

Outcome	Probability
a	.15
b	.20
c	.30
d	.35

Define the events

E = {a, b}
F = {b, c, d}.

Find the probability of each of the following events

(a) E (b) F (c) E′ (d) F′

10. Suppose Pr(E) = .44, Pr(F) = .31 and Pr(E ∩ F) = 0. Find Pr(E ∪ F).

Multiple-Choice Exercise Set 4-3

1. Which of the following probability experiments has a sample spaces with equally likely outcomes?
   (a) Select a college student and record the student's major.
   (b) Select a sophomore and record the GPA.
   (c) Select a card and record the suit.
   (d) Select a person and record the person's annual income.
   (e) none of these

2. If a sample space has 12 equally likely outcomes, then the probability of each outcome is
   (a) 1/12 (b) 1/11 (c) 1/10 (d) 1/2 (e) none of these

3. If a sample space has 10 equally likely outcomes, then the probability of an event with three outcomes is
   (a) 3/12 (b) 3/10 (c) 1/3 (d) 1/2 (e) none of these

4. When a die is rolled and the number of the top face is recorded, Pr(2) =
   (a) 4 (b) 1/4 (c) 4/6 (d) 0 (e) none of these

5. If two coins are flipped, then Pr(exactly one head) =
   (a) 3/12 (b) 3/10 (c) 1/3 (d) 1/2 (e) none of these

6. If a card is selected from a deck and its suit recorded, then the probability of this event is
   (a) 4 (b) 1/4 (c) 4/6 (d) 0 (e) none of these

7. A die is rolled and the number on the top face is recorded. What is Pr(E) where E is the set of even numbers?
   (a) 1/6 (b) 1/4 (c) 1/2 (d) 1/3 (e) none of these

8. If a coin is flipped twice, then Pr(exactly one head) =
   (a) 3/12 (b) 3/10 (c) 1/3 (d) 1/2 (e) none of these

9. If a card is selected from a deck and its value recorded, then Pr(7) =
   (a) 4/48 (b) 1/13 (c) 7/52 (d) 7/13 (e) none of these

10. If a card is selected from a deck, the probability that it is either a 2 or a queen is
    (a) 8/48 (b) 2/13 (c) 14/52 (d) 8/26 (e) none of these

11. If a coin is flipped twice, then the probability of getting two tails is
    (a) 1/3 (b) 1/4 (c) 1/6 (d) 1/2 (e) none of these

12. If a card is selected from a deck, the probability that it is not a king is
    (a) 4/5248 (b) 12/13 (c) 11/12 (d) 44/48 (e) none of these

13. If three coins are flipped, the probability of getting at most four heads is
    (a) 1 (b) 1/2 (c) 1/3 (d) 0 (e) none of these

14. If two dice are tossed, the probability that the sum is less than two is
    (a) 1 (b) 1/2 (c) 1/3 (d) 0 (e) none of these

4-4 THE ADDITION PRINCIPLE

In the previous section some elementary rules allowed us to compute probabilities of various events. However, in all cases the events were mutually exclusive. In this section we will consider events that are not mutually exclusive, that is, they have a nonempty intersection.

For example, suppose that the experiment is to select a card from a deck. Define the two events E and F as follows.

E: a spade is selected
F: a king is selected

To find Pr(E ∪ F) we count the number of outcomes favorable to the event E ∪ F. There are 13 spades and 4 kings, but we do not have 17 *different* cards since the king of spades is counted twice – once as a spade and once as a king. Thus, in E ∪ F there are 13 spades and only 3 additional kings, for a total of 16 cards. Therefore

$$\Pr(E \cup F) = \frac{16}{52} = \frac{4}{13}$$

The Venn diagram in Figure 4–2 helps to explain where the king of spades causes the difficulty, namely in E ∩ F. Hence to find the probability of the union of two events that are not mutually exclusive, the intersection of the events is critical.

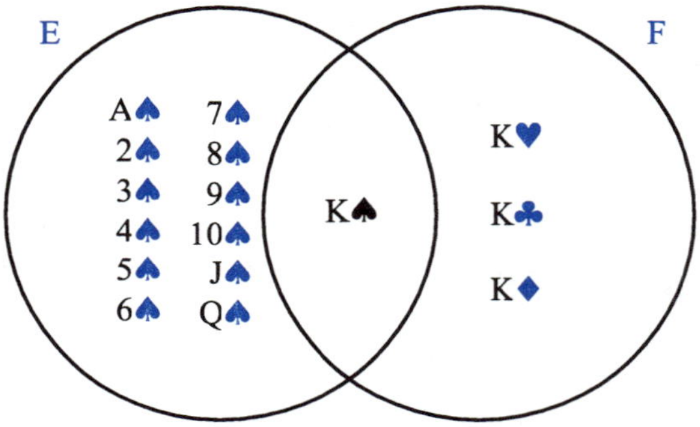

Figure 4–2

The Addition Principle

The example above involving the cards provides the idea needed to find the formula for E ∪ F when the events E and F are not mutually exclusive. When the 13 spades were added to the 4 kings, the resulting answer of 17 was too big because the king of spades had been counted twice. To remedy the situation, the extra king of spades had to be removed.

As pointed out above, the king of spades caused the trouble because it was in the intersection of the two events. In fact, by the same argument, every element in the intersection of E and F is counted twice whenever the number of elements in E is added to the number of elements in F. Therefore, if the intersection of E and F contains, for example, 6 elements, adding the number of elements in E to the number of elements in F will produce a total for E ∪ F that contains 6 elements too many since the intersection is counted twice. Subtracting the number of elements in the intersection from this sum will then result in the correct answer.

This discussion leads to the **Addition Principle**.

The Addition Principle

If E and F are any two events in a sample space,

$$\Pr(E \cup F) = \Pr(E) + \Pr(F) - \Pr(E \cap F)$$

To illustrate how to use the Addition Principle, we consider the above discussion and find $\Pr(E \cup F)$. We have $\Pr(E) = 13/52$, $\Pr(F) = 4/52$ and $\Pr(E \cap F) = 1/52$. Thus

$$\Pr(E \cup F) = \frac{13}{52} + \frac{4}{52} - \frac{1}{52} = \frac{16}{52}$$

The fraction 16/52 can be reduced to 4/13 as was done earlier. However, sometimes it is advantageous to leave the fraction in its unreduced form. In this case, the fraction 16/52 illustrates that 16 of the 52 cards are in $E \cup F$. In general, we will reduce a fraction unless it is more instructive to not reduce. For instance, in the next two examples that further illustrate the Addition Principle, some fractions will be left in their unreduced form because they will be added to other fractions with the same original denominator.

EXAMPLE 1

The experiment is to draw a card from a deck. Define events E, F and G as follows.

E: a heart
F: a face card
G: a red card

Find (a) $\Pr(E \cup F)$ and (b) $\Pr(F \cup G)$.

Solution. There are 13 hearts, 12 face cards and 26 red cards, so $\Pr(E) = 13/52$, $\Pr(F) = 12/52$ and $\Pr(G) = 26/52$.

(a) Since we need to calculate the probability of the event $E \cap F$, it is necessary to first list its elements.

$E \cap F = \{J♥, Q♥, K♥\}$

Therefore, since $E \cap F$ contains three equally likely outcomes, $\Pr(E \cap F) = 3/52$, and

$$\Pr(E \cup F) = \frac{13}{52} + \frac{12}{52} - \frac{3}{52} = \frac{22}{52} = \frac{11}{26}$$

(b) First list the elements in the event $F \cap G$.

$F \cap G = \{J♥, Q♥, K♥, J♦, Q♦, K♦\}$

Therefore, $\Pr(F \cap G) = 6/52$, and

$$\Pr(F \cup G) = \frac{12}{52} + \frac{26}{52} - \frac{6}{52} = \frac{32}{52} = \frac{8}{13}$$

The next example deals with the experiment of tossing two dice and recording the sum of the numbers on the top faces. The probability distribution for this experiment is reproduced as an aid.

EXAMPLE 2

Record Sum of 2 Dice

Outcome	Probability
2	1/36
3	2/36
4	3/36
5	4/36
6	5/36
7	6/36
8	5/36
9	4/36
10	3/36
11	2/36
12	1/36

Toss two dice and record the sum of the top faces. Define the events E, F and G as follows.

E: a sum of 3, 4 or 5 occurs
F: a sum of at most 7 occurs
G: an odd sum occurs

Find (a) Pr(E ∪ G), (b) Pr(F ∪ G)

Solution. We first find Pr(E), Pr(F), and Pr(G). To find Pr(E), add the probabilities of the three sums, 3, 4, and 5, that are in E to get

$$\Pr(E) = \frac{2}{36} + \frac{3}{36} + \frac{4}{36} = \frac{9}{36}$$

Since event F = {2, 3, 4, 5, 6, 7}, we must add together their five respective probabilities to find Pr(F).

$$\Pr(F) = \frac{1}{36} + \frac{2}{36} + \frac{3}{36} + \frac{4}{36} + \frac{5}{36} + \frac{6}{36} = \frac{21}{36}$$

Finally, since G = {3, 5, 7, 9, 11},

$$\Pr(G) = \frac{2}{36} + \frac{4}{36} + \frac{6}{36} + \frac{4}{36} + \frac{2}{36} = \frac{18}{36}$$

(a) The intersection of E and G is given by E ∩ G = {3, 5}, and

$$\Pr(E \cap G) = \frac{2}{36} + \frac{4}{36} = \frac{6}{36}$$

Therefore, from the Addition Principle,

$$\Pr(E \cup G) = \frac{9}{36} + \frac{18}{36} - \frac{6}{36} = \frac{21}{36} = \frac{7}{12}$$

(b) Here, F ∩ G = {3, 5, 7}, and

$$\Pr(F \cap G) = \frac{2}{36} + \frac{4}{36} + \frac{6}{36} = \frac{12}{36}$$

Therefore, from the Addition Principle,

$$\Pr(F \cup G) = \frac{21}{36} + \frac{18}{36} - \frac{12}{36} = \frac{27}{36} = \frac{3}{4}$$

The Addition Principle can be used to find any one of the quantities Pr(E), Pr(F), Pr(E ∪ F), and Pr(E ∩ F) as long as the values of three of them are known.

EXAMPLE 3

(a) Find $Pr(E \cap F)$ if $Pr(E) = .45$, $Pr(F) = .74$, and $Pr(E \cup F) = .9$
(b) Find $Pr(E)$ if $Pr(F) = .25$, $Pr(E \cup F) = .48$, and $Pr(E \cap F) = .21$

Solution. Use the Addition Principle $Pr(E \cup F) = Pr(E) + Pr(F) - Pr(E \cap F)$ to find the requested values.

(a) Substitute the known values into the Addition Principle.

$$.9 = .45 + .74 - Pr(E \cap F) = 1.19 - Pr(E \cap F)$$

Therefore,

$$Pr(E \cap F) = 1.19 - .9 = .29$$

(b) Substitute the known values into the Addition Principle.

$$.48 = Pr(E) + .25 - .21 = Pr(E) + .04$$

Therefore,

$$Pr(E) = .44$$

Since a relative frequency is a type of probability, and since relative frequencies are often given as percentages, percentages can be treated as probabilities and follow the same rules as probabilities. Recall that the word "or" indicates "union," and that the words "and" and "both" indicate "intersection."

EXAMPLE 4

Suppose that in a certain region of the world it rains on 35% of the days, snows on 9% of the days, and either rains or snows on 42% of the days. On what percent of all days does it both rain and snow?

Solution. define the event E to be "it rains on a given day" and event F to be "it snows on a given day." Then we know that $Pr(E) = .35$, $Pr(F) = .09$, and $Pr(E \cup F) = .42$. We then need to use the Addition Principle to find $Pr(E \cap F)$. Substituting into the formula gives

$$.42 = .35 + .09 - Pr(E \cap F) = .44 - Pr(E \cap F)$$

Therefore,

$$Pr(E \cap F) = .44 - .42 = .02$$

Consequently, it both rains and snows on 2% of all days.

An Application to Social Science

The following application from political science provides another example of the addition rule. In an article[6] that studied many uses of probability in the law, Zeisel and Kalvern investigated the perennial debate over the merits of the jury system. The question they raised is: *Would there be a difference between verdicts reached by a jury and verdicts reached alone by a judge in the absence of a jury?* One could argue that if the judge and jury hardly ever differed, the jury would be a wasteful institution; on the other hand, if they differed too often, serious questions would be raised about the rationality of the system, including juries, judges, and the very meaning of justice itself.

Zeisel and Kalvern studied 3576 jury trials in which the judge reported how he or she would have ruled in the absence of a jury. In 33% of the cases studied, the jury acquitted the defendant, but in only 17% of the cases, the judge would have acquitted the defendant. In 14% of the cases they both would have acquitted the defendant.

EXAMPLE 5

A case from the Zeisel and Kalvern data is chosen at random. What is the probability that either the jury acquitted the defendant or the judge would have acquitted the defendant?

Solution. Define the events:

E: the jury acquitted the defendant
F: the judge would have acquitted the defendant

Then $Pr(E) = .33$, $Pr(F) = .17$ and $Pr(E \cap F) = .14$. Using the Addition Principle to find the probability of the desired event, $E \cup F$, yields

$$Pr(E \cup F) = .33 + .17 - .14 = .36$$

Venn Diagrams

In many problems, especially those involving complements, trying to find the answer using formulas such as the Addition Principle only can be very difficult. In these cases, it is usually a good idea to create a Venn diagram. Venn Diagrams employing probabilities are created in much the same way as those using the numbers of elements in sets.

As an example, use the events E and F from Example 3. Start with the overlap of the two circles, and insert the probability of $E \cap F$, in this case 0.14 (Figure 4–3). Then complete the circle for E by making sure that it adds to Pr(E). In this example, $Pr(E) = .33$, therefore, .19 must be inserted in the circle for event E so that the entire

---

[6]Zeisel, Hans, and Kalvern, Harry, "Parking Tickets and Missing Women: Statistics and the Law", *Statistics: A Guide to the Unknown*, Tanur, J. A., ed., Holden-Day, Inc., San Francisco (1972), pp. 102-111. The authors also investigated the 1968 trial of the pediatrician-author Dr. Benjamin Spock for conspiracy to violate the Selective Service act by encouraging resistance to the war in Vietnam. They asked the question, what is the probability that "this total absence of women jurors was an accident?" They computed the probability that it was an accident to be 1/1,000,000,000,000,000,000.

circle adds to .33. Repeat the same process for event F entering .03 on the left so that the circle for event F adds to Pr(F) = .17 (Figure 4–4). Finally, the entire box representing the sample space must add to 1. Therefore, add the three numbers in the circles, and subtract this answer, .36, from 1 to get .64 (Figure 4–5).

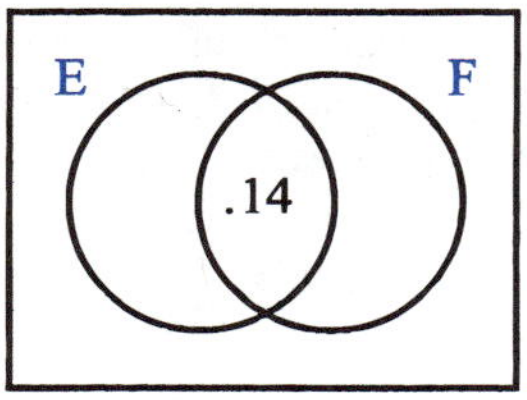

Figure 4–3

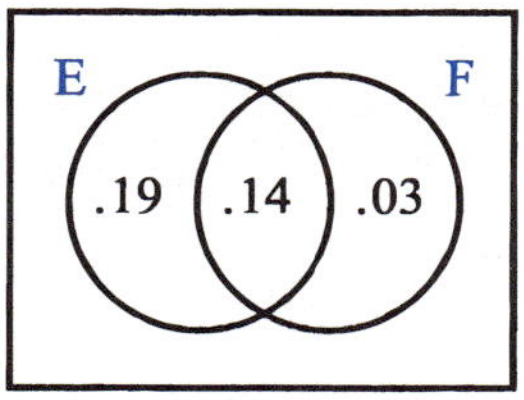

Figure 4–4

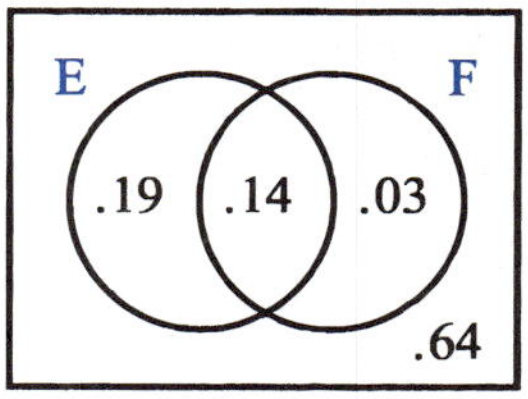

Figure 4–5

A Venn diagram gives another method for finding Pr(E ∪ F), namely adding the three numbers inside the circles. For example, from Figure 4–5,

$$\Pr(E \cup F) = .19 + .14 + .03 = .36$$

which agrees with the answer for Pr(E ∪ F) found in Example 5.

The real benefit of Venn diagrams, though, is that more difficult questions, such as those posed in the following example, can now be answered.

EXAMPLE 6

A case from the Zeisel and Kalvern data is chosen at random. What is the probability that
(a) the jury convicted the defendant and the judge would have acquitted the defendant?
(b) the defendant would have been convicted by both judge and jury?

Solution. Define the events E and F as in Example 5, and refer to Figure 4–5.

(a) Since the jury convicted the defendant, we are looking for the region that is outside of circle E. On the other hand, since the judge would have acquitted the defendant, we are looking for the region that is inside circle F. Since both of these events must be satisfied, we are looking for the intersection of these two regions as shown in Figure 4–6. This region is described as $E' \cap F$. Therefore, $\Pr(E' \cap F) = .03$.

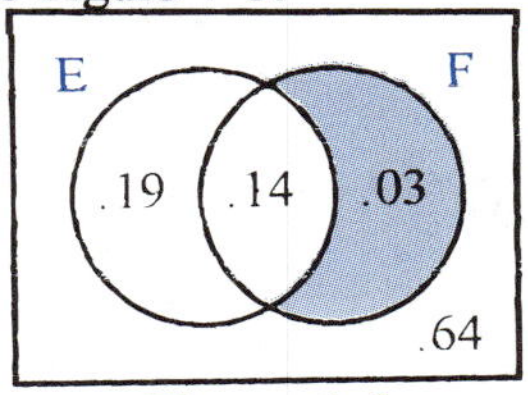

Figure 4–6

(b) The region that is outside of both circles represents the event "the defendant would have been convicted by both judge and jury." Since the region inside the circles represents $E \cup F$, the region outside the circles represents $(E \cup F)'$. Consequently, $\Pr(E \cup F)' = .64$.

Examining the answer to Example 6(a) leads to the fact that in only 3% of the cases would a judge have acquitted the defendant when the jury found the defendant guilty. However, Figure 4–6 shows that in 19% of the cases, the jury reached acquittal when the judge would have found the defendant guilty.

Tables and the Addition Principle

When using tables that place objects into two categories, some care must be taken in determining the probability of the union of two events. For example, suppose that the manager a particular movie theater wants to compare the crowds that were present for *Star Wars* with those that were presents for *Lost World*. The manager sells four kinds of tickets - child (age 12 and under), student (age 13 to 17), adult (age 18 to 64), and senior (65 and over). Below a breakdown of tickets sold by movie and ticket type.

	Star Wars	*Lost World*	Totals
Child	400	250	650
Student	600	900	1500
Adult	1050	1500	2550
Senior	200	100	300
Totals	2250	2750	5000

Table 4–11

Suppose that one ticket is chosen at random. Let C be the event "a child's ticket was sold," and let L be the event "a ticket to *Lost World* was sold." To compute Pr(C), as before, divide the total number of children by the total number of patrons. Therefore, Pr(C) = 650/5000. In a similar manner, Pr(L) = 2750/5000.

To compute probabilities involving intersections and unions, we must know where to look inside the table. The number of outcomes in event $C \cap L$ can be found by looking at the place where the "Child" row intersects the "*Lost World*" column. In other words, $n(C \cap L) = 250$, and hence $Pr(C \cap L) = 250/5000$.

We can now use the Addition Principle to compute $Pr(C \cup L)$.

$$Pr(C \cup L) = Pr(C) + Pr(L) - Pr(C \cap L) = 3150/5000$$

To verify this answer from the table, all numbers in either the "Child" row or the "*Lost World*" column must be considered. The relevant numbers are highlighted in Table 4–12.

	Star Wars	*Lost World*	Totals
Child	400	250	650
Student	600	900	1500
Adult	1050	1500	2550
Senior	200	100	300
Totals	2250	2750	5000

Table 4–12

Adding together the five numbers that are highlighted, gives the number of outcomes in the event $C \cup L$. Therefore,

$$n(C \cup L) = 400 + 250 + 900 + 1500 + 100 = 3150$$

which verifies that $Pr(C \cup L) = 3150/5000$.

It is very important to note that in finding the number of outcomes in $C \cup L$, it is incorrect to simply add the number of child tickets to the number of *Lost World* tickets because then the 250 children who went to *Lost World* are counted twice. This visual depiction of the intersection is very similar to the depiction given by a Venn diagram such as in Figure 4–2 at the beginning of this section.

EXAMPLE 7

Using the data given in Table 4–11, define the following event.

A: an adult ticket was sold
S: a ticket to *Star Wars* was sold

If a ticket is chosen at random, find
(a) $\Pr(A \cap S)$
(b) $\Pr(A \cup S)$

Solution. Note that $\Pr(A) = 2550/5000$ and $\Pr(S) = 2250/5000$
(a) Since there are 1050 people in the intersection of the "adult" row with the "*Star Wars*" column, $\Pr(A \cap S) = 1050/5000$
(b) By the Addition Principle,

$$\Pr(A \cup S) = 2550/5000 + 2250/5000 - 1050/5000 = 3750/5000$$

In this section, we have studied one of the most important formulas in probability, the Addition Principle, in many different ways. Venn diagrams and tables provided very good visual demonstrations that the intersection of two events causes problems when computing the union of these two events. We also saw how to work with the formula for the Addition Principle when faced with a problem where we did not have an actual count of the numbers of objects in each event, but rather had percentages or probabilities.

EXERCISE SET 4-4

In problems 1 to 6 a card is chosen from an ordinary deck. Find $\Pr(E \cup F)$ for the given events E and F.

1. E is the event "a face card is selected"
   F is the event "a black card is selected"
2. E is the event "a red card is selected"
   F is the event "a queen is selected"
3. E is the event "a spade is selected"
   F is the event "an ace is selected"
4. E is the event "a ten is selected"
   F is the event "a diamond is selected"
5. E is the event "a red face card is selected"
   E is the event "a heart is selected"
6. E is the event "a club is selected"
   E is the event "a black five is selected"

In problems 7 to 10 two dice are tossed, and the sum of the top faces is recorded. Find Pr(E ∪ F) for the given events E and F.

7. E is the event "an odd sum occurs"
   F is the event "a sum of 2, 3, or 4 occurs"

8. E is the event "a sum of 6, 7, or 8 occurs"
   F is the event "an even sum occurs"

9. E is the event "a sum of at least 10 occurs"
   F is the event "a sum of 2, 3, or 12"

10. E is the event "a sum of at most 6 occurs"
    F is the event "an even sum occurs"

In problems 11 to 16 determine the indicated probability using the Addition Principle.

11. Find Pr(E ∪ F) if Pr(E) = .2, Pr(F) = .4, and Pr(E ∩ F) = .1

12. Find Pr(E ∪ F) if Pr(E) = .6, Pr(F) = .5, and Pr(E ∩ F) = .2

13. Find Pr(F) if Pr(E) = .35, Pr(E ∪ F) = .7, and Pr(E ∩ F) = .1

14. Find Pr(E) if Pr(F) = .15, Pr(E ∪ F) = .25, and Pr(E ∩ F) = .05

15. Find Pr(E ∩ F) if Pr(E) = .45, Pr(F) = .55, and Pr(E ∪ F) = .75

16. Find Pr(E ∩ F) if Pr(E) = .25, Pr(F) = .2, and Pr(E ∪ F) = .3

17. Suppose that a weatherperson reports that there is a 50% chance of rain on Saturday, a 70 % chance of rain on Sunday, and a 30% chance of rain on both days. What is the probability that it will rain on either Saturday or Sunday?

18. Suppose that a weatherperson reports that there is a 20% chance of snow on Saturday, a 60 % chance of snow on Sunday, and a 5% chance of snow on both days. What is the probability that it will snow on either Saturday or Sunday?

19. If 26% of all cars of a certain model have a cosmetic flaw, 6% have both a mechanical and a cosmetic flaw, and 30% have either a cosmetic or a mechanical flaw, what percent of all of these cars have a mechanical flaw?

20. If 18% of all cars of a certain model have a cosmetic flaw, 7% have a mechanical flaw, and 21% have either a cosmetic or a mechanical flaw, what percent of all of these cars have both a mechanical and a cosmetic flaw?

For problems 21 to 24 assume that Pr(E) = .3, Pr(F) = .5, and Pr(E ∩ F) = .2.

21. Construct a Venn diagram.

22. Use the Venn diagram to find Pr(E ∩ F′).

23. Use the Venn diagram to find Pr(E ∩ F′).

24. Use the Venn diagram to find Pr(E′ ∩ F′).

For problems 25 to 30 assume that Pr(E) = .3, Pr(F) = .5, and Pr(E ∪ F) = .7.

25. Use the Addition Principle to find Pr(E ∩ F)

26. Construct a Venn diagram.

27. Use the Venn diagram to find Pr(E ∩ F′).

28. Use the Venn diagram to find Pr(E′ ∩ F).

29. Use the Venn diagram to find Pr(E ∪ F′).

30. Use the Venn diagram to find Pr(E′ ∪ F′).

31. A car insurance company compiled its records on its customers and classified each customer in two different ways - whether or not they received a ticket for a moving violation within the last three years, and whether or not they were in an accident within the last three years. Below is a table of this data.

	Received Ticket	No Tickets
Accident	65	145
No Accident	15	275

One customer's records are chosen at random.

(a) Find the probability that the customer received a ticket but was not in an accident during this time period.

(b) Find the probability that the customer did not receive a ticket or was not in an accident during this time period.

(c) Find the probability that the customer was not in an accident during this time period.

32. A survey of car owners by a consumers' magazine asked readers several questions. Two of these questions asked of each reader concerned whether the reader's car was bought new or used and whether the car was repaired within the past year or not. The data is given in the table below.

	Bought New	Bought Used
Not Repaired	240	160
Repaired	170	190

One reader's response is chosen at random.

(a) Find the probability that the reader's car was bought new or was not repaired within the past year.

(b) Find the probability that the reader's car was bought used and was repaired within the past year.

(c) Find the probability that the reader's car was bought new.

Problems 33 to 36 refer to the following information.

A computer magazine surveys its readers about their home computer. Two of the questions that were asked involved the type of operating system used (DOS, Windows, or Macintosh), and whether the reader subscribed to an on-line service or not. Below is a table of this data.

	On-line Service	No On-line Service
DOS	20	70
Windows	340	270
Macintosh	60	40

One reader's response is chosen at random, and the events W, M, and S are defined as follows.

W: the reader used the Windows operating system

M: the reader used the Macintosh operating system
S: the reader subscribes to an on-line service

33. Find $Pr(W \cup S)$

34. Find $Pr(M' \cup S)$

35. Find $Pr(W \cup M)$

36. Find $Pr(S')$

Problems 37 to 40 refer to the following information.

Undergraduates in the education department at a university can major in either elementary or secondary education. The table below lists the numbers of students by major and by year in college.

	Elementary	Secondary
Freshman	50	20
Sophomore	41	12
Junior	65	30
Senior	75	25

A student is chosen at random, and the events F, J, and E are defined as follows.

F: the student is a freshman
J: the student is a junior
E: the student is an elementary education major

37. Find $Pr(F \cap E)$

38. Find $Pr(J \cap E')$

39. Find $Pr(J' \cap E)$

40. Find $Pr(F')$

In problems 41 to 44 suppose $Pr(E') = .45$, $Pr(F) = .35$ and $Pr(E \cap F') = .25$.

41. Construct a Venn diagram.

42. Use the Venn diagram to find $Pr(E' \cap F)$.

43. Use the Venn diagram to find $Pr(E \cup F)$.

44. Use the Venn diagram to find $Pr(E \cup F')$.

In problems 45 to 48 suppose $Pr(E') = .65$, $Pr(F') = .45$ and $Pr(E \cup F) = .8$.

45. Construct a Venn diagram.

46. Use the Venn diagram to find $Pr(E' \cap F)$.

47. Use the Venn diagram to find $Pr(E \cap F)$.

48. Use the Venn diagram to find $Pr(E \cup F')$.

Referenced Exercise Set 4-4

1. (Medicine and politics) An article[7] on state-mandated vehicle inspection, studied the effectiveness (as measured by loss of life, personal injury and property damage) of those states that had mandatory vehicle inspection versus those that did not. One measure they considered was the percentage of mortality rates (rate per 100,000 population) by sex. They found that 58% of the deaths were in noninspection states, 24% were female, and 14% were females in noninspection states. Define the events

   E: the accident occurred in a noninspection state
   F: the victim was female

   Find $\Pr(E \cup F)$ and $\Pr(E \cap F')$.

2. The admission process at many universities is a very complex operation, especially in the face of considerable uncertainty of student performance. In an article[8] designed to use probabilistic models to help colleges chose the right admission policy for a particular situation, computer simulation was used to compare student performance of those who were accepted at a major university versus the predicted performance of those who were rejected. Define the events

   E: The applicant was accepted.
   F: The applicant's GPA (Grade Point Average) was or was predicted to be greater than or equal to 2.0

   The event F includes the first year performance of those who were accepted as well as the predicted GPA of those rejected. The data showed that $\Pr(E) = .64$, $\Pr(F) = .60$, and $\Pr(E \cap F) = .50$. Express these three events in words and then find
   (a) $\Pr(E \cup F)$
   (b) $\Pr(E' \cap F)$
   (c) $\Pr(E \cap F')$

Cumulative Exercise Set 4-4

1. Find the sample space of the experiment of rolling a die and recording the number on the top face, and then tossing a coin and recording heads or tails.

2. An experiment consists of choosing a committee of two individuals from a list of four people whose last initials are A, B, C, D. Determine the sample space of the experiment and list the elements of the events E, F and G.

   E: A or C is on the committee
   F: B is not on the committee
   G: C is on the committee but A is not

3. The experiment consists of tossing a coin three times and recording the number of heads. Define the two events:

---

[7]Colton, Theodore, and Buxbaum, Robert, "Motor Vehicle Inspection and Motor Vehicle Accident Mortality," *American Journal of Public Health,* Vol. 58, No. 6, pp. 1090-1099.

[8]Holzman, Albert G. and Johnson, Donald B., "A Simulation Model of the College Admission Process", *Interfaces,* Vol. 5, No. 3, May, 1975.

E: More than one head
F: An even number of heads

Find (a) $E \cup F$, (b) $E \cap F$, (c) E'

4. Find the probability of the events in Problem 3(a), (b), and (c).

5. An experiment has the sample space S = {a, b, c, d, e, f} with the following probability distribution:

Outcome	a	b	c	d	e	f
Probability	0.1	0.1	0.2	0.3	0.2	0.1

Define the events

E = {a, c}, F = {c, d, e}, G = {e, f}

Find (a) $Pr(E \cup F)$, (b) $Pr(E \cup G)$, (c) $Pr(F')$, (d) $Pr(E' \cup G')$.

6. In an introductory business course students were asked if they were marketing majors or not. The survey also recorded the gender of the student. The results are given in the following table.

	Marketing Major	Non-Marketing Major
Men	12	20
Women	8	15

A student is chosen at random.
(a) Find the probability that the student was male and was a marketing major.
(b) Find the probability that the student was female and was not a marketing major.
(c) Find the probability that the student was female or was a marketing major.

7. Two dice are tossed and the sum of the top faces is recorded. Find the probability that
(a) the sum is at most 4
(b) the sum is at least 9
(c) the sum is odd

8. An experiment consists of selecting a marble from an urn containing red, white and blue marbles and recording its color. There are three times as many red marbles as blue marbles and four times as many white marbles as red marbles. Construct the probability distribution and evaluate each of the following probabilities. (The event of selecting a particular color is represented by the first letter of the color.)
(a) Pr(R)
(b) $Pr(R \cup B)$
(c) $Pr(W')$

9. An experiment is conducted with the following probability distribution.

Outcome	a	b	c	d
Probability	.52	.16	.03	p

Define the events

E = {a} and F = {c, d}.

(a) Find p.
(b) Find Pr(E) and Pr(F).

(c) Find $\Pr(E \cup F)$.

10. Suppose $\Pr(E) = .35$, $\Pr(F) = .21$ and $\Pr(E \cap F) = .12$. Find
(a) $\Pr(E' \cap F')$
(b) $\Pr(E' \cup F')$.

11. The following table gives the results of a survey that measured how many people from three levels of income regularly invest in real estate.

		Invest in Real Estate	
		Yes	No
Income level	high	18	6
	medium	21	16
	low	15	24

A person in the survey is chosen at random. Define the events:

H: the person has a high income
L: the person has a low income
Y: the person regularly invests in the stock market

Find the probability of the event (a) $H \cup L$, (b) $H' \cap Y$

12. The odds in favor of the first proposal passing are 8 to 5, the odds of the second proposal passing are 9 to 2 and the odds that both will pass are 4 to 5. What are the odds that at least one proposal will pass?

Multiple-Choice Exercise Set 4-4

In problems 1 to 3 two dice are tossed and the sum of the top faces is recorded. Find $\Pr(E \cup F)$ for the given events E and F.

1. E: a sum of at most 7
F: a sum of 5, 6, or 8
(a) 21/36 (b) 14/36 (c) 26/36 (d) 20/36 (e) 5/36

2. E: a sum of at most 7
F: a sum divisible by 3
(a) 21/36 (b) 12/36 (c) 7/36 (d) 28/36 (e) 5/36

3. E: an odd sum
F: a sum divisible by 3
(a)24/36 (b) 18/36 (c) 12/36 (d) 33/36 (e) 20/36

In problems 4 and 5 a card is chosen from an ordinary deck. Find $\Pr(E \cup F)$ for the given events E and F.

4. E: a black face card
F: an ace or a king
(a) 26/52 (b) 6/52 (c) 14/52 (d) 20/52 (e) 12/52

5. E: a face card that is a club
F: a ten or a king

(a) 11/52 (b) 10/52 (c) 12/52 (d) 8/52 (e) 7/52

In problems 6 to 8 suppose Pr(A) = .3, Pr(B) = .2 and Pr(A ∩ B) = .1.

6. Pr(A ∪ B) =
   (a) 0.3 (b) 0.4 (c) 0.1 (d) 0.2 (e) none of these

7. Pr(A ∪ B′) =
   (a) 0.3 (b) 0.4 (c) 0.1 (d) 0.2 (e) none of these

8. Pr(A′ ∪ B) =
   (a) 0.3 (b) 0.4 (c) 0.1 (d) 0.2 (e) none of these

In problems 9 to 12 an experiment is conducted with the following probability distribution:

Outcome	a	b	c	d	e
Probability	.42	.18	.12	p	.18

Define the events E = {a, b, c}, F = {c, e}, G = {a, c, e}.

9. p =
   (a) .42 (b) .1 (c) .9 (d) .72 (e) .18

10. Pr(E ∪ F) =
    (a) .42 (b) .1 (c) .9 (d) .72 (e) .18

11. Pr(E ∪ G) =
    (a) .42 (b) .1 (c) .9 (d) .72 (e) .18

12. Pr(F′ ∪ G) =
    (a) 0.9 (b) 0.8 (c) 0.1 (d) 0.72 (e) none of these

4-5 CONDITIONAL PROBABILITY

Probability theory is most often used to make predictions about the future, using past performance as a guide. Sometimes the likelihood of an event occurring is altered by new information. In this section we will see how probability handles the situation where a new or different piece of information is included in the problem.

To illustrate consider the experiment of drawing two cards from a deck, and examine the following two problems.

Problem A: Find the probability of observing a king.
Problem B: Suppose the card is known to be a black face card, find the probability that it is a king.

To solve Problem A, note that there are 52 equally likely cards, and since there are four kings, the probability is 4/52.

To solve Problem B note that there is *a restricted sample space*. Since we know the card is a black face card, we are restricted to those six equally likely cards. Two of the black face cards are kings, so the probability is 2/6.

The solution to Problem B is called a **conditional probability**. If E and F are two events, then we write $E|F$ to represent the event "E given that F has occurred" or simply "E given F." For example, in Problem B define the events

E: the card selected is a king
F: the card selected is a black face card

Then $\Pr(E|F) = 2/6$. This expression is a conditional probability and is read "the probability of E given F is 2/6."

Consider for a moment how this answer of 2/6 was computed in Problem B. First we restricted the sample space to the six black face cards, and this number was placed in the denominator. Therefore, the denominator is $n(F)$. Then was looked to see how many of these black face cards were kings and found two cards – the king of clubs and the king of spades. Then "2" was placed in the numerator of the answer.

How is this numerator of "2" related to the events E and F? First we looked at the black face cards, and then we needed to know how many of them were kings, essentially looking at the *intersection* of the two events. In general, the numerator is found by evaluating $n(E \cap F)$. Hence,

$$\Pr(E|F) = \frac{n(E \cap F)}{n(F)}$$

This particular formula works only when the outcomes in the sample space are equally likely and will not work if the outcomes are not equally likely. Therefore a more versatile formula is needed.

If a sample space has N equally likely outcomes, then

$$\Pr(E \cap F) = \frac{n(E \cap F)}{N} \quad \text{and} \quad \Pr(F) = \frac{n(F)}{N}$$

Therefore, the ratio of these two quantities yields the same answer as the formula above.

$$\frac{\Pr(E \cap F)}{\Pr(F)} = \frac{n(E \cap F)/N}{n(F)/N} = \frac{n(E \cap F)}{n(F)} = \Pr(E|F)$$

This formula for conditional probability must be used in problems where there are not equally likely outcomes and in problems when the actual numbers of objects are not present. Recapping the definition gives the following.

DEFINITION

If E and F are two events with Pr(F) $\neq$ 0, then the conditional probability of E given F is

$$\Pr(E|F) = \frac{\Pr(E \cap F)}{\Pr(F)}$$

In cases where the outcomes in the sample space are equally likely, then

$$\Pr(E|F) = \frac{n(E \cap F)}{n(F)}$$

In the following example, we will use the first of these two definitions because the sample space does not have equally likely outcomes.

EXAMPLE 1

Roll two dice and record the sum of the top faces. Define the two events

E: observe an even sum
F: observe a sum of 5, 6 or 7

Find (a) Pr(E| F) and (b) Pr(F | E).

Solution. We will need to compute Pr(E), Pr(F) and Pr(E ∩ F), so it is a good idea to first list the outcomes in each of these events.

$$E = \{2, 4, 6, 8, 10, 12\}$$
$$F = \{5, 6, 7\}$$
$$E \cap F = \{6\}$$

Therefore, using the numbers of ways to get each of these sums,

$$\Pr(E) = \frac{1+3+5+5+3+1}{36} = \frac{18}{36}$$

$$\Pr(F) = \frac{4+5+6}{36} = \frac{15}{36}$$

$$\Pr(E \cap F) = \frac{5}{36}$$

(a) By the definition

$$\Pr(E|F) = \frac{\Pr(E \cap F)}{\Pr(F)} = \frac{5/36}{15/36} = \frac{5}{15} = \frac{1}{3}$$

(b) Since $E \cap F = F \cap E$, by the definition we get

$$\Pr(F\,|\,E) = \frac{\Pr(F \cap E)}{\Pr(E)} = \frac{5/36}{18/36} = \frac{5}{18}$$

Example 1 illustrates a very important point, namely that $\Pr(E\,|\,F)$ and $\Pr(F\,|\,E)$ are almost always different .

Since the outcomes in Example 2 are equally likely, the second form of the equation for $\Pr(E\,|\,F)$ can be used.

EXAMPLE 2

A composition course has 25 students of whom 14 are male and 9 are sophomores. In addition, 5 of the sophomores are male. A student is chosen at random. Find the probability that the student is
(a) a sophomore given that the student is a male
(b) a female given that the student is not a sophomore.

Solution. Define the events

M: the student is male
S: the student is a sophomore

Then we must compute (a) $\Pr(S\,|\,M)$ and (b) $\Pr(M'\,|\,S')$. In order to count the numbers of students in the required events, constructing a Venn diagram (Figure 4–7) for events M and S is a very good idea.

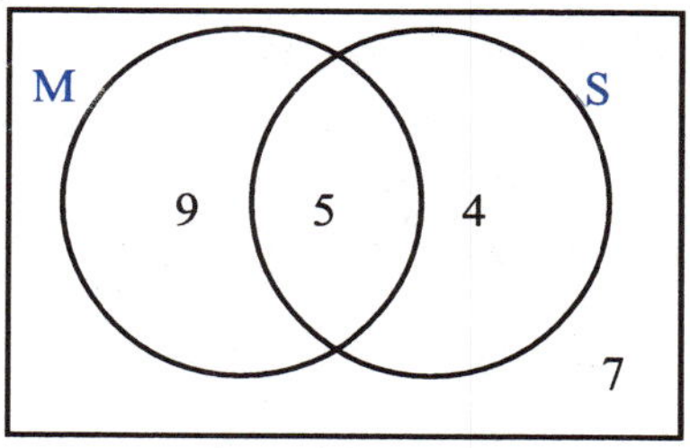

Figure 4–7

(a) From the Venn diagram, we have $n(S \cap M) = 5$ and $n(M) = 14$. Hence

$$\Pr(S\,|\,M) = \frac{n(S \cap M)}{n(M)} = \frac{5}{14}$$

(b) We have $n(M' \cap S) = 4$ and $n(S) = 9$. Hence

$$\Pr(M'\,|\,S') = \frac{n(M' \cap S')}{n(S')} = \frac{7}{16}$$

The next example shows how conditional probability can be used in medicine. The cited article[9] concerns one of the first attempts by the medical profession to distinguish well-designed experiments from poorly-designed experiments. By a well-designed experiment the authors meant a **controlled experiment**, that is, one in which the patients who get the treatment are compared with others who do not. The question of importance was

[9]Gilbert, John P., et. al., "Assessing Social Innovations: An Empirical Base for Policy," *Statistics and Public Policy*, by Fairley and Mosteller, Addison-Wesley, Boston, pp. 185-241, 1977.

"Does a controlled experiment give more reliable results than an uncontrolled experiment?"

The investigators reviewed 53 studies of a particular type of operation, the portavacal shunt, which treats bleeding in the esophagus. They measured the degree of control of the experiment versus the enthusiasm of the physician. They measured whether the study had a good control group, a poor control group or none at all. They compared that measure with whether the physician was very enthusiastic, had moderate enthusiasm or showed no enthusiasm for the operation.

EXAMPLE 3

The following table contains the data of the study of the portavacal shunt operation.

	Degree of Enthusiasm			
Degree of Control	Marked	Moderate	None	Totals
Well-controlled	0	3	3	6
Poorly controlled	10	3	2	15
Uncontrolled	24	7	1	32
Totals	34	13	6	53

Define the events

W: the experiment was well-controlled
P: the experiment was poorly controlled or uncontrolled
M: the physician showed marked enthusiasm
N: the physician showed no enthusiasm

Find (a) $\Pr(M \mid W)$, (b) $\Pr(N \mid W)$, (c) $\Pr(W \mid N)$, (d) $\Pr(P \mid M)$

Solution. Looking at the totals we see that $n(W) = 6$, $n(M) = 34$, and $n(N) = 6$. Since the number of outcomes in the intersection of two events is the number of elements in the overlap of the desired row and column, we have $n(M \cap W) = 0$, $n(N \cap W) = 3$, and $n(P \cap M) = 10 + 24 = 34$. Now we can compute the conditional probabilities using the second formula in the definition of conditional probability.

(a) $$\Pr(M|W) = \frac{n(M \cap W)}{n(W)} = \frac{0}{6} = 0$$

(b) $$\Pr(N|W) = \frac{n(N \cap W)}{n(W)} = \frac{3}{6} = \frac{1}{2}$$

(c) $$\Pr(W|N) = \frac{n(W \cap N)}{n(N)} = \frac{3}{6} = \frac{1}{2}$$

(d) $$\Pr(P|M) = \frac{n(P \cap M)}{n(M)} = \frac{34}{34} = 1$$

The significance of this study is that the investigators found that after four years the survival rates of the patients in the control groups were the same as those who elected to have the operation. Because $\Pr(P \mid M) = 1$, the study shows that uncontrolled and poorly controlled experiments led to far greater enthusiasm than well-controlled experiments.

Later investigation showed, however, that this enthusiasm was unwarranted. On the other hand, since $\Pr(M|W) = 0$, the experimenters that used well-controlled studies had a more accurate opinion of effectiveness of the operation. Consequently, performing poorly-controlled experiments wastes time and can lead to false conclusions.

Product Rule

In applications sometimes a conditional probability is easy to compute but the probability of the intersection is not. In this case the formula in the definition of conditional probability can be altered to give the probability of the intersection. If we multiply each side of the equation in the definition by Pr(F) we obtain the **Product Rule**.

Product Rule

If E and F are any two events then

$$\Pr(E \cap F) = \Pr(F) \cdot \Pr(E|F)$$

The product rule is especially effective when used in conjunction with a tree diagram. The next example shows how tree diagrams can help solve probability problems that occur in stages.

EXAMPLE 4

A utility company submits a rate increase proposal to the state utility commission. The proposal will go to one of two commission members, either Ms Jackson or Mr. Barnes, with Ms Jackson reviewing 70% of all proposals. In the past Ms Jackson has accepted 60% of the company's proposals while Mr. Barnes has accepted 75% of the company's proposals. Determine the probability that the proposal will be accepted.

Solution. First let J be the event "Ms Jackson receives the proposal," let B be the event "Mr. Barnes receives the proposal," let A be the event "the proposal is accepted," and let R be the event "the proposal is rejected." These events then divide the problem into a two stage problem

Stage 1: Either Ms Jackson or Mr. Barnes receives the proposal
Stage 2: The proposal is either accepted or rejected.

This two-stage problem then leads to a tree diagram as in Figure 4–8. However, unlike previous tree diagrams, we can now assign probabilities to each stage. On the branch of the tree leading to J we put .70 since this number is the probability that Ms Jackson receives the proposal. Similarly, since event B is the complement of event J, we have $\Pr(B) = .30$. Therefore .30 is placed on the branch leading to B.

The branch from J to S, represents the proposal being accepted given that Ms Jackson received it. Since she accepts 60% of proposals, we put .60 on this branch. In symbols, we have $\Pr(A|J) = .60$. Since she rejects 40% of her proposals, $\Pr(R|J) = .40$, and this number is placed on the branch from J to R. In a similar manner, on the branch from B to A, we put $\Pr(A|B) = .75$ and on the branch from B to R, we put $\Pr(R|B) = .25$.

There are two paths that correspond to acceptance of the proposal. The first path goes from the start to J and then to A and represents the event Ms Jackson receives the proposal and accepts it," or $J \cap A$. The second path goes from the start to B and then to A and represents the event Mr. Barnes receives the proposal and accepts it," or $B \cap A$. The product rule is used to compute their probabilities by *multiplying along each path.*

$$\Pr(J \cap A) = \Pr(A|J)\cdot\Pr(J) = (.60)(.70) = .42$$

$$\Pr(B \cap A) = \Pr(A|B)\cdot\Pr(B) = (.75)(.30) = .225$$

The event A can occur in one of the two mutually exclusive ways, $A \cap J$ or $A \cap B$. Therefore, using the Addition Principle for Mutually Exclusive Events,

$$\Pr(A) = \Pr(A \cap J) + \Pr(A \cap B) = .42 + .225 = .645$$

Hence there is a 64.5% chance of the proposal being accepted.

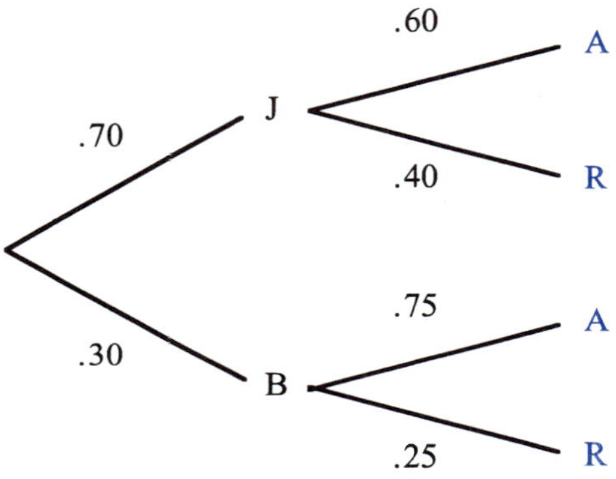

Figure 4–8

The previous example seemed very complicated, but once the tree diagram is constructed, using the formulas is fairly straight-forward. First multiply along all paths that end in the desired event (in this case acceptance of the proposal). Then add all of the answers from the individual paths. In essence, you will be employing the product Rule and the Addition Principle for Mutually Exclusive Events without ever having to write down a single formula!

Independent Events

Each spin of a roulette wheel is a new experiment where the probabilities remain the same. But try to convince gamblers of that! If red has come up 10 straight times, the amount of money placed on black is enormous. Yet the probability that black occurs is still $18/38 = 9/19$. Each spin of the wheel is said to be independent of the others, that is, what happens on one spin does not effect the probability of any other. Two events E and F are said to be independent if the likelihood of one occurring does not affect the chances of the other.

DEFINITION

Two events E and F are **independent** if

$$\Pr(E|F) = \Pr(E)$$

EXAMPLE 5

Two dice are rolled and the sum of the top faces is recorded. Define the three events

E: a sum of 2, 7 or 11
F: an odd number
G: a sum of 4, 7, 10

Determine whether (a) E and F are independent, and (b) F and G are independent.

Solution. First compute the probability of each event.

$\Pr(E) = 9/36 = 1/4$
$\Pr(F) = 18/36 = 1/2$
$\Pr(G) = 12/36 = 1/3$

Then for the conditional probability formula, we need $\Pr(E \cap F)$ and $\Pr(F \cap G)$. List the elements in each event – $E \cap F$ $\{7, 11\}$ and $F \cap G = \{7\}$ – and then compute their probabilities.

$\Pr(E \cap F) = 8/36$
$\Pr(F \cap G) = 6/36$

(a) In order to determine whether E and F are independent or not, we must compare Pr(E) and Pr(E| F). By definition

$$\Pr(E|F) = \frac{\Pr(E \cap F)}{\Pr(F)} = \frac{8/36}{18/36} = \frac{8}{18} = \frac{4}{9}$$

Since $\Pr(E) = 1/4$ and $\Pr(E|F) = 4/9$, $\Pr(E) \neq \Pr(E|F)$. Therefore, E and F are not independent.

(b) In order to determine whether E and F are independent or not, we must compare Pr(F) and Pr(F | G). By definition

$$\Pr(F|G) = \frac{\Pr(F \cap G)}{\Pr(G)} = \frac{6/36}{12/36} = \frac{6}{12} = \frac{1}{2}$$

Since $\Pr(F) = 1/2$ and $\Pr(F|G) = 1/2$, $\Pr(F) = \Pr(F|G)$. Therefore, F and G are independent.

In part (b) of Example 5 we could just as well have shown that $\Pr(G) = \Pr(G|F)$. That is, in the definition of independent events, one equality implies the other. (See Problem 44.) Therefore, the expression "independent events" is not a one-sided relationship. That is, E and F are independent means $\Pr(E|F) = \Pr(E)$ and $\Pr(F|E) = \Pr(F)$.

The definition of independence can be coupled with the Product Rule to form another useful rule. This rule governs when we can multiply the probabilities of independent events.

Suppose E and F are independent. Then $\Pr(E|F) = \Pr(E)$ and so the Product Rule yields

$$\Pr(E \cap F) = \Pr(E|F) \cdot \Pr(F) = \Pr(E) \cdot \Pr(F)$$

The same steps in reverse show that if

$$\Pr(E) \cdot \Pr(F) = \Pr(E \cap F)$$

then E and F are independent. This rule is important so we single it out.

The Product Rule for Independent Events

Two events E and F are independent if and only if

$$\Pr(E) \cdot \Pr(F) = \Pr(E \cap F)$$

For example, in Example 5 we have

$$\Pr(F) \cdot \Pr(G) = \frac{1}{2} \cdot \frac{1}{3} = \frac{1}{6} = \Pr(F \cap G)$$

Thus the Product Rule for Independent Events confirms that F and G are independent.

On the other hand, since $\Pr(E \cap F) = 4/9$

$$\Pr(E) \cdot \Pr(F) = \frac{1}{4} \cdot \frac{1}{2} = \frac{1}{8} \neq \Pr(E \cap F)$$

Therefore, the Product Rule for Independent Events shows that E and F are not independent.

It is very important to only use the Product Rule for Independent Events to determine whether two events are independent or not or when you are certain that two events are independent as in the next example.

EXAMPLE 6

A pair of dice are is rolled twice.
(a) What is the probability of getting "doubles" on both rolls?
(b) What is the probability of getting "doubles" on either the first or the second roll?

Solution. First define the events

E: "doubles" are rolled on the first toss
F: "doubles" are rolled on the second toss

As with the case of the roulette wheel, the outcome of the first toss of the dice has no effect on the probability of the outcome of the second toss. Therefore, events E and F are independent.

(a) The probability of getting "doubles" on both rolls is the same as $\Pr(E \cap F)$. Since the events are independent, we need only multiply $\Pr(E) = 1/6$ by $\Pr(F) = 1/6$ to get the answer.

$$\Pr(E \cap F) = \frac{1}{6} \cdot \frac{1}{6} = \frac{1}{36}$$

(b) The probability of getting "doubles" on either the first or the second roll is the same as $\Pr(E \cup F)$. Using the information from part (a) and the Addition Principle yields

$$\Pr(E \cup F) = \frac{1}{6} + \frac{1}{6} - \frac{1}{36} = \frac{11}{36}$$

Reliability

The Product Rule can be used in the opposite direction also. That is, when two events E and F are known to be independent, $\Pr(E \cap F)$ can be computed by multiplying Pr(E) and Pr(F). An important application of this idea is in reliability. The **reliability** of a machine or process is defined to be the probability that it will not fail during an operation.

For example, suppose the probability of a computer failing is .05. Then its reliability is the probability that it will not fail, which is 1 – .05 = .95. Suppose this computer operates independently with another computer, that is, they both perform the same operation but independently. If the probability of the second computer failing is .02, then the probability of both failing is the intersection of the two independent events that each fails individually. Since we assume that they are independent, to find the probability that both fail we multiply the probabilities and get (.05)(.02) = .001.

This example demonstrates the need for what is often called a *back-up system*. One computer backs up another in the sense that it can handle the load if one fails. A concrete example of such a system is the computer facilities aboard the space shuttle.[10] Each shuttle is equipped with four computers that operate independently. Each has its own tasks but they also serve as back-ups for the others, especially for essential computations. If two do not agree with a calculation the third and fourth are consulted. The historic first flight of the space shuttle Columbia was delayed for several days when the countdown was halted in the last few minutes because the reliability of the computer system was too low.

[10] Lieberman, Gerald J., "Striving for Reliability," *Statistics: A Guide to the Unknown*, by Tanur, Judith, et. al., Holden-Day, San Francisco, pp. 400-406, 1972.

EXERCISE SET 4-5

In problems 1 to 4 a card is chosen from a deck. Define the following three events.

E: a club
F: a black face card
G: a red jack or a black queen

Evaluate each of the following conditional probabilities.

1. Pr(E| F)
2. Pr(F | E)
3. Pr(E| G)
4. Pr(F | G)

In problems 5 to 8 a club has 200 members, 110 of whom are female, and 130 of whom are less than 40 years old. In addition, 60 of the females are less than 40 years old. A member is chosen at random. Find each of the following conditional probabilities.

5. Pr(the person is female given that she is less than 40)
6. Pr(the person is less than 40 given that she is female)
7. Pr(the person is less than 40 given that he is a male)
8. Pr(the person is male given that he is 40 or more)

Problems 9 to 12 refer to the table whose entries are the data from the Kalvern and Zeisel study of criminal jury trials. After the jury gave its verdict the judge was asked what verdict he or she would have given if there had been no jury. The percentages of agreement and disagreement are in the table. One way to analyze the data is by conditional probabilities.

		Jury's Verdict	
		Acquitted	Convicted
Judge's Verdict	Acquitted	14	3
	Convicted	19	64

9. Given that the jury convicted the defendant, what is the probability that the judge would have convicted the defendant?
10. Given that the jury convicted the defendant, what is the probability that the judge would have acquitted the defendant?
11. Given that the judge would have convicted the defendant, what is the probability that the jury acquitted the defendant?
12. Given that the judge would have acquitted the defendant, what is the probability that the jury convicted the defendant?

In problems 13 and 14 a company sends a package by an overnight courier. The package will be picked up by one of 3 delivery vans Van 1 picks up 50% of the packages, Van 2 picks up 30% of the packages, and Van 3 picks up the rest. The problem gives the past record for each van's success in delivering packages on time. Find the probability that the package will be delivered on time.

13. Van 1 delivers the package on time 97% of the time, Van 2 delivers the package on time 95% of the time, Van 3 delivers the package on time 90% of the time.

14. Van 1 delivers the package on time 90% of the time, Van 2 delivers the package on time 93% of the time, Van 3 delivers the package on time 96% of the time.

In problems 15 and 16 Urn 1 and Urn 2 have the indicated number of white and red marbles. The experiment consists of first selecting one of two urns and then selecting a marble from the urn and recording its color. The selection of the urns is decided by a coin flip – heads tells you to pick from Urn 1, while tails tells you to pick from Urn 2. In each problem, find the probability of observing a red marble.

15. Urn 1 has 5 white and 7 red marbles
    Urn 2 has 3 white and 9 red marbles

16. Urn 1 has 9 white and 8 red marbles
    Urn 2 has 2 white and 8 red marbles

In problems 17 to 20 two dice are rolled and the sum of the top faces is recorded. For the given events E and F find Pr(E| F).

17. E: a sum of 2, 3, 4, or 5
    F: a sum of 4, 5, 6, or 7

18. E: a sum of 2, 6, 8, or 12
    F: a sum of 2, 3, 11, or 12

19. E: an odd sum
    F: a sum of 2, 4, or 5

20. E: a sum greater than 5
    F: a sum less than 6

21. Are the events in Problem 17 independent?

22. Are the events in Problem 18 independent?

23. Are the events in Problem 19 independent?

24. Are the events in Problem 20 independent?

In problems 25 to 28 two machines operate independently and have reliability .90 and .95, respectively.

25. Find the probability that they both fail.

26. Find the probability that neither fails.

27. Find the probability that at least one fails.

28. Find the probability that exactly one fails.

29. An urn contains 12 red and 8 white marbles. Two marbles are selected with replacement. Find the probability that (a) both are red, (b) both are white, and (c) one is red and one is white?

30. One percent of all scratch-off lottery tickets are winners. If you buy two of these tickets, what is the probability that (a) both are winners, (b) both are losers, (c) one is a winner and one is a loser?

31. An urn contains 6 red and 9 white marbles. Two marbles are selected without replacement. Find the probability that (a) both are red, (b) both are white, (c) the second is red given that the first is white.

32. An economics class has 40 students, 23 of whom are seniors and 12 of whom are juniors. Five of the juniors and 3 of the seniors are marketing majors, while 4 of the other students are marketing majors. A student is selected at random. Find the probability that the student is (a) a marketing major given that the student is a senior, (b) neither a junior nor a senior given that the student is a marketing major.

33. There are 3 urns: Urn 1 has 1 red and 2 white marbles, Urn 2 has 1 red and 3 white marbles, Urn 3 has 3 red and 2 white marbles. A die is rolled: if 1 occurs then a marble is selected from Urn 1; If 2 or 3 occurs then a marble is selected from Urn 2; if 4, 5 or 6 occurs then a marble is selected from Urn 3. Find the probability that the marble selected is red.

34. For the experiment in problem 31, if a red marble is selected then it is replaced and a second marble is selected from Urn 1, and if the first marble is white a second marble is selected from Urn 2. Find the probability that the second marble is red.

35. A factory has three assembly lines; Line 1, Line 2 and Line 3. The reliability of Line 1 is .90, meaning that 90% of its products are acceptable and 10% are defective. The reliability of Line 2 is .85, and the reliability of Line 3 is .95. If Line 1 makes 60% of the products, Line 2 makes 30% of the products, and Line 3 makes 10% of the products, find the probability that a product selected at random is defective.

36. A survey shows that 10% of a population are left-handed, that 60% of the left-handed people are male, and that 55% of the right-handed people are female. A person is selected at random. What is the probability that this person is (a) a left-handed female (b) a right-handed male

Problems 37 to 40 refer to the table whose entries are the data from a large metropolitan university. Each entry is the percentage of students enrolled in the university that is described by the category heading of its row and column. The categories are full-time and part-time students, and students in the College of Business and students in the College of Arts and Sciences.

College	Full-time	Part-time
Business	25	18
Arts & Sciences	45	12

Find the indicated conditional probability if a student is selected at random.

37. Given that the student is a business major, compute the probability that the student attends full-time.

38. Given that the student is a business major, compute the probability that the student attends part-time.

39. Given that the student attends part-time, compute the probability that the student is an Arts and Science major.

40. Given that the student attends full-time, compute the probability that the student is a business major.

41. Let E and F be events with Pr(E) = .35, Pr(F) = .67 and Pr(E∩F) = .24. Use a Venn diagram to find Pr(E| F ′).

42. Let E and F be events with Pr(E) = .35, Pr(F) = .67 and Pr(E∩F) = .24. Use a Venn diagram to find Pr(E′|F′).

43. Suppose E and F are mutually exclusive events and Pr(F) is not zero. Find Pr(E| F).

44. By definition, if E and F are independent then Pr(E| F) = Pr(E). Show that it is also true that Pr(F | E) = Pr(F).

45. Suppose E and F are independent. Are the events E′ and F′ independent?

46. Suppose E and F are independent. Show that $\Pr(E \cup F) = 1 - \Pr(E')\cdot\Pr(F')$.

47. A family has two children. Suppose that each child is equally likely to be a boy or a girl. What is the conditional probability that both are male given that at least one of the children is a male?

48. For any two events E and F show that Pr(F′ | E) = 1 – Pr(F | E).

Referenced Exercise Set 4-5

1. Brightman and Haslanger[11] studied whether past performance of portfolio managers for mutual funds is a good indication of their ability to perform well in the future. They collected data on 52 portfolio managers over a 6-year period. They ranked each manager's performance according to which quartile they were in (that is, a ranking in the first quartile means the manager was ranked in the top quarter, or top 13). They first measured the performance in the 1971-73 period and then compared that with the performance in the 1974-76 period. The data was given in the following table.

		1971-73 Performance Quartile				
		1	**2**	**3**	**4**	Total
1974-76 Performance Quartile	**1**	3	1	1	8	13
	2	2	3	5	3	13
	3	4	3	5	1	13
	4	4	6	2	1	13
	Total	13	13	13	13	52

[11]Brightman, Haslanger, "Past Investment Performance: Seductive but Deceptive," *Journal of Portfolio Management*, vol. 6, pp. 43-45, 1980.

Find the probability that a manager was ranked

(a) in the first quartile in the 1974-76 period given that the manager was ranked in the first quartile in the 1971-73 period.

(b) in the first quartile in the 1974-76 period given that the manager was ranked in the first or second quartile in the 1971-73 period.

(c) in the third or fourth quartile in the 1974-76 period given that the manager was ranked in the first or second quartile in the 1971-73 period.

2. In one of the first studies[12] of the side effects of oral contraceptive users, Jick, et al., compared the incidence of thromboembolic disease with the user's blood type. A control group of healthy women was studied for comparison. The data is in the following table.

Blood Type	Women with Thromboembolism	Healthy Women
A	32	51
B	8	19
AB	6	5
O	9	70

If a woman is selected at random, find the probability that she

(a) had thromboembolism given that she had blood group A

(b) had thromboembolism given that she had blood group O

(c) had blood group A given that she had thromboembolism.

Cumulative Exercise Set 4-5

1. An experiment consists of tossing a coin five times and recording the number of heads. Find the sample space and determine the elements in the two events:

   E: More than three heads

   F: An even number of heads

   Find (a) $E \cap F$, (b) $E \cup F$, (c) E'

2. Let E and F be events with Pr(E) = .65, Pr(F) = .47 and $Pr(E \cap F) = .14$. Find $Pr(E \cap F')$ and $Pr(E' \cap F')$.

3. A survey of students enrolled in a university revealed the following information about full-time and part-time students, and about students in the College of Business and the College of Arts and Sciences.

	Full-time	Part-time
Business	30	45
Arts & Sciences	45	15

   A student is selected at random. Find

   (a) Pr(the student is full-time and is a business major)

   (b) Pr(the student is part-time or is a business major)

---

[12] Jick, et al., "Venous Thromboembolic Disease and ABO Blood Types", *Lancet*, vol. 1, no. 539, 1969.

4. A card is chosen from a deck. Find the conditional probability for the following events.

   E: a queen or a king
   F: a club or a red card

   (a) Pr(E | F)
   (b) Pr(F | E)

5. The given table has entries that are the data from a study of criminal jury trials.

		Jury's Verdict	
		Acquitted	Convicted
Judge's Verdict	Acquitted	15	5
	Convicted	20	60

   Find the indicated conditional probability.
   (a) Pr(Judge would have convicted | jury acquitted)
   (b) Pr(Judge would have acquitted | jury convicted)

6. Two dice are rolled and the sum of the top faces is recorded. For the given events E and F find Pr(E | F) and determine if E and F are independent.

   E: a sum of 3, 4, 5 or 6
   F: a sum of 4, 5, 6, or 11

7. Two machines operate independently and have reliability .90 and .85, respectively. Find the probability that (a) they both fail, (b) neither fails, and (c) exactly one fails.

8. An accounting class has 50 students, 13 of whom are juniors and 10 of whom are seniors. Five of the juniors, 3 of the seniors, and 9 of the others are marketing majors. A student is selected at random. Find the probability that the student is (a) a marketing major given that the student is a senior, (b) neither a junior nor a senior given that the student is a marketing major.

9. A factory has three assembly lines: Line 1, Line 2 and Line 3. The reliability of Line 1 is .90, meaning that 90% of its products are acceptable and that 10% are defective. The reliability of Line 2 is .80, and the reliability of Line 3 is .85. If Line 1 makes 50% of the products, Line 2 makes 10% of the products, and Line 3 makes 40% of the products, find the probability that a product selected at random is defective.

Multiple-Choice Exercise Set 4-5

1. Let E be the event "a person with lung cancer" and let F be the event "a heavy smoker." The notation for the probability that a person has lung cancer given that the person is a heavy smoker is
   (a) Pr(E | F)   (b) Pr(F | E)   (c) E | F   (d) F | E   (e) none of these

2. If a day of the week is selected at random, E is the event "Saturday," and F is the event "a weekend day" then Pr(E | F) =
   (a) 1/7   (b) 1/5   (c) 1/3   (d) 1/2   (e) none of these

3. If a day of the week is selected at random, E is the event "Friday," and F is the event "a weekend day" then Pr(E | F) =
   (a) 1/7 (b) 1/5 (c) 1/3 (d) 1/2 (e) none of these

4. If a month is selected at random, E is the event "December," and F is the event "ends in 'ber'" then Pr(E | F) =
   (a) 1/3 (b) 1/4 (c) 1/2 (d) 1/5 (e) none of these

5. If a number is selected from {1, 2, …, 9} and F is the event "is an odd number" then Pr(7 | F) =
   (a) 1/3 (b) 1/4 (c) 1/2 (d) 1/5 (e) none of these

6. If Pr(E∩F) = 0.2 and Pr(F) = 0.5 then Pr(E | F) =
   (a) 0.4 (b) 0.04 (c) 0.1 (d) 0.25 (e) none of these

7. If Pr(E | F) = 0.2 and Pr(F) = 0.5 then Pr(E∩F) =
   (a) 0.4 (b) 0.04 (c) 0.1 (d) 0.25 (e) none of these

8. If two dice are rolled, E is the event "the sum is 12," and F is the event "the sum is greater than 10," then Pr(E | F) =
   (a) 1/4 (b) 1/3 (c) 1/2 (d) 1/16 (e) none of these

9. If a card is selected from a deck, E is the event "the card is a queen," and F is the event "the card is a red card," then Pr(E | F) =
   (a) 1/4 (b) 1/3 (c) 1/2 (d) 1/13 (e) none of these

In Problems 11 to 15 a survey of the 40 students in a class revealed that on the night before Test 1, 15 watched TV, 14 talked on the phone, and 10 both watched TV and talked on the phone. A student in the class is selected at random. Let T be the event "watched TV" and P be the event "talked on the phone."

10. Pr(T | P) =
    (a) 14/24 (b) 5/15 (c) 10/14 (d) 4/14 (e) none of these

11. Pr(T ∩ P | T) =
    (a) 14/24 (b) 5/15 (c) 15/29 (d) 4/14 (e) none of these

12. Pr(T | P ′) =
    (a) 5/16 (b) 15/16 (c) 5/26 (d) 1/16 (e) none of these

13. Pr(T ∩ P | T ∪ P) =
    (a) 5/16 (b) 15/16 (c) 5/26 (d) 1/16 (e) none of these

14. Pr(T′ ∩ P ′) =
    (a) 21/40 (b) 31/40 (c) 1/40 (d) 11/40 (e) none of these

CASE STUDY
TRIAL BY JUDGE OR JURY?

As if Rich didn't have problems enough. He was being tried for a crime he didn't commit, which was bad enough, but then his lawyer asked him if he would rather be tried by judge or by jury. Isn't that what you hire a lawyer for, to make those decisions? His lawyer did point out the advantages and disadvantages of each. If your case depends on some technical legal issues that would be difficult for a jury to follow, it is best to select a trial by judge. But if a major part of the defense is emotional, a trial by jury is preferable. Rich's case fit neither description, so it was a toss-up.

The basic question facing Rich and his lawyer is, given that there is no clear reason for choosing one type of trial over the other, which is better for the defendant? And how do you even answer such a question?

Two political scientists, Hans Zeisel and Harry Kalvern, Jr., of the University of Chicago, raised the same issue in an article investigating various ways probability and statistics are used in the law. In one section of the article Zeisel and Kalvern investigated the perennial debate over the merits of the jury system. The question they raised is: Would there be a difference between verdicts reached by a jury and verdicts by a judge if there was no jury? The argument is: if the judge and jury hardly ever differed, the jury would be a wasteful institution; if they differed too often, serious questions would be raised about the rationality of the system, including juries, judges, and the very meaning of justice itself. They studied 3576 jury trials in which the judge reported how he or she would have ruled if there was no jury. The percentages of agreement and disagreement are in the table.

		Judge's Verdict	
		Acquitted	Convicted
Jury's Verdict	Acquitted	14	19
	Convicted	3	64

Table 4–13

View the study as a probability experiment where one of the cases is chosen at random and the decisions of the jury and of the judge are recorded. It is a two stage experiment where the first stage is the jury's verdict (Jur) and the second stage is the judge's verdict (Jud). There are two possibilities at each stage of the experiment, acquit (A) or convict (C). Thus there are four outcomes to the experiment which can be illustrated in the tree diagram shown in Figure 4–9.

Each outcome can also be viewed as an intersection. For example, the outcome "JurA and JudC" can be viewed as the intersection of the event "the jury acquits the defendant" and the event "the judge convicts the defendant". Then the table consists of four numbers (percentages) that represent the probability that a case chosen at random had the given two decisions. Another way to look at the table is:

		Judge's Verdict	
		Acquitted	Convicted
Jury's Verdict	Acquitted	JurA and JudA 14	JurA and JudC 19
	Convicted	JurC and JudA 3	JurC and JudC 64

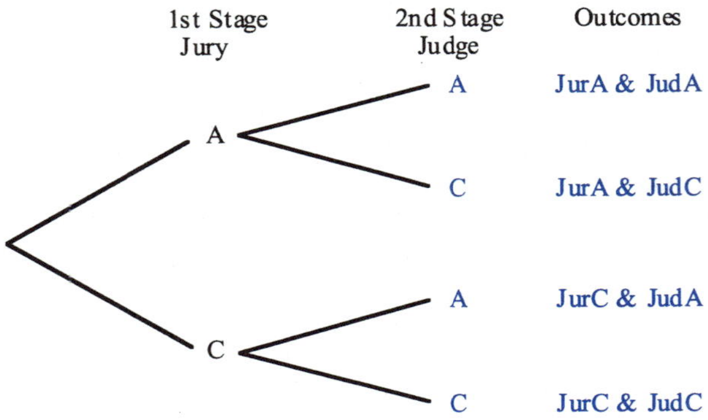

Figure 4–9

The question facing Rich is whether the data yield a definitive answer to his question: Which type of trial is better for him? The answer is not obvious. It requires some critical analysis. The jury and the judge agree in 78% of the cases – in 14% they both acquit and in 64% they both convict, meaning that in most of the cases the evidence was compelling. The more pertinent data are the cases where they disagree. Rich is most interested in when he might make a mistake. Sometimes the key step in critical thinking is to ask the right question. In Rich's dilemma, here is the key question: What is the likelihood of choosing the wrong type of trial? Rich obviously wants to choose the type of trial that has the smaller margin for error. More specifically, Rich makes a mistake if he chooses one type of trial and is convicted but he would have been acquitted if he had chosen the other type. This mistake can happen in two ways:

A) He is convicted by the jury but would have been acquitted by the judge.
B) He is convicted by the judge but would have been acquitted by the jury.

These are conditional events. Expressing them in the language of probability we get:

Event A. He would have been acquitted by the judge given that he was convicted by the jury?
Event B. He would have been acquitted by the jury given that he was convicted by the judge?

Now we compute the probability of each event. If the probability of Event A is larger, he would make a mistake by choosing a trial by jury. We now compute these two probabilities.

$$\Pr(A) = \Pr(\text{JudA} \mid \text{JurC}) = \frac{\Pr(\text{JudA} \cap \text{JurC})}{\Pr(\text{JurC})} = \frac{3}{3+64} \approx .04$$

$$\Pr(B) = \Pr(\text{JurA} \mid \text{JudC}) = \frac{\Pr(\text{JurA} \cap \text{JudC})}{\Pr(\text{JudC})} = \frac{19}{19+64} \approx .23$$

Therefore, in only 4% of all cases where the jury convicted a defendant would the judge have acquitted the defendant. On the other hand, in 23% of all cases where the judge would have convicted the defendant, the jury acquitted the defendant. In other words, there is almost a six times greater chance of making a mistake by choosing a trial by judge.

Rich chose a trial by jury.

CASE STUDY EXERCISE SET

In problems 1 to 6 compute the conditional probabilities from Table 4–13.

1. Pr(JudC| JurA)
2. Pr(JurC| JudA)
3. Pr(JudA| JurA)
4. Pr(JudC| JurC)
5. Pr(JurA| JudA)
6. Pr(JurC| JudC)

Problems 7 to 12 refer to data similar to that in the Zeisel and Kalvern study. From the following table compute the indicated probabilities.

		Judge's Verdict	
		Acquitted	Convicted
Jury's Verdict	Acquitted	10	20
	Convicted	5	65

7. Pr(JudC| JurA)
8. Pr(JurC| JudA)
9. Pr(JudA| JurA)
10. Pr(JudC| JurC)
11. Pr(JurA| JudA)
12. Pr(JurC| JudC)

CHAPTER REVIEW

Key Terms

4-1 Sample Spaces & Events

Frequency
Probability experiment
Outcome (element)
One or the other occurs (union)
Mutually exclusive events (disjoint subsets)
Relative frequency
Sample space (universal set)
Event (subset)
Both occur (intersection)

4-2 Assigning Probabilities

Theoretical probability
Equally likely events
Odds in favor of an event E
Empirical probability
Probability distribution

4-3 Fundamental Probability Properties

Addition Principle for mutually exclusive events
Certain event, Impossible event
Rule for complements
Does not occur (complement)

4-4 The Addition Principle

Addition Principle

4-5 Conditional Probability

Conditional Probability
Independent events
Product rule
Reliability

REVIEW EXERCISE SET

In problems 1 and 2 find the sample space of the experiment.

1. Select a letter in the word "mathematics".
2. Three coins are tossed and the number of heads is recorded.
3. Three urns contain red and blue marbles. The experiment consists of first selecting an urn and then a marble from the urn and recording its color. Define the events:

   E: It is not red
   F: It is red or blue

   Determine the sample space S of the experiment.

   Find (a) $E \cup F$, (b) E'.
4. A marble is selected from an urn containing 4 red, 5 white and 6 blue marbles and its color is recorded. Find (a) Pr(red), (b) Pr(red or blue).

5. A card is drawn from a deck. Define the events E = {king}, F = {black ace}. Find (a) Pr(E), (b) Pr(F), (c) Pr(E ∪ F).

6. A die is rolled and the number on the top face is recorded. Define the two events E = {2, 3} and F = (3, 4, 5}. Find Pr(E ∪ F).

7. Let the sample space be S = {a, b, c, d, e} where each outcome is equally likely. Define the events

   E = {a, b, c}
   F = {b, c, d}

   Find Pr(E ∪ F′)

8. If Pr(E) = .27, Pr(F) = .65 and Pr(E ∪ F) = .78, find Pr(E′ ∪ F′).

9. If Pr(E) = .38, Pr(F) = .68 and Pr(E ∩ F) = .84. Find Pr(E′ ∪ F′).

10. If the odds in favor of E are 6 to 7, find the probability of E occurring.

11. A club has 40 members, 25 of whom are seniors and 18 of whom are female. If 12 of the females are seniors, find the probability that a person selected is a male given that the person is a senior.

5 TOPICS IN PROBABILITY

Section 1. Counting Techniques
Section 2. Bayes' Rule
Section 3. Bernoulli Trials
Section 4. Expected Value
Case Study: Jane Blalock and the LPGA

Chapter Overview

In the previous chapter we studied how probability theory is used to predict the future when events take place randomly and it is not possible to predict what the exact outcome will be. Models were constructed that take the chance likelihood of events into consideration. In this chapter we will study more rigorous models that govern more complicated models of random behavior. The first section builds upon the counting techniques of Chapter 2 and investigates the probability of events when selections are made without replacement. The second section expands the concept of conditional probability. The third section creates a model that studies in detail probability experiments that repeat an action many times. The fourth section gives us a measure of the average value that one can expect to observe in the long run when an experiment is conducted many times.

Case Study Preview

When Jane Blalock, star professional golfer, was banned from participating in tournaments by the Ladies' Professional Golf Association, she sued the organization under the federal anti-trust laws. The court reinstated her but the decision came too late for her to participate in the tournament. She then sued for damages. But how much money would she have earned? She might have won the tournament but she just as well might have finished out of the money. Her court battle was based on probability theory. The case study gives the surprising result of her suit.

5-1 COUNTING TECHNIQUES

In this section we will use the counting techniques introduced in Chapter 3 to solve more complex problems in probability. Their solutions depend upon the equally likely outcomes principle, first covered in Section 4-2. Let us recall it here.

If a sample space S contains N equally likely outcomes and an event E contains m outcomes, then

$$\Pr(E) = \frac{m}{N}$$

In other words, if $n(E)$ is the number of outcomes in E and $n(S)$ is the number of outcomes in S then

$$\Pr(E) = \frac{n(E)}{n(S)}$$

In addition, we will make use of the Multiplication Principle as well as the formulas for permutations and combinations, $P(n,r)$ and $C(n,r)$, respectively. Specifically, we will study problems in which selections are made *without replacement*.

In previous sections we found the probability of selecting only one object from an urn, but now counting techniques will enable us to tackle more complex situations.

EXAMPLE 1

An urn contains 3 red marbles and 9 white marbles. Three marbles are selected at random without replacement. Find the probability of observing:
(a) 3 red marbles
(b) 3 white marbles
(c) 1 white and 2 red marbles (in any order)

Solution. There is one set of 12 marbles, and 3 marbles are being chosen without replacement from this set, so the denominator for each of the probabilities is C(12,3).

To determine each of the numerators, count the number of ways to select the required event. Note that combinations are used since the order in which the marbles are selected is not important.

(a) Since there are 3 red marbles and since all 3 are to be selected, the number of ways to select the 3 red marbles is given by C(3,3) = 1. Hence

$$\Pr(3\text{ red}) = \frac{C(3,3)}{C(12,3)} = \frac{1}{220} \approx .0045$$

(b) In a manner similar to part (a), the number of ways of selecting 3 white marbles is C(9,3) = 84. Hence

$$\Pr(3\text{ white}) = \frac{C(9,3)}{C(12,3)} = \frac{84}{220} = \frac{21}{55} \approx .3818$$

(c) There are C(9,1) ways to select 1 of the 9 white marbles and C(3,2) ways to select 2 of the 3 red marbles. By the Multiplication Principle there are C(9,1)·C(3,2) ways to select 1 white marble and 2 red marbles. Hence

$$\Pr(\text{1 white and 2 red}) = \frac{C(9,1)\cdot C(3,2)}{C(12,3)} = \frac{9\cdot 3}{220} = \frac{27}{220} \approx .1227$$

Many situations that you encounter can be treated as urn problems. The ingredients of an urn problem such as the one in Example 1 are (1) having more than one type of object (the different colors), (2) possibly a different number of each type of object, and (3) selection without replacement. Example 2 looks at applying the concept of marbles in urns to a different situation.

EXAMPLE 2

Fifteen men and 20 women are at a banquet at which 6 door prizes are given out. If no one can win more than one prize, what is the probability that exactly 4 men win prizes?

Solution. This problem can be thought of as an "urn problem," by thinking of men and women as differently colored marbles. Since no one can win more than one prize, we are picking the 6 winners from the group of 35 without replacement. Hence, the denominator is C(35,6).

To determine the numerator, first we must count the number of ways of selecting 4 men from the 15 total men and the number of ways of selecting 2 women from the 20 total women. These numbers are C(15,4) and C(20,2), respectively. Then to determine the number of ways of picking the 6 so that exactly 4 are men, we must multiply these two numbers. Hence

$$\Pr(\text{exactly 4 men}) = \frac{C(15,4)\cdot C(20,2)}{C(35,6)} = \frac{1365\cdot 190}{1623160} \approx .1598$$

Poker Hands

When the big loser inevitably screams "Wait! What beats what!" he is demanding to know which poker hand is better than another. One hand beats a second hand when the probability that it occurs is smaller than the probability that the second occurs. Therefore, the rules of poker are based on probability. To illustrate, we will consider three hands: a pair (two matching cards with three unmatched cards), a full house (three matching cards and a pair of matching cards), and a flush (all the cards the same suit). In all poker hands, the order that the cards appear is immaterial.

EXAMPLE 3

Compute the probability of being dealt the following 5-card poker hands: (a) a flush, (b) a full house, (c) a pair.

Solution. The number of five-card hands is the denominator. Since 5 different cards are being chosen from one set of 52, and since the order in which these cards are selected in unimportant, the denominator is $C(52,5) = 2{,}598{,}960$.

(a) For a flush we need to draw 5 of the 13 cards in one of the 4 suits. Thus, for each suit there are C(13,5) ways to get a flush.[1] Since there are four distinct suits, use the Multiplication Principle to find that there are $4 \cdot C(13,5) = 5148$ ways to draw a flush. Therefore

$$\Pr(\text{a flush}) = \frac{4 \cdot C(13,5)}{C(52,5)} \approx .0020$$

Consequently, approximately 20 out of every 10,000 hands is a flush.

(b) To get 3 matching cards, we must draw 3 of the 4 cards of a particular type. Since there are 13 different types of cards, there are $13 \cdot C(4,3)$ ways of getting 3 matching cards. Now we must also fill out the hand with another pair. There are 12 remaining types of cards. Each type has 4 cards, and we must draw 2 of them. Thus there are $12 \cdot C(4,2)$ ways to fill out the hand. Therefore

$$\Pr(\text{a full house}) = \frac{13 \cdot C(4,3) \cdot 12 \cdot C(4,2)}{C(52,5)} \approx .0014$$

Consequently, approximately 14 out of every 10,000 hands is a full house.

(c) There are 13 different types of cards. Once the type of card is chosen, to get a pair we must draw 2 of the 4 cards of this type, which can be done in C(4,2) ways. Since there are 13 types, there are $13 \cdot C(4,2)$ ways to get a pair.

The other 3 cards must not match. Since there are 12 types of cards remaining, there are C(12,3) ways to select 3 different types. Since each one of these 3 cards can be in any suit, we have 4 choices for the first suit, four for the second, and 4 for the third. Using the Multiplication Principle, there are 4^3 ways of selecting each of the C(12,3) possibilities. Hence there are $C(12,3) \cdot 4^3$ ways to fill out the hand. Therefore

$$\Pr(\text{a pair}) = \frac{13 \cdot C(4,2) \cdot C(12,3) \cdot 4^3}{C(52,5)} \approx .4226$$

Consequently, approximately 4,226 out of every 10,000 hands is a pair.

In Example 3, we have therefore shown that

$$\Pr(\text{full house}) < \Pr(\text{flush}) < \Pr(\text{Pair})$$

Therefore, a full house beats a flush which in turn beats a pair.

Since problems involving poker hands can become very complicated, it is a good idea to "build" each poker hand in some logical order using as many stages as necessary. Table 5–1 outlines how to build a hand for a full house, and consequently how to count the total number of possible full houses.

---

[1] To be precise we should subtract the 40 ways of getting a straight flush. A straight flush is a hand that is both a straight and a flush, for example a hand consisting of the 2, 3, 4, 5, and 6 of diamonds. A straight flush beats every other hand.

Stage	Description	Example	# of Choices
Stage 1	From the 13 types of cards, pick the type of card that will comprise the matching 3.	The matching three cards will be "9's"	13
Stage 2	Once the type of card is chosen, pick 3 of the 4 cards of that type	Choose the 9 of hearts, the 9 of spades and the 9 of clubs.	C(4,3)
Stage 3	From the remaining 12 types of cards, pick the type of card that will comprise the pair.	The pair will be comprised of jack's.	12
Stage 4	Once the type of card for the pair is chosen, pick 2 of the 4 cards of that type	Choose the jack of hearts and the jack of diamonds.	C(4,2)

Table 5–1

In the example in Table 5–1, the hand consists of 9♥, 9‹ , 9♣, J♥, J♦. Using the Multiplication Principle, the product of the numbers in the last column gives the *total* number of full houses possible.

$13 \cdot C(4,3) \cdot 12 \cdot C(4,2) = 3744$ possible full houses

Sampling

One of the most useful applications of probability theory is sampling. When an entire set of data cannot be inspected, usually because of time or cost considerations, a sample of the data is selected and studied. Then an inference about the whole set is made from the sample. For example, when goods are shipped to a firm, often a sample is chosen at random and tested for defectives. If more than a prescribed number of defectives is found, the entire shipment is rejected. The next example illustrates how probability and sampling can be used in accounting. (See Referenced Exercise 2 for another application with data that is more realistic.)

EXAMPLE 4

An auditor generally does not study all the accounts of a firm but rather selects a representative sample. In a certain category the auditor chooses a sample of 4 accounts from a total of 20 accounts to see how many are delinquent. Suppose that, unknown to the auditor, 5 of the 20 accounts are delinquent. If the auditor finds that 1 out of the 4 accounts in the sample are delinquent, the sample will have exactly the same percentage of delinquencies as the entire set – 25%. Knowing that 5 out of 20 accounts are delinquent, if a sample of 4 accounts is selected, what is the probability that exactly 1 account in the sample is delinquent?

Solution. The denominator is given by C(20,4) since there are 20 accounts, and 4 of them are chosen.

To determine the numerator, first count the number of ways of picking 1 delinquent account from the 5 delinquent accounts. This number is given by C(5,1). Then determine the number of ways to pick the other 3 accounts in the sample from the 15 accounts that are not delinquent, yielding C(15,3). Using the Multiplication Principle, the number of ways of choosing the 4 accounts so that exactly 1 of them is delinquent is given by C(5,1)·C(15,3). Therefore

$$\Pr(\text{exactly one delinquency}) = \frac{C(5,1)\cdot C(15,3)}{C(20,4)} = \frac{2275}{4845} \approx .4696$$

Therefore, with a small sample of 4 there is only a 47% (approximately) chance that the sample will yield the correct ratio of delinquent accounts to non-delinquent accounts.

"The Birthday Problem"

In many problems, finding the probability of a given event is too complicated because there are too many computations that need to be performed. Sometimes using the complement of the event will greatly reduce the number of computations needed. Since we know that the probability of an event E and the probability of its complement E′ are related by the formula Pr(E) = 1 – Pr(E′), once the probability of E′ is calculated, we can immediately find the probability of E.

Suppose that we have a sequence of *n* objects, and suppose that event E is "at least two of the objects are the same." Then the complement of this event, or E′, is "all of the objects are different." The event E consists of the outcomes "exactly two are the same," "exactly three are the same," and so on up to "exactly *n* are the same." On the other hand, E′ is comprised of just one outcome.

Though each of the examples that follow seems to consist of an experiment where selection is done with replacement, a large part of each problem will involve selection with replacement.

EXAMPLE 5

An urn contains 10 marbles numbered 1 to 10. Select 3 marbles from the urn with replacement and record the numbers. What is the probability that at least two will have the same number?

Solution. The denominator is given by the total number of ways of selecting the three marbles. Since there are 10 marbles and since replacement is being used, the denominator is given by 10^3.

To determine the numerator, let E represent the event "at least two of the marbles have the same number." Then E′ represents the event "all three marbles have different numbers." Think of the experiment in three stages – at each stage, select a marble, record its number and then replace it. For E′ to occur, the three numbers must be different. Therefore, at the first stage there are 10 possible numbers, at the second stage there are 9 possible numbers, and at the third stage there are 8 possible numbers. (Hence selecting the number of ways that E′ can occur is like picking without replacement.) Using the

Multiplication Principle, the number of ways of picking these three different numbers is given by 10·9·8 = 720. Hence

$$\Pr(E') = \Pr(\text{all 3 numbers are different}) = \frac{10\cdot 9\cdot 8}{10^3} = \frac{720}{1000} = .72$$

To answer the original question of finding the probability that at least two of the numbers are the same, use the formula Pr(E) = 1 – Pr(E′). Hence

$$\Pr(E) = 1 - \Pr(E') = 1 - .72 = .28$$

In calculating the number of ways that E′ could occur, we could have used a permutation instead of the Multiplication Principle since E′ fits the criteria for a permutation.

- we are picking from one set of objects (the 10 marbles)
- no repeats are allowed (all three marbles chosen must be different)
- order counts (for example, picking the marbles in the order 135 produces a different number from picking them in the order 513)

Therefore, the number of ways in which E′ could occur is given by P(10,3) = 720. The use of a permutation can cut down on the number of calculations that you need to perform if many objects are being chosen from the given set. On the other hand, if the number of objects in this set is greater than 69, using a permutation can be impossible on many calculators since many calculators cannot compute 70! or greater.

An interesting related result is one sometimes referred to as "the Birthday Problem." This problem can be stated as "given n people, what is the probability that at least two of them have the same birthday?" A "birthday" will refer to the day of the year only. For example, a person born on January 1, 1954 has the same birthday as a person born on January 1, 1994. To simplify matters, we will ignore birthdays that occur on February 29 because these birthdays do not occur every year.[2]

One way to turn this question around is to ask "how many people are needed so that the probability that at lest two people have the same birthday is p?"

Since there are 365 possible birthdays, 366 people are needed to have a 100% chance that at least two people have the same birthday. To see this fact, imagine that the people come into the room one at a time and that we assign each a birthday. The first person gets January 1, the second January 2, and so on until the 365th person gets December 31. Then we have a group of 365 people with no birthday repeated. However, when the 366th person walks in the room, there are no new birthdays to hand out, so this person must get a birthday that someone else already has.

Since 366 people are needed to have a 100% chance that at least two people have the same birthday, it might seem correct to say that half this many people, or 183 people, are needed for a 50% chance that at least two people have the same birthday. We will see after the next example that 183 people is far more than we actually need.

---

[2]Contrary to popular belief, leap years do not occur "every four years." The actual rule for leap years is that they occur "every four years, except when the year is the beginning of a new century. In these cases, a leap year occurs only if the number given by the first two digits of the year is divisible by 4." For example, there is a leap year in the year 2000 since 20 is divisible by 4, but there is no leap year in the years 1800, 1900 and 2100 since 18, 19, and 21, respectively, are not divisible by 4.

EXAMPLE 6

Seven people are selected at random. What is the probability that at least two of them have the same birthday?

Solution. This problem is similar to Example 5. Here we have 7 stages instead of 3, and at each stage there are 365 possibilities instead of 10. The denominator is given by the total number of ways for the 7 people to have birthdays. Since there are 365 choices for each person's birthday, the denominator is 365^7.

Let E be the event "at least two of these people have the same birthday." Then E′ is the event "all seven have different birthdays." In order for all of the birthdays to be different, there are 365 ways for the first person to have a birthday, 364 ways for the second person, 363 ways for the third, down to 359 ways for the seventh person. Therefore the numerator for E′ is given by 365·364·363·362·361·360·359. Hence

$$\text{Pr(all 7 birthdays are different)} = \frac{365\cdot 364\cdot 363\cdot 362\cdot 361\cdot 360\cdot 359}{365^7} \approx .9438$$

Consequently,

$$\text{Pr(at least two birthdays are the same)} = \Pr(E) = 1 - \Pr(E') \approx 1 - .9438 = .0562$$

Example 6 says that if 7 people are chosen at random, the probability that at least 2 will have the same birthday is just 5.62%. As expected, this number is small, however not quite as small as one might expect for a group of just 7 people. The same process can be used to compute the probability that at least two people have the same birthday if any number of people are selected. Computing the probability for various sizes of the groups of people will allow us to answer the original question of finding the smallest number of people required so that the probability is greater than 50%. Table 5–2 lists the probabilities for some of these group sizes. Let E represent the event "at least 2 people in a group of size *n* have the same birthday."

n	5	10	15	20	22	23	25	30	40	50
Pr(E)	.027	.117	.253	.411	.476	.507	.569	.706	.891	.970

Table 5–2

Therefore, the probability for 22 people is less than .5, but for 23 people the probability is .507, greater than 50%. Notice that by the time the group reaches a size of 50, the probability of at least two people having the same birthday is .97, or almost 1!

EXERCISE SET 5-1

In problems 1 to 8 an urn contains 10 red and 6 white marbles. Five marbles are selected without replacement. Find the probability of each of the following events.

1. All 5 marbles are red
2. All 5 marbles are white
3. 3 marbles are red, and 2 marbles are white
4. 1 marble is red, and 4 marbles are white
5. Exactly 2 marbles are red
6. Exactly 1 marble is white
7. At least 3 marbles are red
8. At most 1 marble is white

In problems 9 to 12, a manufacturer sends a shipment of n automobile batteries, and the customer selects m of them and tests for defects.

9. Suppose that $n = 20$, $m = 4$, and the shipment contains 3 defectives. What is the probability that the sample will contain exactly 1 defective?
10. Suppose that $n = 20$, $m = 6$, and the shipment contains 3 defectives. What is the probability that the sample will contain exactly 1 defective?
11. Suppose that $n = 30$, $m = 5$, and the shipment contains 4 defectives. What is the probability that the sample will contain exactly 2 defectives?
12. Suppose that $n = 40$, $m = 5$, and the shipment contains 4 defectives. What is the probability that the sample will contain exactly 2 defectives?

In problems 13 to 16, an auditor selects a sample of 5 accounts from a total of n accounts and examines them for delinquency. There are r delinquent accounts.

13. Suppose that $n = 20$ and $r = 6$. What is the probability that the sample will contain at most 1 delinquent account?
14. Suppose that $n = 30$ and $r = 4$. What is the probability that the sample will contain at most 1 delinquent account?
15. Suppose that $n = 20$ and $r = 6$. What is the probability that the sample will contain at least 1 delinquent account?
16. Suppose that $n = 30$ and $r = 4$. What is the probability that the sample will contain at least 1 delinquent account?

17. A shipment of 100 engines contains 5 defectives. The quality control inspector selects 3 engines at random to be tested. Find the probability that none of the engines tested is defective.
18. A batch of 50 computers contains 5 defectives. The quality control inspector selects 4 computers at random to be tested. Find the probability that at least one of the computers tested is defective.

In problems 19 to 22 an urn contains n marbles numbered 1 to n. Select m marbles with replacement and record the number. Find the probability that at least 2 of the numbers are the same.

19. $n = 10$, $m = 4$

20. $n = 15$, $m = 4$

21. $n = 20$, $m = 5$

22. $n = 50$, $m = 4$

23. What is the probability that at least 2 people in a group of 6 have the same birthday?

24. What is the probability that at least 2 people in a group of 9 have the same birthday?

25. Four letters are selected with replacement from the alphabet to form a password. Find the probability that the password has at least 1 letter repeated, i.e. at least 2 letters the same.

26. Seven letters are selected with replacement from the alphabet to form a password. Find the probability that the password has at least 1 letter repeated.

27. In poker, what is the probability of getting three-of-a-kind? (Three cards must be of the same type while the other two must be of different types, and must be different from the matching three.)

28. In poker, what is the probability of getting four-of-a-kind?

29. In poker, what is the probability of getting "two pair?" ("Two pair" means that there is a pair of one type, a pair of a different type, and a fifth card whose type is different from both of the other types.)

30. In poker, what is the probability of getting a straight? (The 13 types of cards can be ranked in the order of Ace, 2, 3, 4, 5, 6, 7, 8, 9, 10, Jack, Queen, King, Ace. Note that "Ace" can be both the lowest and highest card. A "straight" consists of any five consecutive types. For example "Ace, 2, 3, 4, 5" and "7, 8, 9 ,10, Jack" are both straights. As always, the order in which the cards are dealt is immaterial.)

Problems 31 to 34 address the fact that many big-jackpot lottery games consist of selecting m balls without replacement from a set of n balls numbered 1 to n. In order to win the numbers on a ticket must match some or all of the balls that are chosen.

31. Suppose that $n = 39$ and $m = 5$. What is the probability that all 5 numbers on a randomly chosen ticket match the balls drawn?

32. Suppose that $n = 48$ and $m = 6$. What is the probability that all 6 numbers on a randomly chosen ticket match the balls drawn?

33. Suppose that $n = 39$ and $m = 5$. What is the probability that at least 4 of 5 numbers on a randomly chosen ticket match the balls drawn?

34. Suppose that $n = 48$ and $m = 6$. What is the probability that at least 5 of 6 numbers on a randomly chosen ticket match the balls drawn?

35. Thirty people are called for jury duty, 18 of whom are women. If the selection of jurors is random, find the probability that at most 3 women are selected for the 12-person jury.

36. Twenty men and 14 women apply for academic scholarships, and 6 are chosen as winners. If the selection of winners is random, what is the probability that at most 2 men win scholarships?

37. There are 15 banks in a city and 8 of them offer students a discount on checking accounts. A student calls 6 banks at random and asks if they offer a discount. Find the probability that at least 3 of the banks questioned offer a discount.

38. A service station has recently repaired 15 domestic and 10 imported cars. If the records of 5 of these repairs are randomly selected, what is the probability that at least 2 of them are for domestic cars?

In problems 39 to 42, suppose that an urn contains 10 red, 1 green, and 3 black marbles, and that 5 marbles are chosen without replacement.

39. What is the probability of selecting 2 red, 1 green and 2 black marbles?

40. What is the probability of selecting 1 red, 1 green and 3 black marbles?

41. What is the probability of selecting exactly 4 red marbles?

42. What is the probability of selecting exactly 3 red marbles?

Referenced Exercise Set 5-1

1. A study[3] of how a bank should best improve waiting time in its drive-in facility found that bank officials define poor service to be "a customer waiting more than 4 minutes." This definition is used because customers tend to base their impatience on the movement of the minute hand between two marks, a movement which represents 5 minutes on a watch. The average service time was just less than 2 minutes, and out of every 14 customers, 3 would require more than 2 1/2 minutes. Poor service would result if two of these people entered the same line. Assume that a bank has 7 teller stations and 14 people are in line, two at each teller station. Suppose 3 of the 14 require more than 2 1/2 minutes. Find the probability that 2 of these 3 are in the same line.

2. In an experiment[4] to test the accuracy of their sampling method of accounting, executives of the Chesapeake and Ohio Railroad compared the actual amount of revenue due them on inter-line freight shipments versus the amount computed by sampling. Using probability, sampling techniques have many checks to insure the sample contains percentages similar to the actual data. In the 6-month period of the study there were 23,000 total bills that were divided into categories according to the size of the bill. Category 3 consisted of 8,000 bills ranging in size from $11 to $20.

---

[3]Foote, B. L., "A Queuing Case Study of Drive-In Banking," *Interfaces*, Vol. 6, No. 4, August, 1976.

[4]Neter, John, "How Accountants Save Money By Sampling," *Statistics: A Guide to the Unknown*, Holden-Day, San Francisco, 1976.

The sample consisted of 20%, or 1,600, of the bills in Category 3. If too many bills in the sample were greater than $18, the sample would give a much larger gross figure than the actual amount. There were 800 bills in Category 3 that were greater than $18. Set up the formula for the probability that the sample had at most 200 bills greater than $18.

Multiple Choice Exercise Set 5-1

1. An urn contains 5 red and 4 green marbles. Three marbles are selected without replacement. What is the probability that all three are red?
   (a) 3/5 (b) 3/9 (c) 5/42 (d) 5/84 (e) none of these

2. An urn contains 5 red and 4 green marbles. Three marbles are selected without replacement. What is the probability that all three are green?
   (a) 3/5 (b) 4/9 (c) 3/4 (d) 1/21 (e) none of these

3. Five marbles are selected without replacement from an urn containing 12 red and 8 black marbles. The number of ways that give 3 red and 2 black marbles is
   (a) C(20, 5) (b) C(12,3)·C(8,2) (c) C(5,3)·C(5,2)
   (d) C(12,3) (e) none of these

4. Seven marbles are selected without replacement from an urn containing 10 red and 10 black marbles. The number of ways that give 2 red marbles is
   (a) C(10,2) (b) $[C(10,2)]^2$ (c) C(7,2) (d) C(10,7) (e) none of these

5. The number of ways to choose three cards from a deck of 52 cards is
   (a) 49 (b) 156 (c) C(52,3) (d) 52·51·50 (e) none of these

6. Using a standard deck of cards, the number of 3-card hands with all aces is
   (a) C(52,3) (b) C(13,3) (c) 4 (d) 13·12·11 (e) none of these

7. Using a standard deck of cards, the number of 3-card hands with exactly 2 aces is
   (a) 288 (b) 72 (c) 192 (d) 48 (e) none of these

8. Using a standard deck of cards, the number of 3-card hands with exactly one ace is
   (a) 4 (b) 384 (c) 1128 (d) 4512 (e) none of these

9. Five balls numbered 1 to 5 are placed in an urn. Three are selected with replacement. The probability that at least two of them have the same number is
   (a) .4 (b) .48 (c) .52 (d) .6 (e) none of these

10. Seven balls numbered 1 to 7 are placed in an urn. Four are selected with replacement. The probability that at least two of them have the same number is approximately
    (a) .14 (b) .35 (c) .57 (d) .91 (e) none of these

11. Six people are in a room. The probability that at least two of them have the same birthday is approximately
    (a) .02 (b) .04 (c) .96 (d) .98 (e) none of these

12. Eight people are in a room. The probability that at least two of them have the same birthday is approximately
    (a) .07 (b) .25 (c) .75 (d) .93 (e) none of these

5-2 BAYES' RULE

The previous section dealt with conditional probability, the probability of an event E occurring given that event F has occurred. In this section the process will be reversed. We will use conditional probability to determine the probability that an earlier event F occurred, given that a later one E has happened. In other words, if we know Pr(E | F), can we find Pr(F | E)?

For example, suppose a medical experiment consists in separating patients into two groups, the treatment group and the control group. Then the treatment is administered to the former group and a placebo is given to the control group. A typical question answered in the last section would be "what is the probability that a patient was cured given that the patient was in the control group?" In this section a typical question would be "what is the probability that a patient was in the control group given that the patient was cured?" The experiments in this section will consist of two stages, and tree diagrams will be useful. Let us look at a simple two-stage problem to illustrate the type of problem and to find its solution.

EXAMPLE 1

The experiment is to choose a marble from either an urn or a hat. Selecting the hat and selecting the urn will be equally likely. The urn contains 1 red and 2 white marbles, and the hat contains 3 red and 2 white marbles. Find the probability the marble came from the urn given that it is red.

Solution. The experiment consists of two stages, where the first stage is to choose either the urn or the hat, and the second is to select a red or white marble. Let U, H, R and W represent selecting the urn, the hat, a red marble, and white marble, respectively. Then in Figure 5–1 we assign the following probabilities: Pr(U) = 1/2, Pr(H) = 1/2, Pr(R | U) = 1/3, Pr(W | U) = 2/3, Pr(R | H) = 3/5, and Pr(W | H) = 2/5. The problem is to find Pr(U | R).

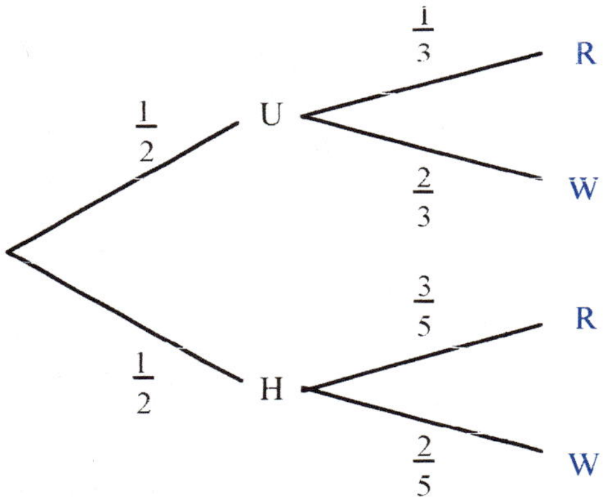

$$\Pr(U)\Pr(R \mid U) = \frac{1}{6}$$

$$\Pr(U)\Pr(W \mid U) = \frac{2}{6} = \frac{1}{3}$$

$$\Pr(H)\Pr(R \mid H) = \frac{3}{10}$$

$$\Pr(H)\Pr(W \mid H) = \frac{2}{10} = \frac{1}{5}$$

Figure 5–1

A red marble, R, can be chosen in two ways, either from the urn, U, or from the hat, H. Therefore there are two paths that correspond to choosing a red marble, namely U–R and H–R. In terms of sets, this correspondence can be expressed as $R = (U \cap R) \cup (H \cap R)$. Because the sets $U \cap R$ and $H \cap R$ are mutually exclusive, we have

$$\Pr(R) = \Pr(U \cap R) + \Pr(H \cap R) \qquad (1)$$

From the definition of conditional probability, we have

$$\Pr(U \mid R) = \frac{\Pr(U \cap R)}{\Pr(R)} \qquad (2)$$

Substitute equation (1) into this equation to get

$$\Pr(U \mid R) = \frac{\Pr(U \cap R)}{\Pr(U \cap R) + \Pr(H \cap R)}$$

The probabilities for the events $U \cap R$ and $H \cap R$ can be computed from the tree diagram by using the product rule. Hence

$$\Pr(U \cap R) = \Pr(U)\Pr(R \mid U) = \frac{1}{2} \cdot \frac{1}{3} = \frac{1}{6}$$

$$\Pr(H \cap R) = \Pr(H)\Pr(R \mid H) = \frac{1}{2} \cdot \frac{3}{5} = \frac{3}{10}$$

Finally, substituting these numbers into equation (2) yields

$$\Pr(U \mid R) = \frac{\frac{1}{6}}{\frac{1}{6} + \frac{3}{10}} = \frac{\frac{1}{6}}{\frac{14}{30}} = \frac{5}{14}$$

Therefore the probability that the marble came from the urn given that it is red is 5/14.

The key to understanding Example 1, is to understand the fact that

$$R = (U \cap R) \cup (H \cap R)$$

and that the two events $U \cap R$ and $H \cap R$ are mutually exclusive. There are two ways to pick a red marble. The first way is to pick a red marble from the urn, and this event is written as $U \cap R$. The second way is to pick a red marble from the hat, written as $H \cap R$. These two events are mutually exclusive since $U \cap R$ involves picking from the urn and $H \cap R$ involves picking from the hat, and since only one selection is being made, both events cannot happen at the same time. The Venn diagram in Figure 5–2 helps to visualize the set R as the union of $U \cap R$ and $H \cap R$.

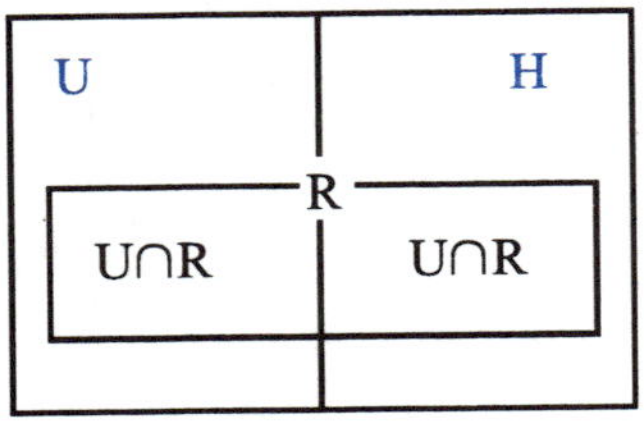

Figure 5–2

Now suppose that an event R intersects three mutually exclusive events, say U, V and H. The counterpart to equation (1) for 3 sets is:

$$\Pr(R) = \Pr(U \cap R) + \Pr(V \cap R) + \Pr(H \cap R) \qquad (3)$$

(See Figure 5–3) Just as we substituted equation (1) into equation (2), the conditional probability formula, so too can we substitute equation (3) equation (2), to find $\Pr(U \mid R)$.

$$\Pr(U \mid R) = \frac{\Pr(U \cap R)}{\Pr(R)} = \frac{\Pr(U \cap R)}{\Pr(U \cap R) + \Pr(V \cap R) + \Pr(H \cap R)}$$

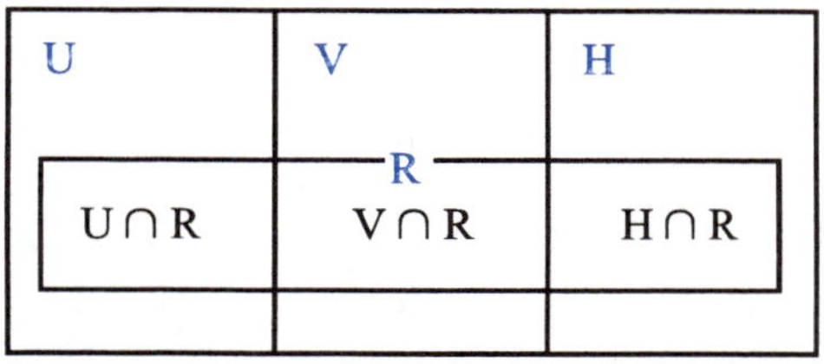

Figure 5–3

In the same way we can extend this result to any finite number of mutually exclusive sets. The principle is called Bayes' Rule, after the Presbyterian minister and part-time mathematician, Rev. Thomas Bayes (1702 - 1736).[5]

Bayes' Rule

Suppose $U_1, U_2, \quad , U_n$, are events whose union is the entire sample space and such that every pair of events is mutually exclusive. Let E be an event with nonzero probability. Then

$$\Pr(U_i \mid E) = \frac{\Pr(U_i \cap E)}{\Pr(U_1 \cap E) + \Pr(U_2 \cap E) + \quad + \Pr(U_n \cap E)}$$

$$= \frac{\Pr(U_i)\Pr(E \mid U_i)}{\Pr(U_1)\Pr(E \mid U_1) + \quad + \Pr(U_n)\Pr(E \mid U_n)}$$

The next example will demonstrate the use of Bayes' Rule in an experiment with three possibilities at the first stage.

EXAMPLE 2

A department store classifies credit card customers according to the maximum amount of credit allowed per month. Preferred customers are allowed over $2000 credit per month, regular customers are allowed a maximum of $2000 per month and basic customers are allowed a maximum of $500 per month. A minimum monthly charge is assessed each account. If the minimum is not paid the account is frozen and no new charges are allowed. Past experience shows that 5% of the preferred accounts are frozen, 10% of the regular accounts are frozen and 20% of the basic accounts are frozen each month. Suppose

[5]Recent evidence indicates that Bayes' Rule was known before Bayes was born. See "Who discovered Bayes' Theorem" by Stephen H. Stigler in *The American Statistician*, Nov. 1983, pp. 290-296.

20% of the accounts are preferred, 50% are regular accounts and 30% are basic accounts. Find the probability that a particular frozen account is a preferred account.

Solution. The first step is to construct a tree diagram as in Figure 5–4. At the first stage let P, R and B represent preferred, regular and basic accounts. At the second stage let F and NF represent frozen and not frozen, respectively. Label each branch with its corresponding probability. The probability that a frozen account is a preferred account is Pr(P | F). Using Bayes' Rule yields

$$\Pr(P \mid F) = \frac{\Pr(F \cap P)}{\Pr(F \cap P) + \Pr(F \cap R) + \Pr(F \cap B)}$$

$$= \frac{\Pr(P)\Pr(F \mid P)}{\Pr(P)\Pr(F \mid P) + \Pr(R)\Pr(F \mid R) + \Pr(B)\Pr(F \mid B)}$$

The probabilities on the tree diagram can be used to get

$$\Pr(P \mid F) = \frac{(.20)(.05)}{(.20)(.05) + (.5)(.1) + (.3)(.2)} = \frac{.01}{.01 + .05 + .06} = \frac{.01}{.12} = \frac{1}{12} \approx .0833$$

Therefore the probability that an account selected at random is preferred given that it is frozen is 1/12. In other words, about 8.3% of the frozen accounts are preferred.

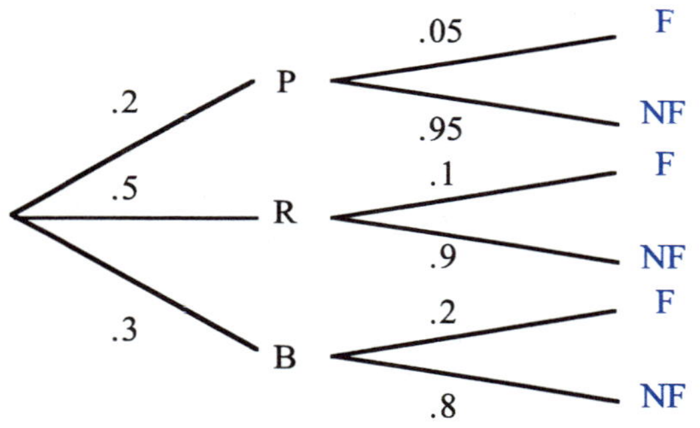

Figure 5–4

The next example illustrates how to use Bayes' Rule when there are more than 3 possibilities at the first stage.

EXAMPLE 3

A television manufacturer purchases cabinets from five different suppliers. The following table gives the percentage of the manufacturer's business that each supplier has as well as the percent of defective cabinets from each supplier.

Supplier	Percent of Business	Percent Defectives
1	.35	.02
2	.25	.04
3	.20	.06
4	.10	.07
5	.10	.09

Table 5–3

Find the probability that a defective cabinet was manufactured by the first supplier.

Solution. Let S_i be the event that the cabinet was manufactured by supplier i. Let D be the event that the cabinet is defective. We want to compute $\Pr(S_1|D)$. In Bayes' Rule, let $n = 5$ and replace U_i by S_1, and E by D. The denominator is the sum of the products of the two columns of percentages. Hence

$$\Pr(S_1 \mid D) = \frac{\Pr(S_1 \cap D)}{\Pr(D)}$$

$$= \frac{\Pr(S_1)\Pr(D \mid S_1)}{\Pr(S_1)\Pr(D \mid S_1) + \quad + \Pr(S_5)\Pr(D \mid S_5)}$$

$$= \frac{(.35)(.02)}{(.35)(.02) + (.25)(.04) + (.2)(.06) + (.1)(.07) + (.1)(.09)}$$

$$= \frac{.007}{.007 + .01 + .012 + .007 + .009} = \frac{.007}{.045} \approx .156$$

Thus about 15.6% of the defectives are made by supplier 1.

The problem in Example 3 can also be solved by a tree diagram as in Figure 5–5. The first stage corresponds to the percent of business of each supplier. The second stage is the percent of defectives for each supplier. The product of the numbers on the branches of each path is then formed and the sum of the products is the solution.

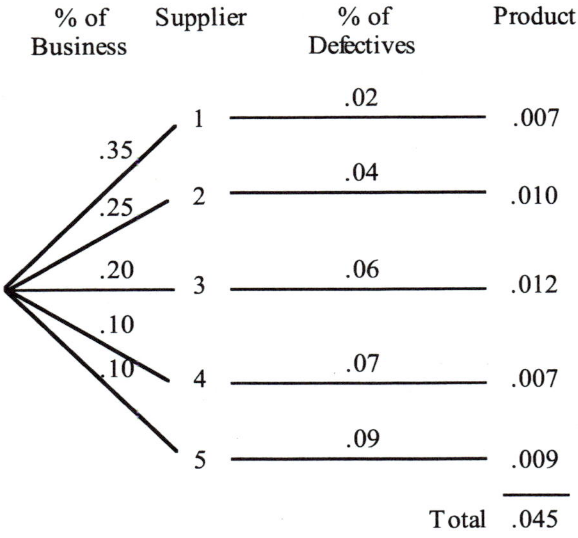

Figure 5–5

EXERCISE SET 5-2

In problems 1 to 4 two mutually exclusive events U_1 and U_2 and their probabilities are given together with the conditional probabilities $Pr(E \mid U_1)$ and $Pr(E \mid U_2)$. Use Bayes' Rule to find $Pr(U_1 \mid E)$.

1. $Pr(U_1) = .55$, $Pr(U_2) = .45$, $Pr(E \mid U_1) = .7$, $Pr(E \mid U_2) = .1$
2. $Pr(U_1) = .35$, $Pr(U_2) = .65$, $Pr(E \mid U_1) = .1$, $Pr(E \mid U_2) = .2$
3. $Pr(U_1) = .3$, $Pr(U_2) = .7$, $Pr(E \mid U_1) = .25$, $Pr(E \mid U_2) = .85$
4. $Pr(U_1) = .59$, $Pr(U_2) = .41$, $Pr(E \mid U_1) = .4$, $Pr(E \mid U_2) = .15$

In problems 5 and 6 three mutually exclusive events U_1, U_2 and U_3 and their probabilities are given together with the conditional probabilities $Pr(E \mid U_1)$, $Pr(E \mid U_2)$ and $Pr(E \mid U_3)$. Use Bayes' Rule to find $Pr(U_1 \mid E)$.

5. $Pr(U_1) = .1$, $Pr(U_2) = .5$, $Pr(U_3) = .4$, $Pr(E \mid U_1) = .9$, $Pr(E \mid U_2) = .2$, $Pr(E \mid U_3) = .1$
6. $Pr(U_1) = .6$, $Pr(U_2) = .2$, $Pr(U_3) = .2$, $Pr(E \mid U_1) = .1$, $Pr(E \mid U_2) = .1$, $Pr(E \mid U_3) = .9$

In problems 7 to 14 find the indicated probabilities by referring to the following tree diagram.

7. Pr(A∩E)
8. Pr(B∩E)
9. Pr(A∩F)
10. Pr(B∩F)
11. Pr(A|E)
12. Pr(B|E)
13. Pr(A|F)
14. Pr(B|F)

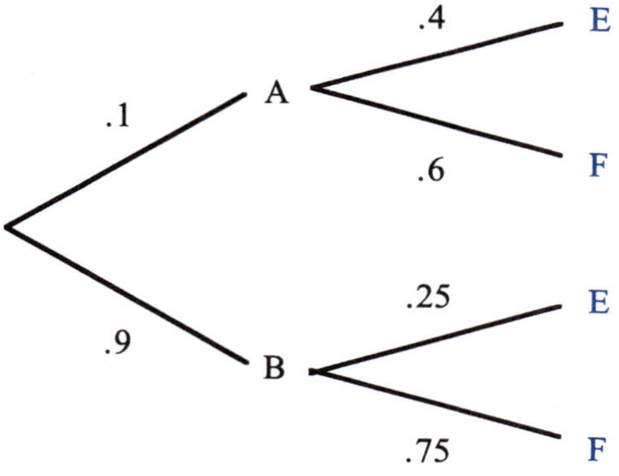

In problems 15 to 22 find the indicated probabilities by referring to the following tree diagram.

15. Pr(A∩E), Pr(B∩E), Pr(C∩E)
16. Pr(A∩F), Pr(B∩F), Pr(C∩F)
17. Pr(A|E)
18. Pr(B|E)
19. Pr(C|E)
20. Pr(A|F)
21. Pr(B|F)
22. Pr(C|F)

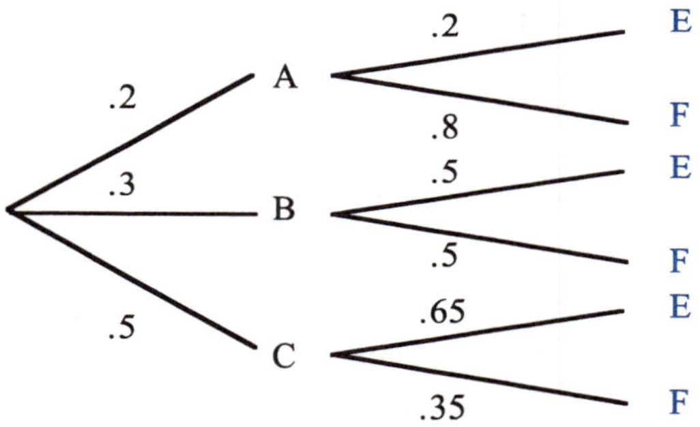

In problems 23 to 28 refer to the Venn diagram. The number of elements in each set is given. For example, $n(A) = 9$ and $n(A \cap E) = 4$. Find the indicated probabilities.

23. Pr(A|E)
24. Pr(B|E)
25. Pr(C|E)
26. Pr(A|E′)
27. Pr(B|E′)
28. Pr(C|E′)

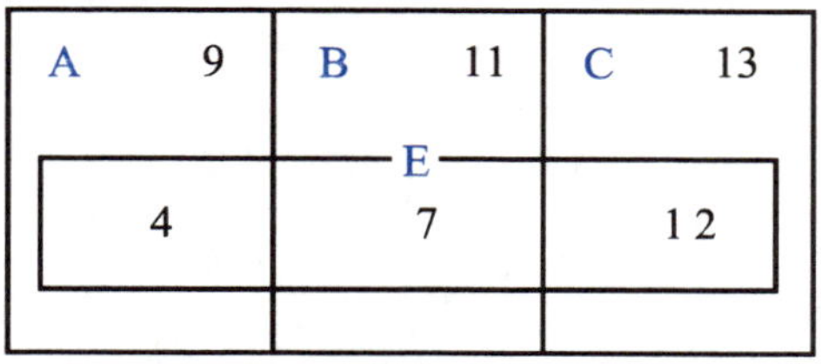

In problems 29 and 30 the experiment is to first choose at random either an urn or a hat, each containing some red and white marbles, and then select a marble. The urn has 4 red and 6 white marbles and the hat has 7 red and 5 white marbles. Find the probability of each event.

29. The urn was chosen given that the marble is red
30. The hat was chosen given that the marble is white

In problems 31 to 34 the experiment is to first choose at random one of three urns, each containing some red and white marbles, and then to select a marble. Urn 1 has 4 red and 8 white marbles, Urn 2 has 3 red and 7 white marbles, and Urn 3 has 7 red and 5 white marbles. Find the probability of each event.

31. Urn 1 was chosen given that the marble is red

32. Urn 2 was chosen given that the marble is red

33. Urn 3 was chosen given that the marble is white

34. Urn 1 was chosen given that the marble is white

35. A firm that specializes in filling out income tax forms has two types of accounts - large (gross income at least $50,000) and small (gross income less than $50,000). They find that 5% of the large accounts and 2% of the small accounts are audited by the IRS. 40% of their accounts are large accounts. Given that an account is audited, what is the probability that it is a large account.

36. Red-green color blind people see these two colors the same. This type of color blindness occurs 4.9% of the general population. and in 8.5% of the male population. If 47% of the population is male, find the probability that a person is male given that he or she is red-green color blind.

37. A survey of skiing accidents at 5 hospitals found that 22% of the fractures at Hospital 1 were severe (required surgery), 12% of the fractures at Hospital 2 were severe, 18% of the fractures at Hospital 3 were severe, 10% of the fractures at Hospital 4 were severe, and 15% of the fractures at Hospital 5 were severe. If Hospital 1 treated 9% of all fractures, Hospital 2 treated 21% of all fractures, Hospital 3 treated 30% of all fractures, Hospital 4 treated 15% of all fractures, and Hospital 5 treated 25% of all fractures, what percentage of all severe fractures occurred at either Hospital 1 or Hospital 2.

38. A college surveyed the members of its administration, faculty and staff inquiring about the means of transportation that individuals use to get to work. 10% of administrators, 18% of faculty and 39% of the staff take public transportation. 8% of all employees are in administration, 57% of all employees are faculty members, and the rest are members of the staff. Given that a person walks to work, what is the probability that he or she is a member of the staff?

39. In order to measure degree of visibility of an ad campaign, consumers are often asked if they can remember seeing certain ads. An advertising firm questioned 1000 TV viewers and asked if they recalled seeing a particular commercial. They were then asked if they had bought the product recently. Of the 110 people who definitely remembered the commercial, 60 had bought the product recently. Of the 330 who vaguely remembered the commercial, 66 had bought the product recently. Of the 560 who did not remember seeing the commercial, 40 had bought the product recently. Given that a person bought the product recently, find the probability that the person definitely remembered seeing the ad.

40. A medical research company conducted a survey of a new ointment that claims to reduce swelling if applied immediately after an injury. They choose 100 people to be in the treatment group and 50 in the control group. Those in the control group were given a similar looking ointment that is known to be ineffective in reducing swelling. Of those in the treatment group, 80 said that the swelling was reduced significantly. Of those in the control group, 15 reported that the swelling was reduced significantly. Find the probability that a person chosen at random was in

the control group given that the person reported that the swelling was reduced significantly.

41. Suppose A, B, C and D are mutually exclusive events in a sample space S where $S = A \cup B \cup C \cup D$. Let E be any event. Show that

$$\Pr(E) = \Pr(E \cap A) + \Pr(E \cap B) + \Pr(E \cap C) + \Pr(E \cap D).$$

42. Suppose A, B and C are mutually exclusive events in a sample space S where $S = A \cup B \cup C$. Let E be any event. Show that

$$\Pr(E) = \Pr(E \mid A)\Pr(A) + \Pr(E \mid B)\Pr(B) + \Pr(E \mid C)\Pr(C).$$

Referenced Exercise Set 5-2

1. In a study[6] of 1405 students, participation in religious activities was compared with the answer to the question, "How happy has your home life been?" C. L. Stone found the following:

	Answer to Question	
Religious Participation	Happy	Not Happy
Not at all	183	25
Very little	406	30
Somewhat	521	24
Very much	203	13

Find the probability that a person who did not have a happy home life participated very little or not at all in religious activities.

7. In an article[7] that studied the effects of seeding hurricanes, Howard et al. used Bayes' Rule extensively. The possibility of mitigating the destructive force of hurricanes by seeding them with silver oxide was suggested as early as 1961, but strong evidence was not obtained until 1969 when Hurricane Debbie was seeded. The researchers defined the events: H_1 - seeding reduces wind speed; H_2 - seeding has no effect; H_3 - seeding increases wind speed. They determined the probabilities $\Pr(H_1) = .49$, $\Pr(H_2) = .49$, and $\Pr(H_3) = .02$. The researchers defined the event "Debbie" to be that wind reduction of at least 31% was observed. They computed $\Pr(\text{Debbie}|H_1) = .6$, $\Pr(\text{Debbie}|H_2) = .25$, and $\Pr(\text{Debbie}|H_3) = .4$. Find the probability that H_1 occurred given that Debbie occurred.

---

[6]Adapted from Stone, C. L., "Church Participation and Social Adjustment of High School and College Youth," *Rural Sociology Series on Youth*, No. 12, Bulletin 550, Agricultural Research Center, Washington State U., 1954.

[7]Howard, Ronald A., et al., "The Decision to Seed Hurricanes," *Science*, vol. 176, No. 4040, (June, 1972), pp. 1191-1202.

Cumulative Exercise Set 5-2

In problems 1 and 2 a coin is tossed n times and the number of heads is recorded. Find the probability of observing the given event E.

1. $n = 4$, E is "exactly 1 head"
2. $n = 8$, E is "exactly 2 heads"

In problems 3 and 4 an urn contains 8 red and 4 blue marbles. Four marbles are selected without replacement. Find the probability of the event.

3. All 4 are red.
4. At least three are red.
5. For a group of 12 people compute the probability that at least 2 of them have the same birthday.
6. There are 30 members in a law firm and 10 agree with the president on whether to hire a new member. The president selects 4 at random and asks their opinion. What is the probability that at least 2 will agree with the president?

In problems 7 and 8 three mutually exclusive events U_1, U_2 and U_3 and their probabilities are given together with the conditional probabilities $Pr(E \mid U_1)$, $Pr(E \mid U_2)$ and $Pr(E \mid U_3)$. Use Bayes' Rule to find $Pr(U_1|E)$.

7. $Pr(U_1) = .3$, $Pr(U_2) = .6$, $Pr(U_3) = .1$, $Pr(E \mid U_1) = .7$, $Pr(E \mid U_2) = .6$, $Pr(E \mid U_3) = .9$
8. $Pr(U_1) = .2$, $Pr(U_2) = .2$, $Pr(U_3) = .6$, $Pr(E \mid U_1) = .2$, $Pr(E \mid U_2) = .6$, $Pr(E \mid U_3) = .1$
9. Urn 1 has 4 blue and 6 red marbles, and Urn 2 has 1 blue and 9 red marbles. An experiment consists of selecting an urn and then selecting a marble from it. Find the probability the selected marble came from Urn 1 given that it was a red marble.
10. In a survey of students' preferences in four different classes, the percentage of students who rated the class "very interesting" was as follows: Class 1 - 10%, Class 2 - 15%, Class 3 - 10%, Class 4 - 5%. Class 1 contained 30 students, Class 2 contained 40 students, Class 3 contained 50 students, and Class 4 contained 40 students. Given that a student rated the class "very interesting," what is the probability that he or she was in Class 3?

Multiple Choice Exercise Set 5-2

In problems 1 and 2 refer to the following probabilities for the two mutually exclusive events U_1 and U_2.

$$Pr(U_1) = .3,\ Pr(U_2) = .7,\ Pr(E \mid U_1) = .1,\ Pr(E \mid U_2) = .3$$

1. $Pr(U_2| E) =$
   (a) .1 (b) .125 (c) .3 (d) .875 (e) none of these
2. $Pr(U_2| E) =$
   (a) .1 (b) .125 (c) .3 (d) .875 (e) none of these

In problems 3 and 4 the experiment is to first choose at random either an urn or a hat, each containing some red and white marbles, and then to select a marble. The urn has 8 red and 2 white marbles, and the hat has 4 red and 6 white marbles.

3. The probability that the urn was chosen given that the marble is red is equal to
(a) 1/2 (b) 1 (c) 2/3 (d) 3/4 (e) none of these

4. The probability that the hat was chosen given that the marble is white is equal to
(a) 1/2 (b) 1 (c) 2/3 (d) 3/4 (e) none of these

Problems 5 to 8 refer to the following problem. A manufacturer purchases parts from five different suppliers. The following table gives the percentage of the manufacturer's business that each supplier has as well as the percent of defective parts from each supplier.

Supplier	Percent of Business	Percent Defective
1	.45	.02
2	.15	.02
3	.10	.07
4	.20	.06
5	.10	.05

A part is chosen at random, and it is found to be defective.

5. The probability that the defective part was manufactured by the first supplier is equal to
(a) 1/4 (b) 10/36 (c) 1/12 (d) 5/36 (e) none of these

6. The probability that the defective part was manufactured by the second supplier is equal to
(a) 1/4 (b) 10/36 (c) 1/12 (d) 5/36 (e) none of these

7. The probability that the defective part was manufactured by the second or the third supplier is equal to
(a) 1/4 (b) 10/36 (c) 1/12 (d) 5/36 (e) none of these

8. The probability that the defective part was not manufactured by the fourth supplier is equal to
(a) 1/3 (b) 2/3 (c) 1/12 (d) 5/36 (e) none of these

In problems 9 to 11, there are three polling places in a township. The first polling place has 400 voters, the second has 500 voters, and the third has 700 voters. Suppose that 200 voters in the first polling place vote to pass a particular ordinance, 225 voters in the second polling place vote to pass the ordinance, and 420 voters in the third polling place vote to pass the ordinance.

Suppose a voter in the township is chosen at random to be interviewed, and that person voted to pass the ordinance.

9. The probability that the voter is from the first polling place is approximately
(a).12 (b) .24 (c) .33 (d) .5 (e) none of these

10. The probability that the voter is from the third polling place is approximately
(a).12 (b) .24 (c) .33 (d) .5 (e) none of these

11. The probability that the voter is not from the second polling place is approximately
(a) .18 (b) .27 (c) .5 (d) .8 (e) none of these

5-3 BERNOULLI TRIALS

In Section 5-1, we concentrated on problems where selection was without replacement. In this section, we will look at those problems where selection is done with replacement.

A common type of problem in probability is one in which an experiment is repeated in the exact same manner several times. To illustrate, consider the following problems.

1. Toss a coin 5 times and record H or T each time.
2. Roll a die 4 times and record the number on the top face on each roll.
3. Select 10 patients and determine if each is cured.

Each repetition is called a trial. In problem 1, the trial is to toss a coin and it is repeated five times. In problem 2, the trial is to roll a die, and it is repeated 4 times. In problem 3, the trial is to select a patient and it is repeated ten times. Each problem has specific characteristics: a physical experiment or trial is repeated several times; the trials are identical; each trial has the same two outcomes. The outcomes are usually called success and failure, or simply S and F. In most experiments deciding which of the two outcomes is labeled "success" is arbitrary. Thus S can be assigned to either outcome, and then F is its complementary event. Therefore, if $\Pr(S) = p$, then $\Pr(F) = 1 - p$.

Bernoulli Trials and Binomial Experiments

Experiments of this kind are called *binomial experiments*, and the trials themselves are called *Bernoulli trials*. Thus a binomial experiment consists of the repetition of Bernoulli trials. In problem 2 the Bernoulli trial is to roll a die, and the binomial experiment is the repetition of the trial 4 times. Let us summarize.

DEFINITION

A **binomial experiment** is a probability problem consisting one experiment repeated several times. Each repetition is called a **Bernoulli trial**. The experiment has the following properties:

1. The experiment in each trial remains the same when it is repeated.
2. The trials are independent.
3. There are two outcomes, called success and failure, or simply S and F.
4. The probabilities of the outcomes remain the same for each trial.

Sometimes a binomial experiment is called a **Bernoulli process**. Example 1 illustrates the derivation of the formula that will be used to find the probability of a binomial experiment.

EXAMPLE 1

Roll a die 4 times. Find the probability of observing
(a) exactly four 1's
(b) exactly three 1's.

Solution. Let us assign S to the event of observing a roll of 1. For each trial Pr(S) = 1/6 and Pr(F) = 5/6. Hence

(a) Pr(exactly four 1's) = Pr(SSSS) = (1/6)(1/6)(1/6)(1/6) = 1/1296.

(b) Here we need to find the probability of getting three successes and one failure. however, before we can compute that value, we need to find *the number of ways* of getting three successes and one failure. Listing the possible number of ways as

SSSF, SSFS, SFSS, and FSSS

we see that there are four ways to get three successes and one failure. One way to determine this number of ways without listing the outcomes is to notice that we are trying to place S in three places out of four with F filling in the remaining place. In other words, we need to calculate the number of ways of picking three places from four. The order in which we choose these three places does not matter, all that matters are the three places with which we are left. Therefore, the number of ways of getting three successes and one failure is given by C(4,3).

In each of these four ways, there are three successes, each with a probability of 1/6, and one failure, with a probability of 5/6. Therefore, the probability that any one of these four occurs is given by

$$\frac{1}{6}\cdot\frac{1}{6}\cdot\frac{1}{6}\cdot\frac{5}{6} = \left(\frac{1}{6}\right)^3\left(\frac{5}{6}\right)$$

Since there are C(4,3) = 4 total ways

$$\text{Pr(exactly three 1's)} = C(4,3)\left(\frac{1}{6}\right)^3\left(\frac{5}{6}\right) = 4\cdot\left(\frac{1}{6}\right)^3\left(\frac{5}{6}\right) = \frac{20}{1296} \approx .0154$$

The second part of Example 1 provides a model for the general case. Suppose a binomial experiment consists of n repetitions of a Bernoulli trial with Pr(S) = p. The problem is to find the probability of observing exactly x successes. One way to observe x successes is for the S's to come first followed by the F's. There would be x S's and $n - x$ F's:

$$\underbrace{(SS\cdots SS)}_{x \text{ times}}\underbrace{(FF\cdots FF)}_{n-x \text{ times}}$$

Since Pr(S) = p, we have Pr(F) = $q = 1 - p$. Hence

$$\Pr(\underbrace{SS\cdots SS}_{x \text{ times}}\underbrace{FF\cdots FF}_{n-x \text{ times}}) = \underbrace{pp\cdots pp}_{x \text{ times}}\underbrace{(1-p)(1-p)\cdots(1-p)(1-p)}_{n-x \text{ times}} = p^x(1-p)^{n-x}$$

The event "exactly x successes" in n trials occurs with any arrangement of the x S's and $n - x$ F's. Since there are C(n,x) such arrangements, we are led to the following general result, where we substitute q for $1 - p$.

The Probability of a Binomial Experiment

In a binomial experiment consisting of n repetitions of a Bernoulli trial with Pr(S) = p and Pr(F) = $1 - p = q$, the probability of x successes is

$$\Pr(\text{exactly } x \text{ successes}) = C(n,x)p^x q^{n-x}$$

The next two examples illustrate the use of the formula. Note that in Example 2, the selection of the marbles is done with replacement.

EXAMPLE 2

An urn contains 4 red and 6 white marbles. Five marbles are drawn with replacement. Find the probability of observing (a) exactly 3 red marbles, and (b) at least 3 red marbles.

Solution. Here we have a binomial experiment with 5 repetitions of the Bernoulli trial of drawing a marble. Let S represent "drawing a red marble" since this event is the one about which we are asking questions. Then p is the probability of drawing a red marble *in one trial*, so p = Pr(S) = 4/10 = .4, and q = Pr(F) = 1 – .4 = .6.

(a) We need to compute the probability of "exactly 3 successes." From the formula with $n = 5$, $x = 3$, $p = .4$, and $q = .6$, we have

$$\Pr(\text{exactly 3 reds}) = C(5,3)(.4)^3(.6)^2 = 10(.064)(.36) = .2304$$

(b) The event "at least 3 red marbles" consists of the three mutually exclusive events "exactly 3," "exactly 4," and "exactly 5" red marbles. As in (a), we find the latter two probabilities from the formula with $n = 5$ and $x = 4$, and $n = 5$ and $x = 5$, respectively.

$$\Pr(\text{exactly 4 reds}) = C(5,4)(.4)^4(.6)^1 = 5(.0256)(.6) = .0768$$

and

$$\Pr(\text{exactly 5 reds}) = C(5,5)(.4)^5(.6)^0 = 1(.01024)(1) = .01024$$

Therefore, if we let "at least 3 successes" be represented by "$x \geq 3$"

$$\Pr(x \geq 3) = .2304 + .0768 + .01024 = .31744$$

Note that in finding p, the probability of success on one trial, the values for n and x *do not come into play*. To calculate p, imagine that you are conducting just *one* trial of the experiment. Hence in Example 2, the probability of picking one marble from the urn and having it be red is 4/10. Consequently, $p = 4/10 = .4$.

Usually, it is easier to write, for example, the expression $\Pr(x = 3)$ to represent the probability of *exactly* 3 successes. Then a similar notation, used in Example 2(b), for the probability of *at least* 3 successes is $\Pr(x \geq 3)$.

EXAMPLE 3

Food companies use what is known as a "triangle taste test" to judge the significance of the taste of one product over another. For example, when the Coca-Cola Co. was preparing to introduce a new brand of Coke they used this test to evaluate the difference in taste between the old brand, later called "Coke Classic", against the new version. In the triangle taste test the taster is given three samples, two of which are alike, and asked to choose the odd food by tasting. Suppose the taster does not have a well-developed sense of taste for this product, and chooses at random. If the taster makes 6 identical taste tests of the product, find the probability that the taster will make (a) no correct choices, (b) exactly 1 correct choice, (c) at least 2 correct choices.

Solution. We have a binomial experiment with 6 repetitions, hence $n = 6$. Let S represent the event "choosing the odd food." Then $\Pr(S) = p = 1/3$ since in *one* taste test, there is a one-in-three chance of picking the odd food. Consequently, $q = 2/3$.
(a) Making no correct choices corresponds to $x = 0$, so using the formula, we have:

$$\Pr(\text{no correct choices}) = \Pr(x = 0) = C(6,0)\left(\frac{1}{3}\right)^0\left(\frac{2}{3}\right)^6 = \frac{64}{729} \approx .088$$

(b) Making exactly one correct choices corresponds to $x = 1$, so using the formula, we have:

$$\Pr(\text{exactly one correct choice}) = \Pr(x = 1) = C(6,1)\left(\frac{1}{3}\right)^1\left(\frac{2}{3}\right)^5 = \frac{192}{729} \approx .263$$

(c) One way to calculate the probability of at least two correct choices is to calculate the probabilities for exactly 2, 3, 4, 5, and 6 correct choices and then add them together as in Example 2(b). However, the complement of exactly 2 correct choices is comprised of just the two outcomes "exactly one correct choice" and "no correct choices." Therefore, using the complement will involve fewer calculations. Therefore,

$$\Pr(x \geq 2) = 1 - [\Pr(x = 0) + \Pr(x = 1)] = 1 - [.088 + .263] = 1 - .351 = .649$$

Thus there is only about a 64.9% chance that a taster without a well-developed sense of taste for this product will get at least 2 of 6 choices correct.

When the percentage of the population that falls into a certain category is known, this number can be used for the value of p.

EXAMPLE 4

If 9% of the population is left-handed, what is the probability that in a random sample of 18 people, at least one will be left-handed?

Solution. A random sample of the population constitutes a binomial experiment. Since 18 people are selected, this problem is considered to have 18 trials, hence $n = 18$. Since the probability that *one* randomly selected person is left-handed is .09, then letting S be the event "a person is left-handed," $\Pr(S) = p = .09$. Therefore, $q = .91$. The question asks for the probability that "at least one person is left-handed," therefore, the complement of this event is "none of the 18 is left-handed." Consequently,

$$\Pr(x \geq 1) = 1 - \Pr(x = 0) = 1 - C(18,0)(.09)^0(.91)^{18} \text{ ª } 1 - .183 = .817$$

Therefore, there is an 81.7% chance that at least one person in this group of 18 is left-handed.

EXERCISE SET 5-3

In problems 1 to 4 the experiment is to toss a coin 6 times and record the sequence of heads and tails. Find the probability of the event.

1. Exactly 4 heads.
2. Exactly 5 heads.
3. At least 4 heads.
4. At least 5 heads.

In problems 5 to 8 the experiment is to toss a coin 8 times and record the sequence of heads and tails. Find the probability of the event.

5. Exactly 2 heads.
6. Exactly 4 heads
7. At least 6 heads.
8. At least 7 heads.

In problems 9 to 12 an urn contains 5 red and 15 white marbles.

9. Find the probability of selecting exactly 4 red marbles if 5 marbles are selected with replacement.
10. Find the probability of selecting exactly 3 red marbles if 5 marbles are selected with replacement.
11. Find the probability of selecting exactly 5 white marbles if 6 marbles are selected with replacement.
12. Find the probability of selecting exactly 5 white marbles if 8 marbles are selected with replacement.

In problems 13 to 16 an urn contains 5 red and 20 white marbles.

13. Find the probability of selecting at least 4 white marbles if 5 marbles are selected with replacement.
14. Find the probability of selecting at least 3 white marbles if 5 marbles are selected with replacement.

15. Find the probability of selecting at most 2 red marbles if 6 marbles are selected with replacement.

16. Find the probability of selecting at most 1 red marble if 8 marbles are selected with replacement.

In problems 17 to 20 a taste test described in Example 3 is conducted.

17. Find the probability that the taster makes exactly 2 correct choices in 5 taste tests.

18. Find the probability that the taster makes exactly 1 correct choice in 5 taste tests.

19. Find the probability that the taster makes exactly 3 correct choices in 7 taste tests.

20. Find the probability that the taster makes exactly 4 correct choices in 8 taste tests.

In problems 21 to 24 a taste test described in Example 3 is conducted. The taster makes the given number n of taste tests of the product. Find the probability of the taster making the given number of correct choices.

21. Find the probability that the taster makes at most 2 correct choices in 5 taste tests.

22. Find the probability that the taster makes at most 1 correct choice in 5 taste tests.

23. Find the probability that the taster makes at least 5 correct choices in 7 taste tests.

24. Find the probability that the taster makes at least 7 correct choices in 8 taste tests.

In problems 25 to 28 a survey of voters is conducted in an election where in fact 40% of the voters favor a particular candidate.

25. If 10 voters are surveyed, what is the probability that exactly 4 favor the candidate?

26. If 20 voters are surveyed, what is the probability that exactly 8 favor the candidate?

27. If 10 voters are surveyed, what is the probability that at least 2 favor the candidate?

28. If 12 voters are surveyed, what is the probability that at least 2 favor the candidate?

29. It is known that 80% of the population in a ward favor an upcoming bond issue. Eight people are selected. Find the probability that (a) exactly 6 of them will favor the bond issue, (b) at least 6 of them will favor the bond issue.

30. It is known that 75% of the population in a ward favor an upcoming bond issue. Ten people are selected. Find the probability that (a) exactly 8 of them will favor the bond issue, (b) at most 8 of them will favor the bond issue.

31. A student takes a 10 question true-false test and guesses at the answers. What is the probability of getting (a) exactly 7 correct, (b) at least 7 correct?

32. A student takes a 20 question true-false test and guesses at the answers. What is the probability of getting (a) exactly 15 correct, (b) at least 18 correct?

33. The past performance of a machine shows that it produces 0.5% defectives. What is the probability that in a run of 50 items (a) exactly 1 will be defective, (b) at least 1 will be defective?

34. Each year a firm selects several undergraduates to participate in a co-op program. In the past the firm has hired 40% of the co-op workers to full-time employment after

graduation. If 15 students participate this year, find the probability that (a) exactly 5 will be hired after graduation, (b) at least 5 will be hired after graduation

35. A die is rolled 8 times. What is the probability of observing (a) exactly five 6's, (b) at most five 6's?

36. A die is rolled 10 times. What is the probability of observing (a) exactly eight 6's, (b) at least eight 6's?

37. A part of a final exam consists of 10 questions that the student has not studied. If the questions are multiple choice, each with 5 answers, and the student guesses, find the probability that the student gets (a) exactly 3 correct, (b) at least 3 correct.

38. A part of a final exam consists of 5 questions that the student has not studied. If the questions are multiple choice, each with 4 answers, and the student guesses, find the probability that the student gets (a) exactly 3 correct, (b) at most 3 correct.

39. If a baseball player has a batting average of .300 and gets to bat 5 times in the next game, find the probability the player will get (a) exactly 3 hits, (b) at most 3 hits.

40. If a basketball player has a shooting percentage of 60% find the probability the player will make at least two of her next six shots.

41. In a small city it is expected that 30% of the population will contract the latest strain of influenza. If a company has 15 employees, find the probability that at least 3 employees will contract the disease.

42. A company advertises that its new toothpaste will prevent tartar buildup. It claims that 40% of people will have less tartar after three months of using the toothpaste. To justify the claim a clinical experiment is run with 20 people using the toothpaste for three months. Find the probability that at least 4 of them will have less tartar.

Referenced Exercise Set 5-3

1. In an attempt to determine if performance on ESP tests is affected by the person's belief in ESP, Ryzl[8] scored people according to their relative ability on an ESP test. Forty percent of the people performed well on the ESP test. Suppose 10 are selected at random. Find the probability that at least 9 of them performed well.

---

[8]Ryzl, M., "Precognition Scoring and Attitude Toward ESP", *Journal of Parapsychology*, 32:1-8, 1968. (Ryzl found that 96% of the people who scored well on the test also believed in ESP. Do you think there is a connection between scoring well on the ESP test and believing in ESP?)

Cumulative Exercise Set 5-3

1. An experiment consists of tossing two dice and recording the sum of the top faces. Define the events E and F as follows

   E: the sum is greater than 9
   F: the sum is even

   Find (a) $Pr(E \cup F)$, (b) $Pr(E \cap F')$, and (c) $Pr(E | F)$

2. There are 50 members of a state legislature, and 30 will vote to pass the next bill. If three are selected at random and asked how they will vote what is the probability that at least one will vote to pass the bill?

3. Urn A has 3 blue and 7 red marbles, and Urn B has 2 blue and 8 red marbles. The experiment is to select an urn and then select a marble and record its color. Find the probability of observing a red marble.

4. There are 100 patients in a hospital, 40 of which are in outpatient care and 35 which are being treated for influenza. If 10 of the outpatients are being treated for influenza and if a patient is selected at random, find the probability that the patient is an outpatient given that the patient is not being treated for influenza.

5. In a survey of students' attitudes in five different classes, the percentage of students who rated the professor the best they ever had was as follows: class 1 - 20%, class 2 - 15%, class 3 - 10%, class 4 - 25%, class 5 - 5%. Assuming that there were the same number of students in each class, find the probability that a student selected at random was selected from class 1 or class 2.

6. A bank has three types of checking accounts, regular, deluxe and executive class. Past experience shows that 15% of regular accounts are assessed an overdraft charge, 10% of deluxe accounts are charged an overdraft charge, and 5% of executive accounts are charged an overdraft charge. Suppose 60% of the accounts are regular, 30% are deluxe, and 10% are executive class. Find the probability that an account that was charged for an overdraft was an executive class account.

7. A die is rolled ten times. What is the probability of observing (a) exactly seven 6's, (b) at most seven 6's?

8. It is known that 60% of the population in a ward favor an upcoming bond issue. Ten people are selected. Find the probability that (a) exactly 8 of them will favor the bond issue, (b) at least 8 of them will favor the bond issue.

9. A student takes a 10 question multiple-choice test with four choices per question and guesses at the answers. What is the probability of getting (a) exactly 1 correct, (b) at most 2 correct?

10. A firm is going to introduce a new product and wants to predict how well it will do. Past records show that similar products break even 70% of the time. If there are 10 sales regions find the probability that the product will break even in at least 8 regions.

11. There are two used-car lots. The small lot has 6 overpriced cars and 4 underpriced cars, and the large lot has 15 overpriced cars and 5 underpriced cars. (a) If a car is selected at random from the small lot, what is the probability that it is underpriced?

(b) If a lot is selected at random and a car is selected from it at random, what is the probability that it is underpriced?

12. There are two used-car lots. The small lot has 20 overpriced cars and 5 underpriced cars and the large lot has 35 overpriced cars and 15 underpriced cars. (a) If a lot is selected at random and three cars are selected from it at random, what is the probability that all three are underpriced? (b) If a car is selected at random and found to be underpriced, what is the probability that it was selected from the small lot?

Multiple Choice Exercise Set 5-3

1. Which of the following is a Bernoulli trial?
   (a) Select a card from a deck and record the suit
   (b) Select a card from a deck and record the color
   (c) Select a card from a deck and record the value
   (d) Select a card from a deck and record the suit or the color
   (e) None of these.

2. Which of the following is a Bernoulli trial?
   (a) Toss two dice and record the sum
   (b) Toss two dice and record the difference
   (c) Toss two dice and record whether the numbers match
   (d) Toss two dice and record the higher value
   (e) None of these.

3. If one die is rolled three times, the probability of getting three 4's is

   (a) $1/3^6$ (b) $1/6^3$ (c) $1/3^4$ (d) $1/4^3$ (e) none of these

4. If one die is rolled four times, the probability of getting four 5's is

   (a) $1/4^5$ (b) $1/5^4$ (c) $1/36$ (d) $4/36^2$ (e) none of these

5. If one die is rolled three times, the probability of getting exactly two 4's is

   (a) 3/8 (b) 3/16 (c) 1/4 (d) 15/216 (e) none of these

6. If one die is rolled four times, the probability of getting exactly three 5's is

   (a) $12/6^4$ (b) $20/6^4$ (c) $1/6^4$ (d) $4/6^4$ (e) none of these

7. If one die is rolled five times, the probability of getting exactly two 4's is

   (a) $C(4,2)\left(\frac{1}{6}\right)^2\left(\frac{5}{6}\right)^3$ (b) $C(5,2)\left(\frac{1}{6}\right)^3\left(\frac{5}{6}\right)^2$ (c) $C(4,2)\left(\frac{1}{6}\right)^3\left(\frac{5}{6}\right)^2$

   (d) $C(5,2)\left(\frac{1}{6}\right)^2\left(\frac{5}{6}\right)^3$ (e) none of these

8. If one die is rolled 15 times, the probability of getting exactly seven 4's is

(a) $C(15,7)\left(\frac{1}{6}\right)^{15}\left(\frac{5}{6}\right)^{7}$ (b) $C(15,7)\left(\frac{1}{6}\right)^{15}\left(\frac{5}{6}\right)^{8}$ (c) $C(15,7)\left(\frac{1}{6}\right)^{7}\left(\frac{5}{6}\right)^{8}$

(d) $C(15,7)\left(\frac{1}{6}\right)^{8}\left(\frac{5}{6}\right)^{7}$ (e) none of these

9. If one die is rolled eight times, the probability of getting exactly four 1's is

(a) $C(8,4)\left(\frac{1}{6}\right)^{8}$ (b) $C(8,4)\left(\frac{1}{2}\right)^{4}\left(\frac{1}{2}\right)^{4}$ (c) $C(6,4)\left(\frac{1}{6}\right)^{6}\left(\frac{5}{6}\right)^{4}$

(d) 1/2 (e) none of these

10. If one die is rolled 20 times, the probability of getting exactly ten 1's is

(a) $C(20,10)\left(\frac{1}{6}\right)^{10}\left(\frac{5}{6}\right)^{10}$ (b) $C(20,10)\left(\frac{1}{2}\right)^{10}\left(\frac{1}{2}\right)^{10}$ (c) $C(5,4)\left(\frac{1}{6}\right)^{10}\left(\frac{5}{6}\right)^{10}$

(d) 1/2 (e) none of these

11. $C(8,5)\left(\frac{1}{6}\right)^{5}\left(\frac{5}{6}\right)^{3}$ can represent the probability of rolling

(a) exactly five 3's when one die is rolled five times
(b) exactly three 5's when one die is rolled five times
(c) exactly five 3's when one die is rolled eight times
(d) exactly three 5's when one die is rolled eight times
(e) none of these

12. $C(12,5)\left(\frac{1}{6}\right)^{5}\left(\frac{5}{6}\right)^{7}$ can represent the probability of rolling

(a) exactly five 3's when one die is rolled five times
(b) exactly three 5's when one die is rolled five times
(c) exactly five 3's when one die is rolled 12 times
(d) exactly three 5's when one die is rolled 12 times
(e) none of these

13. $C(8,5)\left(\frac{1}{2}\right)^{8}$ can represent the probability of tossing

(a) exactly five pairs of heads when two coins are tossed five times
(b) exactly five heads when a coin is tossed eight times
(c) exactly five pairs of heads when two coins are tossed eight times
(d) exactly eight heads when a coin is tossed five times
(e) none of these

14. $C(6,3)\left(\frac{1}{2}\right)^{6}$ can represent the probability of tossing

(a) exactly three pairs of heads when two coins are tossed six times
(b) exactly three tails when a coin is tossed six times
(c) exactly three pairs of heads when six coins are tossed two times
(d) exactly six heads when a coin is tossed three times
(e) none of these

5-4 EXPECTED VALUE

An important topic in many applications of probability is that of the "expected value" of an experiment. The expected value is very similar to the concept of "average value." In the next chapter we will show the close connection between the average and expected values.

As an intuitive example of expected value, consider the experiment of tossing a coin 100 times. How many times would we *expect* heads to appear? While we might not get exactly 50 heads each time the experiment is run, we would certainly expect the number of heads to be close to 50, since Pr(H) = 1/2 and (1/2)·100 = 50.

Similarly, if a die were rolled 600 times, about how many 2's would appear? Since Pr(2) = 1/6, we would expect about (1/6)·600 or one hundred 2's. Therefore, for the binomial experiment of rolling a die 600 times and recording the number of 2's, the expected value is 100.

These examples suggest the following result concerning the expected value of a binomial experiment.

> If a binomial experiment has n repetitions with Pr(S) = p then the expected value E is
>
> $$E = np$$

The expected value is a useful property for any probability distribution whose outcomes are numbers. If the outcomes are not numbers, such as the experiment to toss a coin and record H or T, it makes no sense to speak of the "average" outcome. Therefore, in this section, all probability distributions have numerical outcomes.

To illustrate how to compute the expected value, look at a simple experiment whose outcomes are not equally likely. Suppose that an urn contains 10 marbles such that one marble has the number "1" marked on it, three marbles have the number "2" marked on them and six have the number "3" marked on them. A marble is selected from the urn and its number is recorded. What is the expected value? If the experiment were done many times, the number "1" would be expected to appear one-tenth of the time, "2" would appear three-tenths of the time and "3" would appear six-tenths of the time. Therefore, the probability distribution is given by

outcome, x_i	1	2	3
probability, p_i	.1	.3	.6

This table implies that if the experiment were repeated 100 times, we would expect to observe ten 1's, thirty 2's and sixty 3's. The average of these numbers is

$$\frac{1(10) + 2(30) + 3(60)}{100} = \frac{250}{100} = 2.5$$

Another way to view this computation is to divide each term in the numerator by 100 to get

$$\frac{1(10)}{100} + \frac{2(30)}{100} + \frac{3(60)}{100} = 2.5$$

Dividing each numerator and denominator by 10 gives

$$1\cdot\frac{1}{10}+2\cdot\frac{3}{10}+3\cdot\frac{6}{10}=1\cdot(.1)+2\cdot(.3)+3\cdot(.6)=2.5$$

This expression demonstrates what we call **the expected value**. It is the sum of the products of the outcomes and their respective probabilities. It is independent of how many times the experiment is actually repeated. If the experiment were done 1000 times we would expect the average of the 1000 outcomes, that is, their sum divided by 1000, to be the same value, 2.5. We can now formally state the definition.

DEFINITION

If an experiment has numerical outcomes and the probability distribution is:

x_i	x_1	x_2	...	x_n
p_i	p_1	p_2	...	p_n

then the **expected value**, E, of the experiment is

$$E = x_1p_1 + x_2p_2 + \cdots + x_np_n$$

The following examples illustrate the use of this definition. Example 1 discusses the concept of a "fair game", meaning one in which each person has an equal chance to win. That is, if the game were played many times, each player would expect to win about the same amount of money.

EXAMPLE 1

Two players, A and B, play the following game. A draws a card from a shuffled deck. If a face card or an ace is drawn, A wins \$2. If any other kind of card is drawn, B wins \$1. Is the game fair?

Solution. We compute the expected value and then interpret its meaning. There are two outcomes: A either wins \$2 or loses \$1, so the two outcomes are 2 and –1. The outcome 2 occurs if one of the face cards or aces is drawn. The probability that one of these cards is drawn is 4/13. Hence Pr(2) = 4/13. On the other hand, A loses a dollar if a value of two through ten, inclusive, occurs. Since there are 36 such cards, Pr(–1) = 36/52 = 9/13. Using the notation in the definition, we have $x_1 = 2$, $p_1 = 4/13$, $x_2 = -1$ and $p_2 = 9/13$. Thus the probability distribution is:

x_i	2	–1
p_i	4/13	9/13

Hence the expected value of Player A's winnings is given by

$$E = x_1p_1 + x_2p_2 = 2\cdot\frac{4}{13}+(-1)\cdot\frac{9}{13}=\frac{8}{13}-\frac{9}{13}=-\frac{1}{13}$$

The game is not fair because, in the long run, A will likely lose \$1 for every 13 plays.

Since E is a negative number in Example 1, Player A likely loses money in the long run. On the other hand, if E were positive, then Player A would likely win money in the long run. The previous example motivates the following definition.

DEFINITION

A game is **fair** if the expected value is zero.

EXAMPLE 2

A fair die is rolled, and a player receives an number of dollars equal to the number on the top face of the die if a roll of 1, 3, 4 or 5 occurs. If a roll of 2 or 6 occurs, the player loses \$2 or \$6 respectively. Is the game fair?

Solution. We compute the expected value by first finding the probability distribution. There are six outcomes: 1, –2, 3, 4, 5 and –6. Each outcome has probability 1/6 because each one corresponds to one of the six equally likely faces of a die. Thus the probability distribution is:

x_i	1	3	4	5	–2	–6
p_i	1/6	1/6	1/6	1/6	1/6	1/6

Hence the expected value E is

$$E = 1\cdot\frac{1}{6} + 3\cdot\frac{1}{6} + 4\cdot\frac{1}{6} + 5\cdot\frac{1}{6} + (-2)\cdot\frac{1}{6} + (-6)\cdot\frac{1}{6} = \frac{5}{6}$$

Therefore, the game is not fair because the expected value is not zero.

Note that in Example 2, the player will likely *win* an average of 5/6 of a dollar, or about \$0.83, per roll of the die. In other words, the player can expect to win about \$5 for every 6 plays in the long run. However, even though the player can expect to win money in the long run, the game is not fair because in order for Player A to win money, someone else has to lose money. Remember that a fair game is one in which *all* players have an equal chance to win.

EXAMPLE 3

In the game described in Example 2, change the value that the player wins when a roll of "1" occurs in order to make the game fair.

Solution. A fair game has expected value 0. We must replace the outcome 1 by a variable outcome, say a, compute the expected value of this probability distribution, and then set it equal to 0 and solve for a. Thus the probability distribution is:

x_i	a	3	4	5	–2	–6
p_i	1/6	1/6	1/6	1/6	1/6	1/6

Hence the expected value E is

$$E = a\cdot\frac{1}{6}+3\cdot\frac{1}{6}+4\cdot\frac{1}{6}+5\cdot\frac{1}{6}+(-2)\cdot\frac{1}{6}+(-6)\cdot\frac{1}{6}=\frac{a}{6}+\frac{4}{6}=\frac{a+4}{6}$$

Now let E = 0 and solve for a.

$$\frac{a+4}{6}=0$$

$$a+4=0$$

$$a=-4$$

Therefore the game is fair if the player loses \$4 whenever a roll of 1 occurs.

The next two examples show how the expected value is often used in business applications. When someone starts a new venture, a good knowledge of the prospect for success is necessary. Often the expected value can give a good indication of the likelihood for success.

EXAMPLE 4

A person has applied for a donut shop franchise. There are two locations that look profitable. Each has a good traffic pattern indicating that a heavy volume of customers is possible. Site 1 has most of its traffic in the morning and afternoon: 50% of the customers will come in the morning, 40% in the afternoon and 10% in the evening. Site 2 has a more homogeneous clientele: 30% will come in the morning, 30% in the afternoon and 40% in the evening. Each site has the same projected volume – about 1000 customers per day. The parent company has determined that the average customer spends \$5 in the morning, \$4 in the afternoon and \$3 in the evening. Which site has the larger expected profit?

Solution. The problem consists of finding the expected value for two locations: Site 1 and Site 2. The probability distributions for these locations are

Time of Day	Expenditure per Customer	Percentage of Volume Site 1	Site 2
Morning	\$5	.5	.3
Afternoon	\$4	.4	.3
Evening	\$3	.1	.4

The expected value for Site 1 is

$$E = 5(.5) + 4(.4) + 3(.1) = 2.5 + 1.6 + .3 = 4.4$$

The expected value for Site 2 is

$$E = 5(.3) + 4(.3) + 3(.4) = 1.5 + 1.2 + 1.2 = 3.9$$

Hence Site 1 has a better expected value of \$4.40 per customer than Site 2 which has the expected value of \$3.90 per customer. Since the volumes are assumed to be equal, Site 1 has a larger expected profit.

To interpret the meaning of the expected values in Example 4, assume that there is a volume of 1000 customers per day. Then the owner can expect a revenue of about \$4,400 per day from Site 1 and only about \$3,900 per day from Site 2.

EXAMPLE 5

A real estate company is bidding on rights to develop two properties, Property A and Property B. The company estimates that a profit of \$525,000 will be made on Property A, while Property B will realize a profit of \$400,000. Legal and other costs of bidding amount to \$9,000 for Property A and \$5,000 for Property B. The probability of being awarded the contract if the company bids for Property A is .45, while this probability is .6 for Property B. For which property should the company bid? (Note: Profits already take bidding into consideration.)

Solution. The problem consists of finding the expected value for each of the two properties: Property A and Property B. The probability distributions for these properties are

	Property A	Property B
Profit	\$525,000	\$400,000
Cost of bidding	\$9,000	\$5,000
Probability of success	.45	.6

Therefore, for Property A, the probability of making \$600,000 is .45, and the probability of *losing* \$9,000 is therefore 1 – .45 = .55. Hence, the expected value for Property A is

$$E = 525{,}000(.45) + (-9000)(.55) = 231{,}300$$

Similarly, the expected value for Property B is

$$E = 400{,}000(.6) + (-5000)(.4) = 238{,}000$$

Hence Property B has a better expected value and is therefore the property for which the real estate company should bid.

EXERCISE SET 5-4

In problems 1 to 4 find the expected value of the binomial experiment.

1. Toss a coin 50 times and record the number of heads.
2. Toss two coins 200 times and record the number of times two heads occurs.
3. Roll a die 12 times and record the number of times 6 occurs on the top face.
4. Roll two dice 10 times and record the number of times the sum of 2 occurs.

In problems 5 and 6 find the expected value of the probability experiment where a marble is selected from the urn and its number is recorded.

5. An urn contains 10 marbles. One marble has the number 1 marked on it, four marbles have the number 2 and five have the number 3.

6. An urn contains 20 marbles. One marble has the number 2 marked on it, eight marbles have the number 4 and eleven have the number 6.

In problems 7 to 10 find the expected value for the games. Is the game fair? (A game is fair if the expected value is 0.)

7. Player A chooses a card from a deck of 52 cards and wins \$4 if a face card appears and loses \$1 if it is not a face card.

8. Player A chooses a card from a deck of 52 cards and wins \$13 if a face card appears and loses \$3 if it is not a face card.

9. Player A rolls 2 dice and records the sum of the top faces. A wins \$9 if 7 or 11 appears and loses \$2 otherwise.

10. Player A rolls 2 dice and records the sum of the top faces. A wins \$3 if 2, 7 or 11 appears and loses \$1 otherwise.

11. A game consists of rolling a fair die and the player receives an amount of dollars equal to twice the number turning up on the top face if the numbers 1, 2, 3 or 4 occur. If 5 or 6 occur the player loses \$9. Is the game fair?

12. A game consists of rolling a fair die and the player receives an amount of dollars equal to twice the number turning up on the top face if the numbers 1, 2, 3 or 4 occur. If 5 or 6 occur the player loses \$10. Is the game fair?

In problems 13 and 14 consider the game described in problem 11. Change the value that the player wins when the given number is rolled in order to make the game fair.

13. Change the amount the player wins when "2" is rolled.

14. Change the amount the player wins when "4" is rolled.

In problems 15 and 16 determine which site is predicted to produce the larger revenue for a proposed franchise. Assume that the volumes at the two sites are equal.

15.

Time of Day	Expenditure per Customer	Percentage of Volume Site 1	Site 2
Morning	\$6	.1	.25
Afternoon	\$14	.35	.25
Evening	\$20	.5	.25
Late Night	\$11	.05	.25

16.

Time of Day	Expenditure per Customer	Percentage of Volume Site 1	Site 2
Morning	\$11	.4	.55
Afternoon	\$5	.3	.15
Evening	\$8	.15	.2
Late Night	\$7	.15	.1

In problems 17 to 20 a real estate company is bidding on rights to develop two properties, Property A and Property B. The estimated profit for each property is given along with the costs of bidding and the probabilities of being awarded the contract if the company bids for Property A and for Property B. For which property should the company bid?

17.

	Property A	Property B
Profit	\$200,000	\$100,000
Cost of bidding	\$8,000	\$3,000
Probability of success	.4	.5

18.

	Property A	Property B
Profit	\$400,000	\$500,000
Cost of bidding	\$8,000	\$9,000
Probability of success	.7	.6

19.

	Property A	Property B
Profit	\$300,000	\$500,000
Cost of bidding	\$12,000	\$9,000
Probability of success	.3	.4

20.

	Property A	Property B
Profit	\$500,000	\$500,000
Cost of bidding	\$10,000	\$9,000
Probability of success	.3	.4

In problems 21 to 25 find the expected value of the binomial experiment.

21. It is known that 70% of the population in a ward favor an upcoming bond issue. Five people are selected at random and each is asked whether the person is in favor of the issue. The number in favor is recorded.

22. A student takes a 20 question true-false test and guesses at the answers. The number of questions answered correctly is recorded.

23. The past performance of a machine shows that it produces 2% defectives. Ten items are selected at random ,and it is recorded whether each item is defective or not.

24. A die is rolled 8 times and the number of times the top face is even is recorded.

25. A part of a final exam consists of 5 questions that the student has not studied. The questions are multiple choice, each with 4 answers, the student guesses each answer and the number of correct responses is recorded.

26. Find the expected value for the probability distribution.

x_i	2	3	4	5	6
p_i	.05	.25	.35	.15	.20

27. An urn contains 3 red and 7 white marbles. Ten marbles are selected with replacement and the number of red marbles is recorded. Find the expected value of the experiment.

28. A raffle offers a first prize of $5000, two second prizes of $1000, four third prizes of $500 and ten prizes of $100. One thousand tickets are sold at $20 each. Find the expected value for a person who buys one ticket.

29. A lottery consists in a player choosing a three-digit number. Repetitions are allowed. Each ticket costs $1 and it pays $500 to win. Compute the expected earnings of a player who purchases one ticket.

30. Compute the expected earnings of the lottery player in problem 29 if repetitions are not allowed.

31. In a lottery a player purchases a ticket with a three digit number for $2. Repetitions are not allowed. If the number is drawn, the player wins $400. If the number drawn is not the player's, but the digits are the same except for the order, the player wins $200. Find the expected earnings.

32. The annual premium for a $6,000 insurance policy against the loss of jewelry is $100. If the probability that the jewelry will be lost is .005, what is the expected gain or loss on the insurance policy for the insurance company?

33. An urn contains 2 red and 3 white marbles. An experiment consists of selecting two marbles without replacement. If the experiment were conducted 100 times, what is the expected number of times that both marbles would be red?

34. A professional tennis player is asked what he expects to win each week. Of course, it is impossible to determine how he will perform in any given week, but it is possible to calculate the player's expected winnings from past performances. Below is a table that gives the likelihood that the tennis player finishes the next tournament in one of the first five places or in a lower position. It also gives the average prize for that place. Find his expected winnings for the next tournament.

Place	1	2	3	4	5	>5
Probability	.02	.04	.08	.12	.2	.54
Prize (in $1,000's)	50	25	12	5	2.5	1

In problems 35 to 38 the outcomes x_i in a game are given along with their probabilities, p_i. Find the expected value.

35.

x_i	−1	0	2	3
p_i	.3	.1	.5	.1

36.

x_i	−1	−.5	1	10
p_i	.2	.3	.4	.1

37.

x_i	−10	2	3	4
p_i	.15	.75	.08	.02

38.

x_i	−10	1	2	3
p_i	.15	.7	.1	.05

Referenced Exercise Set 5-4

1. There is a difference between the game of roulette played in America versus Europe. The American roulette wheel has 38 slots, numbered 1 to 36 with one slot labeled 0 and another 00. Of the slots numbered 1 to 36, 18 are red and 18 are black. The other two are green. The European wheel does not have the 00 slot and the betting differs. The differences can be compared via the expected value to see which version gives the player the better advantage. Consider first the version played in America. If you bet \$1 on red, you have two possibilities: you may win \$1 (in addition to your initial bet of \$1) with probability 18/38 = 9/19, or you may lose your bet of \$1 with probability 20/38 = 10/19. Here is the probability distribution.

Outcome, x_i	Probability, p_i	$x_i \cdot p_i$
1	9/19	9/19
–1	10/19	–10/19

In European roulette there is a third outcome besides the same two for the American version - if the ball lands on green, you do not necessarily lose your bet. Your bet is carried to the next spin where there are again three possibilities: if the ball lands on red, you get your \$1 bet back, so you break even; if it lands on black, you lose \$1; and if it lands on green again, you get back \$.50, which means you lose \$.50 of your initial bet. There are 4 outcomes: 1 (or win \$1), 0 (or break even), –.5 (or lose \$.50), and –1 (or lose \$1). Construct the probability distribution for European roulette and find the expected value for each game.

Cumulative Exercise Set 5-4

1. An experiment consists of tossing two dice and recording the sum of the top faces. Find the sample space and determine the elements in the two events:

   E: the sum is less than 6
   F: the sum is 4, 5 or 6

   Find (a) $\Pr(E \cup F)$, (b) $\Pr(E' \cap F)$

2. Work problem 1 assuming that three dice are tossed.

3. There are 40 members in a union and 10 agree to take a pay cut. If three are selected at random and asked their opinion, what is the probability that at least one will agree to take a pay cut?

4. Urn 1 has 5 white and 7 red marbles, and urn 2 has 4 white and 8 red marbles. The experiment is to select an urn and then select a marble and record its color. Find the probability of observing a red marble.

5. A portfolio has 50 accounts, 22 of which are executive accounts and 10 of which are delinquent. Five of the executive accounts are delinquent. If an account is selected at random, find the probability that the account is an executive account given that it is not delinquent.

6. In a survey of biking accidents at 5 hospitals, the incidence of fractures was as follows: Hospital 1 – 60% of the injuries were fractures; Hospital 2 – 55%; Hospital

3 – 45%; Hospital 4 – 50%; Hospital 5 – 30%. Assuming that there were the same number of biking accidents at each hospital, given that a fracture occurred, find the probability that the fracture occurred at Hospital 1 or Hospital 2.

7. Find the expected value of the binomial experiment of tossing two coins 200 times and recording the number of times two heads occurs.

8. Determine which site is predicted to produce the larger revenue for a proposed franchise.

Time of Day	Expenditure per Customer	Percentage of Volume Site 1	Site 2
Morning	$14	.4	.6
Afternoon	$10	.3	.1
Evening	$20	.2	.2
Late Night	$11	.1	.1

9. A student takes a 10 question true-false test and guesses at the answers. What is the probability of getting (a) exactly 8 correct, (b) at least 6 correct?

10. The annual premium for a $9,000 insurance policy against the loss of jewelry is $80. If the probability that the jewelry will be lost is .005, what is the expected gain or loss on the insurance policy for the insurance company?

11. An urn contains 3 red and 2 green marbles. An experiment consists of selecting two marbles without replacement. (a) What is the probability that both marbles are red? (b) If the experiment were conducted 200 times, what is the expected number of times that both marbles would be red?

12. The outcomes of a game are given along with their probabilities. Find the expected value of the game.

x_i	-1	0	2	3
p_i	.2	.1	.4	.3

13. Traffic police monitor cars parked in a two-hour parking zone by marking with chalk on the tires the hour of the day rounded to the next largest hour. Thus there are twelve possible marks corresponding to the twelve hours on a clock face. For example, if it is 11:45 they would put a chalk mark at the top of the tire to indicate 12:00. Suppose a car drives away and returns to the same parking spot. (a) What is the probability the mark will be in the same spot on the tire? (b) Suppose the officer marks both curbside tires and suppose the tires rotate at different speeds, making the events of the hour the chalk mark on the front tire makes independent of the chalk mark on the back tire. What is the probability the marks will come to rest in the original locations?

14. A city has three tests it administers to drivers to test for alcohol intoxication: a breathalyzer, walking a straight line, and touching index fingers by swinging them from an outstretched sideways position. A type I error is made when the driver is judged intoxicated when in fact the driver is sober. Suppose the probability of a type I error for the breathalyzer is .01, for the walk it is .1, and for the finger test is .15. Suppose for a sober person the tests are independent. What is the probability that a type I error was made on all three tests?

Multiple Choice Exercise Set 5-4

1. The expected value of the binomial experiment of tossing a coin 200 times and recording the number of heads is
(a) 200 (b) 100 (c) 50 (d) 2 (e) none of these

2. The expected value of the binomial experiment of rolling a die 600 times and recording the number of times "1" occurs on the top face is
(a) 600 (b) 100 (c) 50 (d) 1/6 (e) none of these

3. The expected value of the binomial experiment of rolling two dice 500 times and recording the number of times the sum of six occurs is
(a) 500 (b) 100 (c) 50 (d) 6 (e) none of these

4. An urn contains 20 marbles. One marble has the number 2 marked on it, eight marbles have the number 4 marked on them, and eleven have the number 6 marked on them. The expected value of the probability experiment where a marble is selected from the urn and its number is recorded is
(a) 20 (b) 100 (c) 5 (d) 1/20 (e) none of these

5. A person chooses a card from a deck of 52 cards and wins \$6 if a red card appears and loses \$2 if it is not a red card. The expected value of the game is
(a) 52 (b) 1/4 (c) 4 (d) 3 (e) none of these

6. A game consists of rolling a fair die, and the player receives an amount of dollars equal to twice the number turning up on the top face if the numbers 1, 2, 4 or 5 occur. If 3 or 6 occurs the player loses \$10. The expected value is
(a) 2/3 (b) 1/4 (c) 6 (d) –2/3 (e) none of these

7. It is known that 55% of the population in a ward favor an upcoming bond issue. Ten people are selected at random and each is asked whether he or she is in favor of the issue. The number in favor is recorded. The expected value is
(a) 5 (b) 5.5 (c) 55 (d) 10 (e) none of these

8. A student takes a 50 question true-false test and guesses at the answers. The number of questions answered correctly is recorded. The expected value is
(a) 50 (b) 5 (c) 25 (d) 10 (e) none of these

9. A die is rolled 100 times, and the number of times the top face is even is recorded. The expected value is
(a) 50 (b) 5 (c) 25 (d) 2 (e) none of these

10. An urn contains 4 red and 6 white marbles. Twenty marbles are selected with replacement and the number of red marbles is recorded. The expected value is
(a) 20 (b) 10 (c) 8 (d) 2 (e) none of these

11. A raffle offers a first prize of \$2000, four second prizes of \$1000, five third prizes of \$500 and ten prizes of \$100. One thousand tickets are sold at \$20 each. The expected value is
(a) 11,000 (b) –11 (c) 8600 (d) 0 (e) none of these

CASE STUDY
THE EXPECTED VALUE IN COURT

In a celebrated court case in 1974, Jane Blalock sued a group of her peers for damages, claiming they had deprived her of her means of income. Blalock joined the Ladies Professional Golf Association (LPGA) in 1969 and soon became a star performer. By May, 1972 she had become the year's leading money winner on the tour. Then, during the final round of the Bluegrass Invitational Tournament, she was disqualified from the tournament for an alleged infraction of the rules. A committee of her competitors, selected from the LPGA, also suspended her from the next tournament, the Ladies Carling Open. Blalock sought relief in court under the Sherman Anti-trust Act which holds that individuals cannot be suspended from their profession by their peers because it would tend to lessen competition. This legal precedent appeared justified because her peers would clearly benefit by her absence from the tournament. Blalock received an injunction enabling her to play in the Carling Open, but it was issued too late for her to join the tournament.

Besides being denied the chance to earn prize money in the Carling Open, Blalock had incurred a large legal fee, so she again sued the LPGA for damages under the Sherman Act. The only legal point the court could rule on was her loss of potential earnings from being denied the right to participate in the tournament. This issue is very sticky in most Anti-trust cases because the court must decide what the earnings of the claimant, if any, would have been had the competitors not interfered, in this case by suspending the claimant. Thus her problem was to convince the court that, had she been allowed to play, she would most likely have scored low enough to win prize money (remember that in golf the lower the score the better). In addition, she had to convince them of the amount of money that she would have won.

Therefore she had to estimate her score in a tournament in which she never played – and most importantly, her estimate, as well as her means of obtaining this estimate, had to convince a jury that it was valid

One of the key issues for Blalock was to use a measure that would not only give a clear indication of her most probable score but would also be easy for a jury to understand. She used the expected value.

Some fairly complicated statistical measures were used to measure the probability of Blalock attaining a particular score. A wide range of statistics was gathered, and her legal team decided to use the nine tournaments before the disqualification to compare Blalock's performance with that of her competitors.

We summarize the results in Table 5–4. The possible scores, or outcomes, are the integers from 209 (that being the lowest since it was the winning score in the tournament) to 232 (that being the highest because anyone with a higher score did not win any money). The data becomes more manageable if we limit the 24 outcomes to 8, so we will let the outcome 210 represent the three possible scores, 209, 210 or 211. The corresponding probability, .07, is therefore the probability that she would score either 209, 210 or 211. Thus the probability that she would score between 212 and 214, represented by the outcome 213, is .16. In Table 5–4 the possible outcomes are the x_i's, and the probabilities are represented by p_i's. In the third column, we compute $x_i \cdot p_i$. Then the sum of the $x_i \cdot p_i$'s is the expected value.

Outcome, x_i	Probability	$x_i \cdot p_i$
210	.07	14.70
213	.16	34.08
216	.23	49.68
219	.24	52.56
222	.17	37.74
225	.09	20.25
228	.03	6.84
231	.01	2.31
	Total	218.16

Table 5–4

Thus the expected value is 218.16, or a score of about 218. Blalock reasoned that this was the best estimate of her score for the tournament in which she was disqualified and thus she should be awarded a judgment equal to the prize money given to the person(s) who scored 218 in the tournament. Three women shot 218 and tied for 5th place, each earning $1,427.50. If the jury ruled in her favor, these damages would be tripled and her legal fees included under the provisions of the Sherman Act. This argument was so overwhelmingly convincing to the jury that they decided that Blalock's chances of not only winning the prescribed amount were very good but the likelihood of her winning the tournament and garnering the 1st place prize was great enough that they awarded her the 1st place money – $4500.

This award was then tripled, and her legal expenses were paid.

CASE STUDY EXERCISE SET

In problems 1 to 6 refer to Table 5–4 and find the probability of Jane Blalock attaining the described score.

1. A score less than or equal to 220.
2. A score less than or equal to 226.
3. A score between 218 and 226, inclusive.
4. A score between 221 and 229, inclusive.
5. A score of at least 227.
6. A score of at least 215.

In problems 7 and 8 assume that Jane Blalock's probability of attaining a particular outcome in a tournament is given in the table. As in Table 5–4, assume that the outcomes represent her probability of her getting that score, the one lower or the one higher. Thus the outcome of 222 represents the probability of her getting 221, 222 or 223. Find the expected value in each problem.

7.

Outcome, x_i	Probability
210	.05
213	.20
216	.25
219	.22
222	.15
225	.08
228	.03
231	.02

8.

Outcome, x_i	Probability
210	.10
213	.15
216	.20
219	.25
222	.22
225	.04
228	.03
231	.01

CHAPTER REVIEW

Key Terms

5-1 Counting Techniques
Urn Problems
Probability experiment
Poker Hands
"At least two the same"

5-2 Bayes' Rule
Conditional probability
Venn Diagrams
Tree Diagrams
Bayes' Rule

5-3 Bernoulli Trials
Binomial experiments
Success & failure
Bernoulli process

5-4 Expected Value
Expected value
Fair games

REVIEW EXERCISE SET

1. A true-false test consists of 12 questions. If the student guesses at each answer, find the probability the number correct is at least 10.

2. An urn contains 5 red, 3 white and 4 blue marbles. Four marbles are selected without replacement. Find the probability that
   (a) all 4 are red,
   (b) at least 3 are red.

In problems 3 to 5 the experiment is to first choose at random one of three urns, each containing some red and white marbles, and then select a marble. Urn 1 has 2 red and 8 white marbles, urn 2 has 3 red and 7 white marbles and urn 3 has 6 red and 4 white marbles. Find the probability of each event.

3. Urn 1 was chosen given that the marble is red

4. Urn 2 was chosen given that the marble is white

5. Urn 3 was chosen given that the marble is white

6. A die is rolled twenty times. What is the probability of observing
   (a) exactly seven 6's
   (b) at most seven 6's?

7. It is known that 40% of the population in a ward favor an upcoming bond issue. Ten people are selected. Find the probability that
   (a) exactly 6 of them will favor the bond issue
   (b) at least 6 of them will favor the bond issue.

8. A student takes a 10 question multiple-choice test with five choices per question and guesses at the answers. What is the probability of getting
   (a) exactly 1 correct
   (b) at most 2 correct?

9. A firm is going to introduce a new product and wants to predict how well it will do. Past records show that similar products at least break even 60% of the time. If there are 12 sales regions find the probability that the product will break even in at least 8 regions.

In problems 10 and 11 find the expected value of the probability experiment where a marble is selected from the urn and its number is recorded.

10. An urn contains 15 marbles. One marble has the number 1 marked on it, four marbles have the number 2 and ten have the number 3.

11. An urn contains 20 marbles. One marble has the number 2 marked on it, eight marbles have the number 4 and eleven have the number 6.

6 STATISTICS

Chapter Overview

Statistics is concerned with presenting, describing, and interpreting sets of data. This chapter begins by placing probability distributions in a statistical setting. It shows how histograms present various kinds of distributions in a graphic manner. Next, five units of measure are introduced. The mean, median, and expected value describe the "average" of a set of numbers, while the variance and standard deviation describe the spread from the average. The normal curve is introduced to interpret data from the binomial experiments discussed in the preceding chapter.

Statistics is one of the most useful areas of mathematics. This chapter considers applications to such diverse fields as marketing, law, medicine, sociology, ecology, agriculture, industrial accidents, gambling, drug studies, and the over-booking of reservations by airlines.

Case Study Preview

The two Case Studies examine two entirely different issues in two entirely different revolutions in two entirely different countries.

The first Case Study takes place in Russia. The revolutionary leader V. I. Lenin is well-known for his political savvy and cunning. A recent study concludes that he was an accomplished statistician as well. Lenin's early studies, in which comparative statistics support his argument about the worsening plight of Russian peasants, reflect his command of the subject.

The second Case Study takes place in the USA The 13 colonies in America joined forces from 1776 to 1783 to fight for their independence from Great Britain. But then a family squabble broke out as the independent colonies wrestled with their own self-governance. The writing of a Constitution was finally completed in September of 1787. The next step was its ratification by the colonies, and one of the most instrumental documents supporting it was *The Federalist* papers.

As we all know, the Constitution was ratified unanimously. However, for almost 200 years the identity of the authors of some of *The Federalist* papers remained unknown. Conclusive evidence was published in 1964. The statistical methods described in this chapter will enable you to understand the results of that historic study.

6-1 PROBABILITY DISTRIBUTIONS

This section bridges the gap between probability and statistics. First some topics from probability are presented in a statistical setting. Then it is shown how certain graphs called histograms represent data geometrically. Finally the section demonstrates how histograms are related to probability. Histograms can be seen in most daily newspapers, especially in the business section.

Triskaidekaphobia

Let us begin by constructing a histogram using familiar data that may produce a surprising result. Many superstitions are associated with Friday the 13th. Hollywood has exploited a fear of this date by filming several horror movies with the same title, yet the thirteenth of a month is more likely to fall on a Friday than any other day. Let us explain.

The calendar repeats itself every 400 years. There are 4800 months in this time period since every year has 12 months. Table 6–1 lists the days of the week that the 13th falls on for all 4800 months. It shows that the 13th occurs on a Friday more frequently than any other day.[1]

Table 6–1 is called a **frequency distribution**. The sample space consists of the seven days of the week. Since numerical outcomes are more convenient to handle, we assign integers 1 through 7 to the days of the week. They are denoted by the letter X. The number of months that the 13th falls on each day is called the **frequency**. The frequencies are listed in the column under f.

A **random variable** is a rule that assigns a real number to each outcome in the sample space. The term "random variable" refers to the unpredictable nature of the outcomes. In the example above, the variable X is a random variable.

Day	X	f
Sunday	1	687
Monday	2	685
Tuesday	3	685
Wednesday	4	687
Thursday	5	684
Friday	6	688
Saturday	7	684
Total		4800

Table 6–1

In Table 6–1 the frequency of X = 6 is 688, meaning that the 13th of a month falls on a Friday in 688 of the 4800 months. The proportion of these occurrences is called the **relative frequency**. The relative frequency for X = 6, rounded to 4 decimal places, is 688/4800 = 0.1433. Since the relative frequency can be interpreted as a probability, we write

---

[1] V. Frederick Rickey, "Mathematics of the Gregorian Calendar," *The Mathematical Intelligencer*, Volume 7, Number 1, 1985, pp. 53-56.

Pr (X = 6) = 0.1433

In general, Pr (X = x) is read "the probability that the random variable X is equal to x." Table 6–2 is formed from Table 6–1 by omitting the names of the days and by adding a column of relative frequencies under the letter p (for probability). It is called a **probability distribution**.

Example 1 shows how to find the probability that a value of X falls between two numbers a and b. The notation for this is Pr ($a \le X \le b$).

EXAMPLE 1

What is the probability that the 13th of a month falls on a weekday?

Solution. Weekdays are assigned integers between 2 and 6 inclusive. Therefore the problem is to find Pr ($2 \le X \le 6$). It is obtained by summing the relative frequencies.

X	f	p
1	687	.1431
2	685	.1427
3	685	.1427
4	687	.1431
5	684	.1425
6	688	.1433
7	684	.1425
Total	4800	.9999

Table 6–2

$$\begin{aligned} \Pr(2 \le X \le 6) &= \Pr(X = 2) + \Pr(X = 3) + \Pr(X = 4) + \Pr(X = 5) + \Pr(X = 6) \\ &= .1427 + .1427 + .1431 + .1425 + .1433 \\ &= .7143 \end{aligned}$$

Therefore the probability of the 13th of a month falling on a weekday is .7143.

Example 2 applies these concepts to coin-tossing experiments that will occur repeatedly in this chapter. The notation Pr ($X \ge c$) refers to the probability that "X is greater than or equal to the number c."

EXAMPLE 2

Toss 3 coins and let X represent the number of heads that occur. Construct the probability distribution and find Pr ($X \ge 1$).

Solution. The sample space consists of 8 outcomes, each of which is a sequence of 3 H's or T's. The random variable X assumes the values 0, 1, 2, and 3 for the number of heads that can occur. The probability distribution is given in Table 6–3.

X	f	p
0	1	1/8
1	3	3/8
2	3	3/8
3	1	1/8
Total	8	8/8 = 1

Table 6–3

The notation Pr $(X \geq 1)$ stands for the probability of tossing at least one head. It can be obtained directly.

$$\Pr(X \geq 1) = \Pr(X = 1) + \Pr(X = 2) + \Pr(X = 3)$$

Alternately it can be obtained from the rule for complementary probabilities:

$$\Pr(X \geq 1) = 1 - \Pr(X < 1) = 1 - \Pr(X = 0) = 1 - 1/8 = 7/8$$

Histograms

Statistics makes vital use of various graphing techniques to analyze data. A **histogram** represents data by means of a bar graph. Figure 6–1 shows the histogram for the days on which the 13th of a month occurs. The horizontal axis represents the days of the week and the vertical axis represents the frequencies. The jagged line at the origin indicates that an interval of data has been omitted, resulting in a graph whose true depth is distorted.

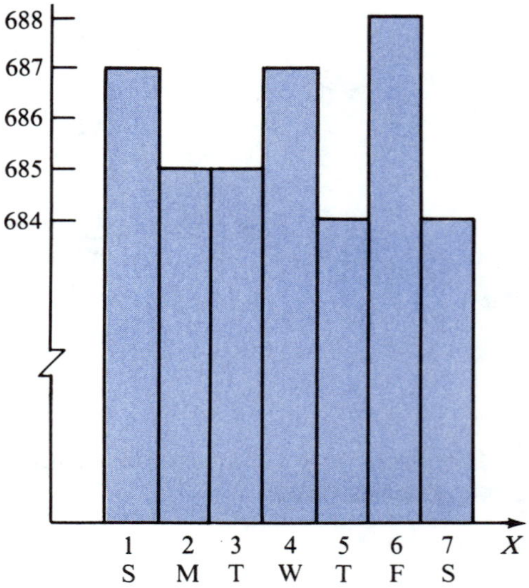

Figure 6–1

A set of data can have a histogram for its frequency distribution and a histogram for its probability distribution. In both cases the units along the horizontal axis are the same. However, the units along the vertical axis are either the frequencies or the relative frequencies. The shapes of the two histograms are identical, but the histogram with relative frequencies has a direct connection with probability. When the widths of the bars are equal to one unit and the heights of the bars are the relative frequencies, the probability of each outcome is equal to the area of that bar. In this way the probability of an event E is equal to the sum of the areas of the bars of all outcomes in E. Example 3 illustrates this property.

EXAMPLE 3

Construct a histogram for the probability distribution in Example 2. Which bars represent the probability of tossing at least one head, and what is the probability of tossing at least one head?

Solution. The histogram for the probability distribution is given in Figure 6–2.
The shaded portion represents the region where $X \geq 1$. Consequently, $\Pr(X \geq 1) = 7/8$ because the total area of the shaded rectangles equals

$$3/8 + 3/8 + 1/8 = 7/8.$$

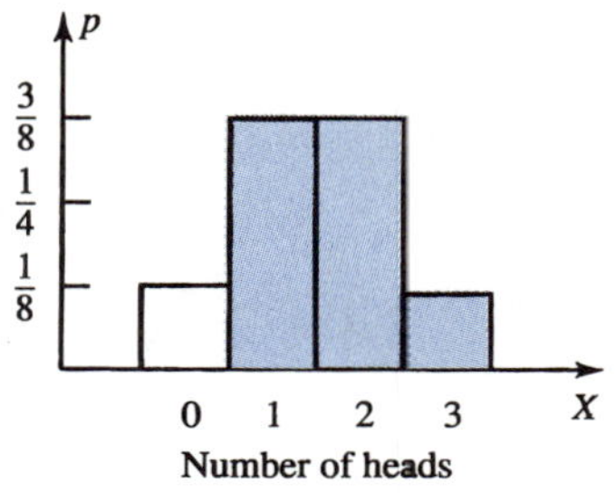

Figure 6–2

Histograms are often used to compare different sets of frequencies of the same sample space. The Contraceptive Drug Study, which examined the possible side effects of oral contraceptives on blood pressure, made use of such a comparison. Table 6–4 presents the results of the study for women aged 17 to 24 who were neither pregnant nor were taking hormonal medication. It gives one frequency distribution for women who used the Pill and another frequency distribution for women who did not. The random variable X represents systolic blood pressure (in millimeters) and the relative frequency represents the percentage of women with the given blood pressure.

Non-users		Users	
X	f	X	f
100	14	100	9
110	26	110	21
120	33	120	27
130	18	130	25
140	7	140	14
150	2	150	4

Table 6–4

EXAMPLE 4

Draw the histograms for each frequency distribution in Table 6–4. Shade the portion that represents blood pressures that are 130 or lower.

Solution. The histograms are drawn in Figure 6–3.

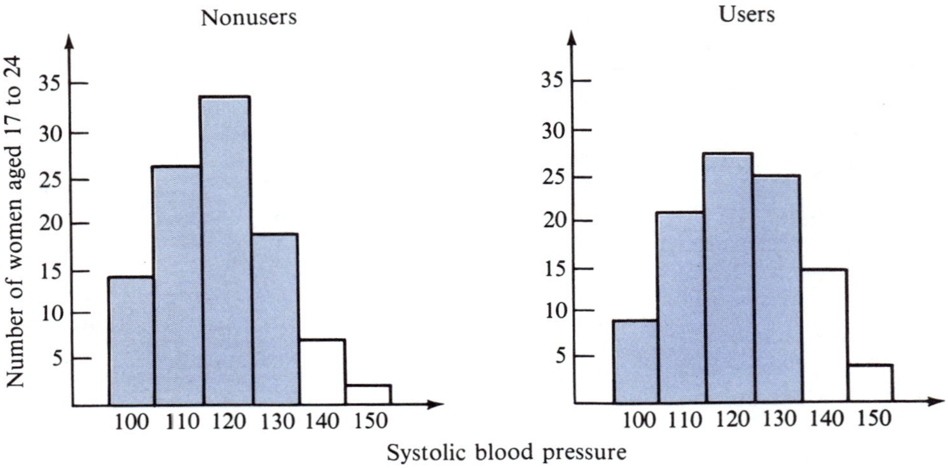

Figure 6–3

Intervals

Sometimes intervals of numbers are tabulated instead of the numbers themselves. For statistical purposes the entries become the midpoints of the intervals. Table 6–5 gives the weights of the 1988 Super Bowl champion team, the Washington Redskins. Example 5 shows how to draw the histogram for this data.

weight	frequency
170 - 179	4
180 - 189	5
190 - 199	3
200 - 209	2
210 - 219	3
220 - 229	5
230 - 239	5
240 - 249	3
250 - 259	1
260 - 269	2
270 - 279	6
280 - 289	1
290 - 299	3

Table 6–5

EXAMPLE 5

Construct the frequency distribution and draw the histogram for the weights of the Washington Redskins.

Solution. The random variable X assumes values that are midpoints of the intervals. The interval 170 - 179 includes all players who weigh between 170 and 180, so the midpoint is 175. The remaining entries are 185, 195, etc. This leads to the accompanying frequency distribution. The histogram is drawn in Figure 6–4.

X	*f*
175	4
185	5
195	3
205	2
215	3
225	5
235	5
245	3
255	1
265	2
275	6
285	1
295	3

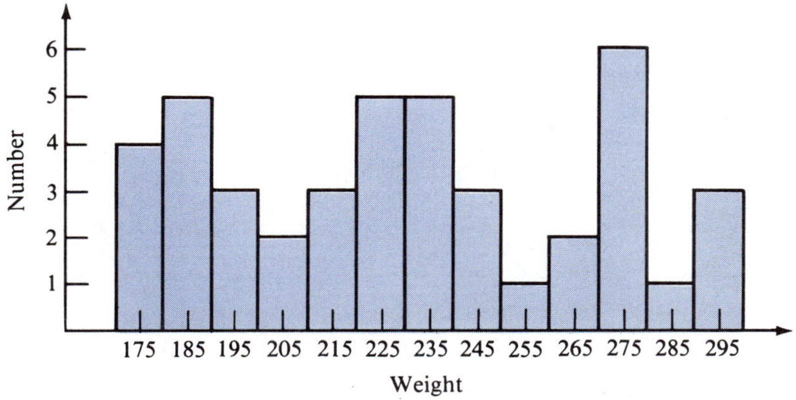

Figure 6–4

The histogram in Figure 6–4 is fairly level. A histogram that is perfectly level is called a **uniform distribution**. Contrast this histogram with the one in Figure 6–2. The latter histogram is more "mounded" and is closer to what is called a **normal distribution** (or a **bell-shaped curve**).

For the general population the distribution of weight is a normal distribution. It is somewhat surprising that the distribution of weights on an NFL team is not a normal distribution. We certainly expect the weights to be higher than the general population, but we do not expect them to be so uniform.

The distribution of income is an example of a **skewed distribution** – one where most of the data occurs at either the low end or the high end. of the x-range. This type of histogram will be illustrated in the exercises.

EXERCISE SET 6-1

Problems 1 to 4 refer to Example 1. Find

1. Pr (the 13th of a month occurs on a weekend).
2. Pr (the 13th of a month does not occur on a Friday).
3. Pr $(X \geq 5)$.
4. Pr $(X \leq 6)$.

In problems 5 to 8 toss three coins and let X be the number of heads that appear. Find

5. Pr $(X = 1)$
6. Pr $(X = 3)$
7. Pr $(X \geq 2)$
8. Pr $(X \leq 2)$

In problems 9 to 12 draw the histogram for the experiment of tossing three coins, and shade the stated portion.

9. Exactly one head.
10. Exactly three heads.
11. At least two heads.
12. At most two heads.

In problems 13 and 14 toss two coins and let X be the number of heads that appear.

13. Construct the frequency distribution and draw its histogram.
14. Construct the probability distribution and draw its histogram.

15. Draw the histogram for the ages of students in a marketing course using the data in the following frequency distribution.

age	frequency
17 - 19	10
20 - 22	14
23 - 25	7
26 - 28	2
29 - 31	1

16. Draw the histogram for the scores on an exam using the data in the following frequency distribution.

age	frequency
50 - 59	4
60 - 69	5
70 - 79	6
80 - 89	5
90 - 99	5

In problems 17 to 22 using a probability distribution and/or a histogram, find the stated probability.

17. Toss four coins and let X be the number of heads that occur. Find Pr $(X \geq 1)$.

18. A jar contains four vanilla cookies and three chocolate cookies. Two cookies are selected at random. Let X be the number of vanilla cookies. Find Pr $(X \geq 1)$.

19. Toss two dice and let X be the sum of the top faces. Find Pr $(6 \leq X \leq 8)$.

20. Toss two dice and let X be the sum of the top faces. Find Pr $(7 \leq X \leq 10)$.

21. Select three cards from an ordinary deck without replacement and let X be the number of hearts. Find Pr $(X \leq 2)$.

22. Select two cards from an ordinary deck without replacement and let X be the number of kings. Find Pr $(X \geq 1)$.

Problems 23 to 28 refer to the following table of scores on two quizzes for a class of 24 students.

	frequencies	
Score	Quiz #1	Quiz #2
10	7	2
9	7	3
8	5	8
7	1	6
6	1	1
5	3	4

23. Find Pr $(X \geq 8)$ for Quiz 1.

24. Find Pr $(X \geq 8)$ for Quiz 2.

25. Find Pr $(6 \leq X \leq 8)$ for Quiz 1.

26. Find Pr $(6 \leq X \leq 8)$ for Quiz 2.

27. Draw the histogram for each quiz.

28. Which quiz scores are higher? Do the histograms drawn in Problem 31 convey this visually?

Problems 33 to 36 refer to Figure 6–5.

29. Find Pr (X ≥ 4).

30. Find Pr (2 ≤ X ≤ 4).

31. Find Pr (3 ≤ X ≤ 5).

32. Find Pr (X ≥ 5).

Problems 37 to 40 refer to Figure 6–6.

33. Find Pr (X ≤ 2).

34. Find Pr (2 ≤ X ≤ 4).

35. Find Pr (1 ≤ X ≤ 3).

36. Find Pr (X ≤ 0).

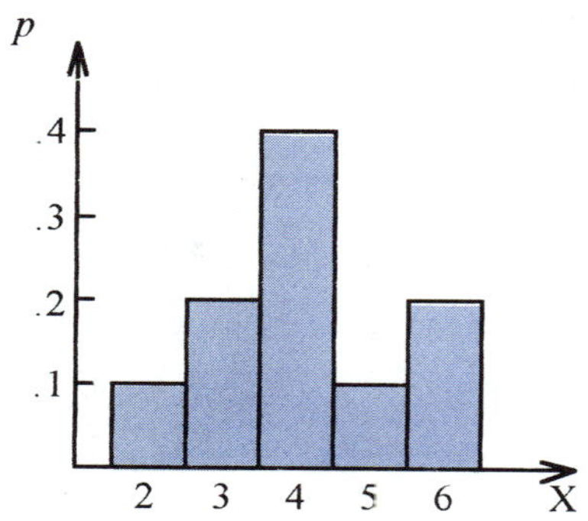

Figure 6–5

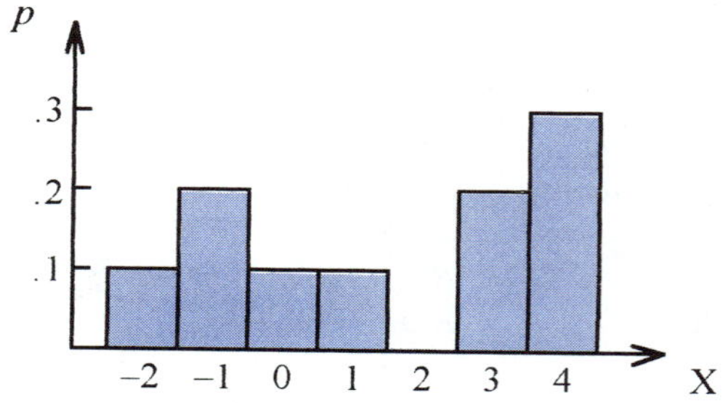

Figure 6–6

Referenced Exercise Set 6-1

1. Table 6–6 lists the total attendance (in millions) and average ticket price (in dollars) for Broadway theater over a 12-year span.[2] Draw the histograms for total attendance and ticket price. What conclusion can be drawn from comparing the histograms? (Note that years in the first column do not represent a range of values, but rather indicate that a Broadway season lasts from near the end of one year to the middle of the next year in much the same way that a basketball or hockey season runs.)

---

[2]Source: The League of American Theatres and Producers.

season	attendance	ticket price
1974-1975	6.6	8.63
1975-1976	7.3	9.72
1976-1977	8.8	10.56
1977-1978	9.6	11.87
1978-1979	9.6	13.95
1979-1980	9.6	15.20
1980-1981	11.0	17.90
1981-1982	10.1	22.07
1982-1984	8.4	24.88
1983-1985	7.9	28.73
1984-1986	7.4	28.78
1985-1975	6.5	28.96

Table 6–6

2. Draw the histogram for Table 6–7, which lists the distribution of income of families in the United States by percentages.[3] What kind of distribution is it?

Income level	Percent
0 - 5,000	4
5,000-10,000	13
10,000-15,000	17
15,000-20,000	20
20,000-25,000	16
25,000-30,000	12
30,000-35,000	7
35,000-40,000	4
40,000-45,000	3
45,000-50,000	2
50,000-55,000	1

Table 6–7

Problems 3 to 5 refer to Table 6–8, which lists the weights, in intervals, of the Denver Broncos, the losing team in the 1988 Super Bowl.

weight	frequency
170 - 179	4
180 - 189	5
190 - 199	9
200 - 209	2
210 - 219	2
220 - 229	3
230 - 239	5
240 - 249	3
250 - 259	2
260 - 269	8
270 - 279	2

Table 6–8

---

[3] Source: The Current Population Survey

3. Construct a frequency distribution with the midpoints of the intervals as entries.
4. Draw the histogram.
5. Does the histogram resemble a uniform distribution?
6. The following table traces the size of Hershey's standard milk chocolate bar from its introduction in 1950 to its discontinuation in 1974.[4] Draw its histogram with the horizontal axis the years from 1950 to 1974.

Common Retail Price: 10¢	Ounces
1950	2
July 1951	1.875
February 1954	1.625
August 1954	1.5
July 1955	1.75
February 1958	1.5
June 1960	2
August 1963	1.75
August 1965	2
June 1966	1.25
February 1968	1.5
March 1970	1.375
January 1973	1.26
Discontinued January 1, 1974	

7. The histogram in Figure 6–7 is taken from a study that contrasted several forecasting models for estimating the expected cash receipt of a company.[5] Does it represent a probability distribution? Why?

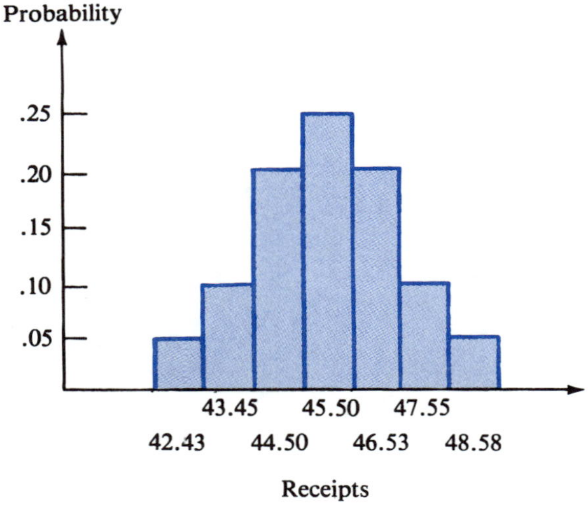

Figure 6–7

[4]H. J. Maidenberg, "Misadventures of Cocoa Trade," *New York Times*, Feb. 25, 1979, and Hershey Foods Corp.

[5]James L. Pappas and George P. Huber, "Probabilistic Short-term Financial Planning," *Financial Management*, August 1973, pp. 36-44.

8. Table 6–9 lists the height of men and women in the United States. Draw the histogram for each distribution. What kind of distributions are they?

Men Height (inches)	Men f (1000's)	Women Height (inches)	Women f (1000's)
60	1	53	0.5
61	5	54	0.5
62	9	55	0.2
63	17	56	0.2
64	37	57	10
65	35	58	13
66	70	59	38
67	63	60	45
68	94	61	83
69	54	62	104
70	63	63	73
71	32	64	90
72	28	65	47
73	11	66	44
74	6	67	14
75	1	68	12
		69	1
		70	1

Table 6–9

9. What does the word "triskaidekaphobia" mean?

Multiple-Choice Exercise Set 6-1

1. What kind of distribution is the table below?

x	f
–1	2
0	3
1	4
2	1

(a) normal (b) frequency (c) probability (d) uniform (e) none of these.

2. What kind of distribution is the table below?

x	p
–1	0.2
0	0.3
1	0.4
2	0.1

(a) normal (b) frequency (c) probability (d) uniform (e) none of these.

Problems 3 to 8 refer to the table below.

x	f
0	1/8
1	3/8
2	3/8
3	1/8

3. Pr (X ≥ 2) =
(a) 3/8 (b) 1/2 (c) 7/8 (d) 1/8 (e) none of these

4. Pr (X ≤ 2) =
(a) 3/8 (b) 1/2 (c) 7/8 (d) 1/8 (e) none of these

5. Pr (X < 2) =
(a) 3/8 (b) 1/2 (c) 7/8 (d) 1/8 (e) none of these

6. Pr (X > 2) =
(a) 3/8 (b) 1/2 (c) 7/8 (d) 1/8 (e) none of these

7. Pr (X = 2) =
(a) 3/8 (b) 1/2 (c) 7/8 (d) 1/8 (e) none of these

8. Pr (X = 0) =
(a) 3/8 (b) 1/2 (c) 7/8 (d) 1/8 (e) none of these

9. A histogram is
(a) a unit of measure in the metric system
(b) a medicine for sinus relief
(c) the name of a labor party in Israel
(d) a bar graph
(e) none of these

10. The shapes of the histograms for a frequency distribution and its corresponding probability distribution are
(a) inverses (b) unrelated (c) bell-shaped curves
(d) identical (e) none of these

Problems 11 to 16 refer to the probability distribution below.

x	p
–1	0.2
0	0.3
1	0.4
2	0.1

11. Pr (X ≥ 1) =
(a) 0 (b) 0.2 (c) 0.3 (d) 0.5 (e) none of these

12. Pr (X ≤ 0) =
(a) 0 (b) 0.2 (c) 0.3 (d) 0.5 (e) none of these

13. Pr (X < 0) =
(a) 0 (b) 0.2 (c) 0.3 (d) 0.5 (e) none of these

14. Pr (X > 0) =
(a) 0 (b) 0.2 (c) 0.3 (d) 0.5 (e) none of these

15. Pr (X = –1) =
(a) 0 (b) 0.2 (c) 0.3 (d) 0.5 (e) none of these

16. Pr (X = 0) =
(a) 0 (b) 0.2 (c) 0.3 (d) 0.5 (e) none of these

17. Flip three coins and let X be the number of tails that occur. Then Pr ($X \geq 1$) =
(a) 3/8 (b) 1/2 (c) 7/8 (d) 1/8 (e) none of these

18. Flip four coins and let X be the number of tails that occur. Then Pr ($X < 1$) =
(a) 3/16 (b) 1/2 (c) 7/16 (d) 1/16 (e) none of these

19. Toss two dice and let X be the sum of the top faces. Then Pr (X = 7) =
(a) 1/6 (b) 15/36 (c) 21/36 (d) 7/36 (e) none of these

20. Toss two dice and let X be the sum of the top faces. Then Pr ($X \leq 11$) =
(a) 1/36 (b) 33/36 (c) 35/36 (d) 11/36 (e) none of these

6-2 MEASURES OF CENTRAL TENDENCY

Section 6-1 showed a histogram provides a quick overall impression of a set of data, giving it a form and structure that is sometimes missing from the mere listing of the individual entries. A histogram reveals other characteristics of the data as well, like its average. This section examines three meanings of the term "average": mean, median, and expected value.

Average

The League of American Theatres and Producers reported that the average price for a ticket to a Broadway show during the 1985-1986 season was $28.96. By the term "average" the League meant the mean, which was obtained by dividing the total ticket revenue by the total attendance. However, another type of "average" is the median ticket price which is directly in the middle of all ticket prices, so that half the tickets cost more and half the tickets cost less than the median price.

The mean and median are numerical values which represent an entire collection of data. Since averages measure the center of a set of data, they are called "measures of central tendency." The discussion of Broadway theater ticket prices illustrates these two measures of central tendency, the mean and the median. It is important to state precisely which one is being used. The exercises examine a federal court ruling in a Civil Rights discrimination suit that revolved around the distinction between these two terms.

The *mean* of a set of data, denoted by $\bar{x}$ is defined in the following display.

DEFINITION

The mean of a set of n numbers $x_1, x_2, \ldots, x_n$, is

$$\bar{x} = \frac{x_1 + x_2 + \ldots + x_n}{n}$$

As an example, the mean of five test grades 82, 84, 78, 76, 95 is

$$\bar{x} = \frac{82 + 84 + 78 + 76 + 95}{5} = \frac{415}{5} = 83$$

The **median** of a set of data is the middle number when the entries are arranged in order. Curiously there is no special notation for this concept. To find the median, first list the elements of the set in either ascending or descending order. Next, determine if there is an odd or even number of elements in the set. If there is an odd number of elements in the set then the median is the middle element. If there is an even number of elements in the set then the median is the mean of the middle two numbers. Thus the median grade on tests 82, 84, 78, 76, 95 is 82 because when these grades are listed in order, 82 is the middle grade.

$$\underset{\underset{\text{median}}{\uparrow}}{76,\ 78,\ 82,\ 84,\ 95}$$

Notice that for this set of grades the mean is 83, while the median is 82, a different value. In most cases, the mean and the median will give different values.

Restricting attention to the first four exams only, we see that the median grade is equal to the mean of the middle two values of 78 and 82, yielding a median of 80.

$$76, \underbrace{78, 82}_{\substack{\text{middle}\\\text{two}}}, 84$$

Computing the mean of these four numbers yields $\bar{x} = 80$, and therefore for this set of data, the mean and median are equal.

Calculating the Median

1. List the elements of the set in either ascending or descending order.
2. (a) If there is an odd number of elements in the set, then the median is the middle element.
   (b) If there is an even number of elements in the set then the median is the mean of the middle two numbers.

Example 1 shows how to compute the mean and median for sets with repeated entries.

EXAMPLE 1

A branch of a bank has 8 tellers who earn \$15,000 per year, 2 managers who earn \$35,000, a vice-president who earns \$70,000, and a president who earns \$100,000.
(a) What is the mean salary?
(b) What is the median salary?

Solution.
(a) The mean salary is equal to the total of the salaries divided by the number of employees. Since the numbers are rather large, it is best to work with the data in units of \$1000. In other words, the number 15 will correspond to a salary of \$15,000. Rather than adding the numbers one by one, it is easier to use multiplication along with addition to find the total of the salaries.

$$\begin{aligned}\text{total} &= \underbrace{15+15+15+15+15+15+15+15}_{8\times 15} + \underbrace{35+35}_{2\times 35} + \underbrace{70}_{1\times 70} + \underbrace{100}_{1\times 100}\\ &= 8(15) + 2(35) + 1(70) + 1(100)\\ &= 360\end{aligned}$$

The number of employees is given by

$$8 + 2 + 1 + 1 = 12$$

Therefore,

$$\bar{x} = \frac{8(15)+2(35)+1(70)+1(100)}{8+2+1+1} = \frac{360}{12} = 30$$

So the mean salary is $30,000.

(b) Since there are 12 employees, the median salary is the mean of the sixth and seventh numbers:

$$15, 15, 15, 15, 15, \underbrace{15, 15}_{\text{middle two}}, 15, 35, 35, 70, 100$$

Therefore the median salary is $15,000.

In Example 1, the median is probably a better description of the "average employee" than the mean is. The mean can be affected greatly by just a relatively few numbers that are much larger or much smaller than the bulk of the list, as is the case in this problem with the vice-president's and the president's salaries.

Example 1 also suggests the derivation of an alternate formula for the mean in terms of frequencies.

Formula for the Mean of Frequency Distributions

If a set has f_1 entries equal to x_1, f_2 entries equal to x_2, ..., and f_n entries equal to x_n, then the mean is given by

$$\bar{x} = \frac{f_1x_1 + f_2x_2 + \ldots + f_nx_n}{f_1 + f_2 + \ldots + f_n}$$

A convenient way to evaluate $\bar{x}$ from this formula is to make a frequency table with a third column for the products of each entry x and its frequency f.

x	f	xf
15	8	120
35	2	70
70	1	70
100	1	100
Totals	12	360

Consequently,

$$\bar{x} = \frac{360}{12} = 30$$

The next example uses this alternate formula to compute the mean from a frequency distribution.

EXAMPLE 2

A psychology professor enters the 100 scores on a ten-question test into a grade book. To find the mean, the grades are grouped in the table below. What is the mean?

Number correct	10	9	8	7	6	5	4	3	2	1	0
Number of students	11	16	20	15	13	10	10	4	1	0	0

Solution. Write the entries x and their frequencies f in a table, with a third column consisting of the products xf.

x	f	xf
10	11	110
9	16	144
8	20	160
7	15	105
6	13	78
5	10	50
4	10	40
3	4	12
2	1	2
Totals	100	701

Consequently,

$$\bar{x} = \frac{701}{100} = 7.01$$

The mean number of correct answers is slightly more than 7.

Expected Value

The alternate formula for the mean has an analog formula for probability distributions. In the alternate formula for $\bar{x}$ set $N = f_1 + f_2 + \cdots + f_n$. Then

$$\begin{aligned}\bar{x} &= \frac{f_1x_1 + f_2x_2 + \cdots + f_nx_n}{f_1 + f_2 + \cdots + f_n} \\ &= \frac{f_1x_1 + f_2x_2 + \cdots + f_nx_n}{N} \\ &= \left(\frac{f_1}{N}\right)x_1 + \left(\frac{f_2}{N}\right)x_2 + \cdots + \left(\frac{f_n}{N}\right)x_n\end{aligned}$$

Each fractional part f_i/N is a relative frequency, so it can be interpreted as a probability. The mean of a probability distribution is called the *expected value*, denoted by E(X). By setting $p_i = f_i/N$, the formula for the expected value becomes

$$E(X) = p_1x_1 + p_2x_2 + \cdots + p_nx_n$$

We encountered expected value in an earlier chapter, where it referred to the average number of outcomes that occur when a probability experiment is repeated many times. Both the mean and the expected value describe the average of a set of numbers, but the mean is computed from a frequency distribution whereas the expected value is computed from a probability distribution.

The **expected value** of the random variable X is given by

$$E(X) = p_1x_1 + p_2x_2 + \ .\ + p_nx_n$$

To compute the expected value, extend the probability distribution of a random variable X to a third column containing the products X*p*. The formula for E(X) shows that it is equal to the sum of the numbers in the X*p* column.

Example 3 finds the expected value for the sum of two dice.

EXAMPLE 3

Toss two dice and let X be the sum of the faces. Find E(X).

Solution. The probability distribution given below the distribution previously used for tossing two dice.

X	*p*	X*p*
2	1/36	2/36
3	2/36	6/36
4	3/36	12/36
5	4/36	20/36
6	5/36	30/36
7	6/36	42/36
8	5/36	40/36
9	4/36	36/36
10	3/36	30/36
11	2/36	22/36
12	1/36	12/36
	Total	252/36

The expected value is then E(X) = 252/36 = 7.

Example 3 means that if you toss two dice many times and form the sum each time, the mean of all the sums will be very close to 7. Example 4 computes another expected value directly from the probability distribution.

EXAMPLE 4

Suppose that the New Jersey Turnpike Commission reports the number of accidents during the Friday evening rush hour to be 0, 1, 2, or 3 with probabilities .92, .03, .02, and .03, respectively. What is the expected number of accidents during the Friday evening rush hour?

Solution. The phrase "expected number" refers to the expected value E(X), where the random variable X represents the number of accidents. Form the expanded probability distribution and total the X*p* column.

x	*p*	X*p*
0	.92	.00
1	.03	.03
2	.02	.04
3	.03	.09
	Total	0.16

Therefore E(X) = .16. Since the expected number of accidents is approximately 1/6 per Friday evening rush hour, one accident will occur about every six weeks.

EXERCISE SET 6-2

In problems 1 to 8 find the mean and median.

1. 2, 5, 7, 9, 2, 1, 13, 12, 50, 60
2. 2, 5, 7, 9, 2, 1, 13, 12, –50, –60
3. .06, .07, .02, .10, .05, .02
4. 5.2, 6.1, 10.8, 11.2, 6.3, 0.5, 8.7, 7.2
5.

x	0	1	2	3
f	2	5	3	4

6.

x	1	2	3	4
f	8	2	1	5

7.

x	3	5	7	9
f	2	1	0	4

8.

x	11	12	13	14	15
f	3	5	2	0	1

In problems 9 to 12 find the expected value from the probability distribution.

9.

X	2	3	5	7
p	1/8	1/2	1/4	1/8

10.

X	2	4	6
p	1/2	1/3	1/6

11.

X	2	3	4
p	0.2	0.3	0.5

12.

X	1	3	5	7
p	0.2	0.4	0.1	0.3

In problems 13 and 14 find the mean from the given histogram.

13.

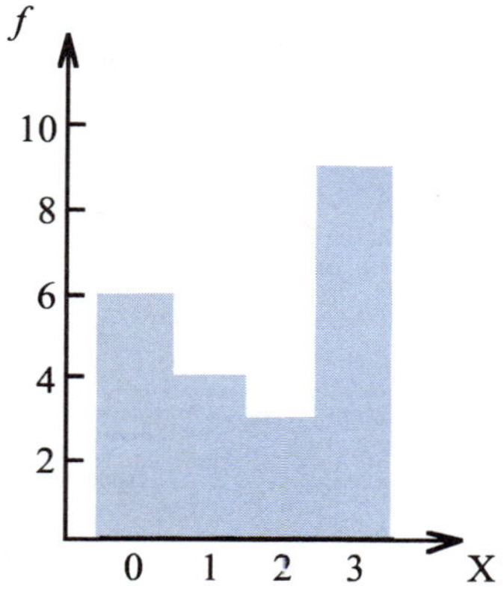

14.

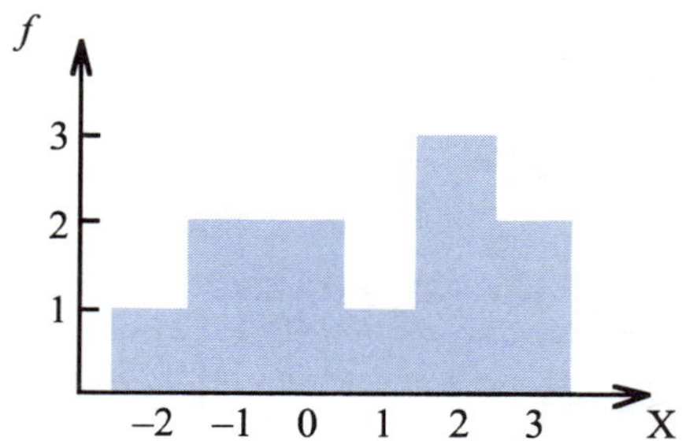

In problems 15 and 16 find the expected value from the given histogram.

15.

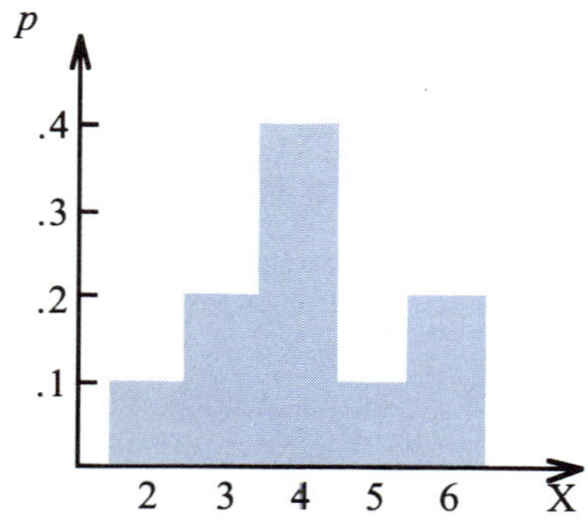

16.

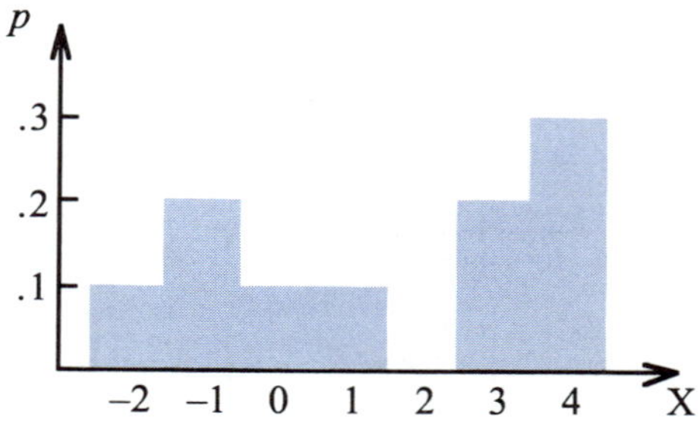

In problems 17 to 22 find the mean and median.

17. −3, 2, −3, 1, −3, 1, −1, 0, 0, 1

18. 3, −3, 2, 2, 2, −3, −1, 0, 0, −2

19.

x	4.4	5.1	6.4	7.3	8.2
f	3	4	1	5	7

20.

x	0.5	0.6	0.7	0.9
f	9	5	2	4

21.

x	−3	−2	−1	0	2	3
f	2	1	1	2	3	1

22.

x	−3	−1	0	1	2
f	3	1	2	3	1

In problems 23 to 26 find the expected value from the probability distribution.

23.

X	−3	−1	0	1	3
p	0.2	0.2	0.1	0.4	0.1

24.

X	−2	−1	0	1	2
p	0.2	0.1	0.4	0.2	0.1

25.

X	−10	−5	0	5	10
p	0.2	0.4	0.1	0.2	0.1

26.

X	−5	−3	0	1	4
p	0.1	0.3	0.1	0.2	0.3

27. Toss three coins and let X be the number of heads that occur. Find E(X).

28. Toss four coins and let X be the number of heads that occur. Find E(X).

29. Find the mean and median weights of the Washington Redskins from the data given in Example 5 of Section 6-1.

30. What is the median grade on the psychology test described in Example 2?

In problems 31 and 32 the data are tabulated in intervals. Find the mean using the midpoints of the intervals as the data.

31.

interval	0-2	2-4	4-6	6-8	8-10
frequency	4	5	3	0	1

32.

interval	0-5	5-10	10-15	15-20	20-25	25-30
frequency	16	14	10	8	2	1

33. A set of data has five entries. Each entry is either 0, 1, or 2.
(a) What is the set if the mean is 0?
(b) What is the set if the mean is 2?
(c) Can the mean be 3?

34. A set of data has five entries. Each entry is either 10, 11, or 12. What is the set if the mean is 10.8 and the median is 10?

35. A psychologist studies whether reward or punishment is a more effective method of inducing learning by running 11 mice through two sets of mazes. Subtracting the times (in seconds) in the punished trials from the times in the rewarded trials yields the following set of data:

    −1.1, 2.0, 1.3, −0.9, −1.9, 0.8, 1.2, −1.2, 4.0, 0.2, −0.1

    (a) Find the mean of these rates.
    (b) Which method is more effective?

36. An entomologist studying the morphological variation in a species of mosquito selects ten samples and measures their body lengths (in centimeters):

    1.2, 1.4, 1.3, 1.6, 1.0, 1.5, 1.7, 1.1, 1.2, 1.3

    What are the mean and median body lengths?

In problems 37 and 38 find the mean and the median of the data in the distribution.

37.

x	0	2	4	6	8
f	4	8	10	8	10

38.

x	10	15	20	25	30
f	5	4	6	3	2

39. Find the expected value for the following probability distribution.

X	−1	0	1	2	3	4
p	0.2	0.1	0.1	0.3	0.1	0.2

40. Find the mean for the following frequency distribution.

interval	0-2	3-5	6-8	9-11	12-14
frequency	8	18	10	10	4

Referenced Exercise Set 6-2

1. The 1964 Civil Rights Act prohibits discrimination by race, sex, or national origin in hiring and promotions. In 1981, 20 people filed suit that they were discriminated against when they were not hired in 1975 by the Federal Reserve Bank.[6] The court agreed. The attorneys for the two sides disagreed on the amount of money the bank should pay each plaintiff. The table below presents a sample of 319 employee records revealed this pertinent information about the job turnover rate, where M denotes the number of months worked before termination and E denotes the number of employees.

M	0.5	1	2	3	4	5	8	10	11	14	18	20	24	30	35	40	60
E	15	21	27	33	42	9	17	32	18	12	22	26	11	9	6	8	11

(a) Find the mean number of months worked.

[6]Source: Data from EEOC vs. *Federal Reserve Bank* 30 FEP Cases 1138-1169.

(b) Find the median number of months worked.
(c) Which side argued for which average?
(d) Each of the 20 plaintiffs was to be compensated at a rate of seven (years) times \$9500 (the yearly salary) divided by the average number of months worked. What is the difference in the amount of money the bank would have to pay according to whether the average is interpreted as the mean or the median?

2. J. E. Kerrich described a coin-tossing experiment while interned during World War II.[7] He tossed a coin 1000 times and recorded the number of heads. He conducted the experiment 10 times, obtaining these results:

   502, 511, 497, 529, 504, 476, 507, 528, 504, 529

   What are the mean and median number of heads?

3. Sociologist David Phillips designed a study to determine whether some individuals have a will that is strong enough to prolong life.[8] He recorded the number of months between the birthday and days of death of 1251 famous people, reasoning that famous people anticipate being honored on their birthday and might ward off death for a period of time. Table 6–10 lists the results, where B = 0 is the month in which the person was born, B = –1 is one month before the "birth month," B = 1 is one month after the birth month, etc. and where D is the number of deaths. In relation to his/her birth month, when does the average famous individual die?

B	–6	–5	–4	–3	–2	–1	0	1	2	3	4	5
D	90	100	87	96	101	86	119	118	121	114	113	106

Table 6–10

4. Table 6–7 in Section 6-1 describes income of families in the United States[9]. What is the median income?

5. Table 6–9 in Section 6-1 lists the height of women in the United States.[4] What are the mean and median heights?

6. Toss n coins and let X be the number of heads that occur. Find E(X).

7. The probability distribution for the number of games played in each World Series from 1922 to 1987 is given below.[10] What is the expected number of games played?

x	4	5	6	7
p	0.167	0.197	0.197	0.439

[7] J. E. Kerrich, *An Experimental Introduction to the Theory of Probability*, Copenhagen, Belgisk Import Co.

[8] Source: Data from J. M. Tanur et al. (eds.), *Statistics: A Guide to the Unknown*, San Francisco, Holden-Day Inc., 1978, pp. 71-85.

[9] Source, The Current Population Survey

[10] Source: Craig Carter (ed.), *The Complete Baseball Record Book*, St. Louis, The Sporting News Publishing Co., 1988, p. 323.

8. Let X represent the number of children of a president of the United States. The probability distribution is given for all presidents from George Washington to Ronald Reagan in the accompanying table.[11] What is the expected number of children of a president of the United States based on this information?

x	0	1	2	3	4	5	6	7	8	10	15
p	.154	.026	.179	.179	.179	.077	.102	.026	.026	.026	.026

9. Professors at the University of Wisconsin devised a five-question multiple-choice vocabulary test whose results were expected to be close to what would happen if the students guessed each question.[12] There are five possible answers for each question on the test.
   (a) What is the expected number of correct answers if a student guesses on each question?
   (b) What is the expected number of correct answers based on the accompanying table, which shows the results from one class that took the test? Here x is the number of correct answers and f is the frequency.

x	0	1	2	3	4	5
f	6	11	6	1	1	0

   (c) Do the answers to parts (a) and (b) suggest that the students in the class probably knew some of the answers?

Multiple-Choice Exercise Set 6-2

1. Which of the following is not a synonym for the term "average?"
   (a) expected value (b) mean (c) median
   (d) standard deviation (e) none of these

2. The notation for the mean of a set of data is
   (a) $\bar{x}$ (b) Σx (c) $\in$ (d) σ (e) none of these

3. The notation for the median of a set of data is
   (a) $\bar{x}$ (b) Σx (c) $\in$ (d) σ (e) none of these

4. The mean of the numbers 1, 2, 3, 6 is
   (a) 2.5 (b) 3 (c) 4 (d) 5 (e) none of these

5. The median of the numbers 1, 2, 3, 6 is
   (a) 2.5 (b) 3 (c) 4 (d) 5 (e) none of these

6. The median of the numbers 3, 1, 2, 5, 9 is
   (a) 2.5 (b) 3 (c) 4 (d) 2 (e) none of these

---

[11] Source: Joseph N. Kane, *Facts About the Presidents* (4th edition), New York, H. W. Wilson Co., 1981, pp. 348-349.

[12] Ken Kundert, "Student-generated data in elementary statistics," *Mathematics Teacher*, April 1990, pp. 322-325.

7. The mean of the numbers 1, 2, 3, 3, 6 is
(a) 2.5 (b) 3 (c) 4 (d) 5 (e) none of these

8. The median of the numbers 1, 2, 3, 3, 6 is
(a) 2.5 (b) 3 (c) 4 (d) 5 (e) none of these

9. If the mean of six positive numbers is 3 then the mean of the opposite of these six numbers is
(a) –6 (b) 6 (c) –3 (d) 3 (e) indeterminable

10. If the mean of six positive numbers is 3 then the median of those numbers is
(a)–6 (b) 6 (c) –3 (d) 3 (e) indeterminable

Problems 11 and 12 refer to the frequency distribution below.

x	1	2	3	4
f	5	2	1	2

11. The mean of the data is
(a) 1 (b) 1.5 (c) 2 (d) 3 (e) none of these

12. The median of the data is
(a) 1 (b) 1.5 (c) 2 (d) 3 (e) none of these

13. The expected value of a probability distribution corresponds to
(a) the median of a frequency distribution
(b) the mean of a normal distribution
(c) the mean of a frequency distribution
(d) the median of a uniform distribution
(e) none of these

14. The notation for the expected value of a probability distribution is
(a) E(X) (b) $\sigma(x)$ (c) $\Sigma(X)$ (d) $\bar{x}$ (e) none of these

15. The expected value of an experiment with outcomes 0, 1, 2 with corresponding probabilities 1/4, 1/2, 1/4 is
(a) 0 (b) 1 (c) 2 (d) 1.5 (e) none of these

16. The expected value of an experiment with outcomes 0, 1, 2 with corresponding probabilities 0.4, 0.2, 0.4 is
(a) 0 (b) 1 (c) 2 (d) 1.5 (e) none of these

17. The expected value of the probability distribution below is

x	1	2	3	4
p	0.5	0.2	0.1	0.2

(a) 1 (b) 2 (c) 2.5 (d) 3 (e) none of these

18. The expected value of the probability distribution below is

x	1	2	3	4
p	0.1	0.3	0.1	0.5

(a) 1 (b) 2 (c) 2.5 (d) 3 (e) none of these

19. If you toss two dice many times and form the sum each time, the mean of all the sums will be very close to
(a) 36 (b) 12 (c) 2 (d) 7 (e) none of these

20. If you flip two coins many times and count the number of heads each time, the mean of all the numbers will be very close to
(a) 1 (b) 2 (c) 1.5 (d) 2.5 (e) none of these

21. Find the expected value of the distribution from the table below.

x	0.5-1.5	1.5-2.5	2.5-3.5	3.5-4.5	4.5-5.5	5.5-6.5
p	0.2	0.1	0.1	0.3	0.1	0.2

(a) 36 (b) 3.6 (c) 0.36 (d) 0.036 (e) none of these

22. Find the expected value of the distribution from the table below.

x	0.5-1.5	1.5-2.5	2.5-3.5	3.5-4.5	4.5-5.5	5.5-6.5
p	0.1	0.1	0.1	0.1	0.1	0.5

(a) 38 (b) 3.8 (c) 0.38 (d) 0.038 (e) none of these

23. If a set of data has ten entries, each 0 or 1, then the expected value is
(a) 0 (b) 1 (c) 0.5 (d) 0.25 (e) indeterminable

24. If a set of data has ten entries, each 0 or 1, and the expected value is 0.7, the number of entries equal to 1 is
(a) 5 (b) 6 (c) 7 (d) 8 (e) indeterminable

CASE STUDY
LENIN AS STATISTICIAN

Vladimir Ilich Lenin (1870-1924) is well-known throughout the world for his pivotal role in the revolution that toppled the Czarist regime in Russia and transformed it into a communist state. It is not widely known that Lenin was an accomplished statistician as well as a highly successful political leader. His role in statistics was recently evaluated by two prominent statisticians.[13] Each April, numerous articles about his life appear in Russian publications to commemorate his birth. Many such articles citing his contributions to statistics have appeared in *Vestnik Statistiki*, the official journal of the Central Statistical Association of the Soviet Union. These articles reflect Lenin's standing with the leadership of the country at the time of their writing.

Thus there was less praise for Lenin's work at the height of Stalin's regime after World War II (1946-1953) than during the Khrushchev era and the early years of Brezhnev's era (1954-1970). The adulation diminished slightly during the peak period of Brezhnev's consolidation of power (the later 1970s and early 1980s). According to Kotz and Seneta, "At present, in the first years of Gorbachev's era Lenin's [standing] remains intact and his pronouncements have not as yet been challenged even in the avant-garde, pro-reform and pro-democratic, liberal press in the Soviet Union."

In 1894 Lenin wrote an article called "What are 'the Friends of the People' and how do they fight with the Social Democrats?" The aim of the article was largely to attack "anti-Marxist liberal *narodniks*," whom Lenin scathingly accused of giving an erroneous picture of the peasant economy. To provide a more accurate picture he analyzed the budgets of 24 peasant households within a county in Voronezh. It was crucial for Lenin's argument that his sample space be typical of peasant households in that entire region since otherwise his findings would not be taken seriously. To achieve this the 24-year old revolutionary compared four measures with the corresponding measures from an extensive study that had been gathered, analyzed, and published by the eminent Soviet statistician, F. A. Shcherbina, in 1887. Lenin called Shcherbina's study "extraordinarily complete." Others regarded it as the definitive study of economics in that area.

Table 6–11 contains the items from Shcherbina's study that Lenin used for comparison. Shcherbina's original table lumped together households with two or three draft animals into one category and lumped together households with four or more draft animals into another category. We have adjusted his figures to be consistent with his findings. The corresponding entries from Lenin's study are given in Table 6–12.

Draft Animals	Households	Land Allotment	Rented Land	Family Size	Cattle
0	8,728	6.2	0.2	4.6	0.7
1	10,510	9.4	1.3	5.7	3.0
2	6,763	13.9	2.1	7.7	5.1
3	4,428	14.0	3.5	8.5	6.9
4	3,152	21.3	12.3	11.2	14.3

Table 6–11 Shcherbina's study of peasant households

[13] Samuel Kotz and Eugene Seneta, "Lenin as a statistician: A Non-Soviet View," *Journal of the Royal Statistical Society* **A 153** (1990) (Part 1), pp. 73-94. We are indebted to Professor Kotz for sending us a copy of this fascinating paper.

Strata Class	Households	Land Allotment	Rented Land	Family Size	Cattle
Farm laborers	2	7.2	0.0	4.5	0.5
Poor peasants	5	8.7	3.9	5.6	2.8
Middle peasants	11	9.2	7.7	8.3	8.1
Prosperous	6	22.1	8.8	7.8	13.5

Table 6–12 Lenin's study of peasant budgets

Notice that the strata classes in the two studies are different. Shcherbina's study was based on the legal classification of draft animals, which were mostly horses. Lenin chose an entirely different classification. His aim was to dramatize the deteriorating condition of Russian peasants so his strata were organized along socioeconomic lines. It was important, however, that the averages (more specifically, the means) of each column in the studies be as close as possible. We will carry out two of these calculations in Example 1.

EXAMPLE 1

The land allotment was measured in *dessiatina*. Each *dessiatina* is equal to roughly 2.7 acres. Calculate the mean number of *dessiatina*s that were allotted per household in both the Lenin and Shcherbina studies.

Solution. Consider Shcherbina's study first. (See Table 6–11.) The random variable is the number of *dessiatina* in each strata class. The frequency is the number of households in each class. Thus the relevant table is as follows.

x	f	xf
6.2	8,728	54,113.6
9.4	10,510	98,794.0
13.9	6,763	94,005.7
14.0	4,428	61,992.0
21.3	3,152	67,137.6
Totals	33,581	376,042.9

The mean is $\bar{x} = \dfrac{376,042.9}{33,581} \approx 11.20.$

The relevant table for the Lenin study (See Table 6–12) is

x	f	xf
7.2	2	14.4
8.7	5	43.5
9.2	11	101.2
22.1	6	132.6
Totals	24	291.7

The mean is $\bar{x} = \dfrac{291.7}{24} \approx 12.15.$

Example 1 shows that the land allotments were remarkably similar for the two groups of households. In the exercises you will be asked to verify that the mean family sizes per household were just as remarkably similar, and that the mean rented land per household and the mean number of cattle owned per household were very similar.

Kotz and Seneta wrote, "We can imagine Lenin's elation on seeing the agreement." Although as modern statisticians they view the agreement as an impressive coincidence and *not* a statistically significant fact, mainly due to the small sample size within each of Lenin's strata classes, they admit that he derived great benefit from his data.

Lenin used his study for political ends. He concluded that the legal classification used by Shcherbina concealed huge socioeconomic differences which his own analysis revealed. He went on to describe how the plight of the peasant in the Soviet Union had worsened under the Czar, a conviction that remained one of his fundamental tenets for the rest of his life, showing that Lenin was not beyond turning statistical analysis to political ends.

CASE STUDY EXERCISE SET

1. Compare the mean of the family sizes per household in the Lenin study with the mean of the family sizes per household in the Shcherbina study.
2. Compare the mean amount of rented land (in *dessiatina*) per household in the Lenin study with the mean amount of rented land per household in the Shcherbina study.
3. Compare the mean number of cattle owned per household in the Lenin study with the mean number of cattle owned per household in the Shcherbina study.
4. Based on the results obtained in Example 1 and Problems 1, 2, and 3, do you agree that the 24 households in Lenin's study of peasant budgets are typical of the peasant households in Shcherbina's study? Explain your answer.

In 1889 Shcherbina published a detailed study of peasant budgets based on 66 households. The relevant data are given in Table 6–13. As before, Shcherbina's numbers have been altered slightly so that the final computations agree with his figures. Problems 5 and 6 refer to Table 6–13.

Draft Animals	Households	Land Allotment	Rented Land	Family Size	Cattle
0	12	5.9	0.0	4.08	0.8
1	18	7.4	1.5	4.94	2.6
2	17	12.7	3.0	8.23	4.9
3	9	18.5	5.6	13.0	9.1
4	5	22.9	14.2	14.2	12.8
5	5	23.5	27.4	16.0	19.5

Table 6–13 Shcherbina's 1889 study of 66 peasant households

5. (a) Calculate the mean land allotment (in *dessiatina*) per household in this study.
   (b) Calculate the mean rented land (in *dessiatina*) per household.
   (c) Calculate the mean family size per household.
   (d) Calculate the mean number of cattle owned per household.
6. Based on the results obtained in Example 1 and Problems 1, 2, 3, and 5, do you agree that the 24 households in Lenin's study of peasant budgets are typical of the peasant households in Shcherbina's study of 66 households? Explain your answer.

6-3 STANDARD DEVIATION

The mean of a set of data by itself is not enough information about the set to make meaningful conclusions. The mean gives an "average" value for the set, but does not indicate whether data is bunched near the mean or whether it is spread out greatly. For example, in manufacturing pharmaceuticals, it is necessary that all pills have an amount of the active ingredient that is extremely close to the mean. On the other hand, if everyone had a score on the Verbal portion of the SAT that was very near the mean, little could be concluded about which students were better prepared to enter college. This case is one where a spread of data is necessary.

In this section we discuss the standard deviation, which measure the spread of the data from the mean. Then in later sections, using knowledge of *both* the mean and the standard deviation of a set of data, many conclusions will be made.

Measures of Dispersion: Lists

Consider these three sets of data:

set A: 6, 5, 5, 4, 5, 3, 5, 7, 5, 5
set B: 6, 9, 7, 6, 5, 5, 2, 1, 4, 5
set C: 6, 2, 10, 9, 4, 1, 0, 1, 7, 10

The mean of each set is 5, but as you can see from their histograms in Figure 6–8, their spreads are quite different. The entries in set C are more dispersed than the entries in set B, and the entries in set B are more dispersed than the entries in set A.

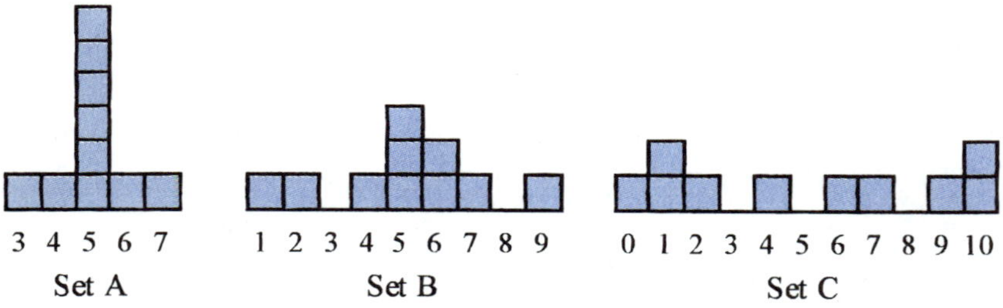

Figure 6–8

Though the difference in spread of the three sets is obvious from the histograms, in order to work with the data, a quantitative measure of this dispersion is required. There are several ways that the spread of data can be quantitatively measured, however, the most common measure of dispersion is called the standard deviation. The standard deviation is usually calculated using a two step process. First a quantity called the variance is calculated, then the standard deviation is found by taking the square root of the variance.

Essentially, the variance measures the differences between the mean of a set of data and the individual elements in the set of data. Unfortunately, however, if we were to merely subtract the mean from every point and add up the resulting numbers, we would always get an answer of 0. The reason for this answer is that the mean is essentially a "balancing act," with the numbers greater than the mean exactly canceling out the numbers less than the mean. Notice in Table 6–14 that the sum of the column labeled $x - \bar{x}$ is 0.

One way around this problem of the differences canceling each other out would be to take the absolute value of these differences and then add. While this approach would yield a meaningful number, the resulting statistic is very difficult with which to work.

The variance resolves both problems by *squaring the differences between the data points and the mean, and then adding*. Finally, this number is then divided by n - 1 where as usual n is the number of elements in the set.

To recap, finding the variance, usually denoted s^2, is a four step process

1. subtract the mean from every element in the set of data
2. square each of the numbers in step 1
3. add the resulting numbers in step 2
4. divide the answer from step 3 by n - 1

As is the case with computing most of the statistics encountered so far, creating a table makes the work proceed much more easily. To illustrate, we calculate the variance of the data in Set B. Recall that in Set B, the mean is 5.

	x	$x-\bar{x}$	$(x-\bar{x})^2$
	6	1	1
	9	4	16
	7	2	4
	6	1	1
	5	0	0
	5	0	0
	2	–3	9
	1	–4	16
	4	–1	1
	5	0	0
Totals	50	0	48

Table 6–14

Computing the variance now is a simple matter of dividing the sum of the last column by 10 – 1 = 9. Therefore the variance is given by

$$s^2 = \frac{48}{9} = \frac{16}{3}$$

As mentioned earlier, the standard deviation, denoted s, is now found by taking the square root of the variance. Therefore the standard deviation is given by

$$s = \sqrt{\frac{16}{3}} \approx 2.3$$

The rather ominous-looking formula for variance is given below. However, keep in mind that the formula is simply a mathematical description of the table method.

DEFINITION

For a set of data with n entries $x_1, x_2, \ldots, x_n$ and with mean $\bar{x}$, the **variance** is

$$s^2 = \frac{(x_1-\bar{x})^2 + (x_2-\bar{x})^2 + \ldots + (x_n-\bar{x})^2}{n-1}$$

The **standard deviation** is then given by $s = \sqrt{s^2}$

Most calculators can compute the standard deviation, however, there are two types of standard deviation, the sample standard deviation, which is the standard deviation outlined above, and the population standard deviation. Since most calculators are capable of computing both of these standard deviations, refer to your manual so that you know the difference between the two.

Example 1 looks at the data Set C and quantitatively shows that it is more dispersed than the data in Set B.

EXAMPLE 1

Find the standard deviation of set C:

6, 2, 10, 9, 4, 1, 0, 1, 7, 10

Solution. The table below contains all of the information needed to calculate the variance. Again, recall that the mean of this set of data is 5, and that $n = 10$.

	x	$x - \bar{x}$	$(x - \bar{x})^2$
	6	1	1
	2	–3	9
	10	5	25
	9	4	16
	4	–1	1
	1	–4	16
	0	–5	25
	1	–4	16
	7	2	4
	10	5	25
Totals	50	0	138

$$s^2 = \frac{138}{9} = \frac{46}{3}$$

Consequently, standard deviation is given by

$$s = \sqrt{\frac{46}{3}} \approx 3.9$$

Since the standard deviation of Set C is larger than the standard deviation of Set B, we can quantitatively see that Set C is more dispersed than Set B.

An Alternate Formula

While the method for calculating the variance that has been presented works well in the preceding examples, that particular method can be very tedious in cases where the mean is not an integer an in cases involving frequency distributions

An alternate method for finding the variance of a list is outlined using the data in Set A which contains the elements

6, 5, 5, 4, 5, 3, 5, 7, 5, 5

First create a table containing two columns – one, labeled x, for the values in the list and a second containing the squares of the values in the first column. Add the numbers in each of these columns.

	x	x^2
	6	36
	5	25
	5	25
	4	16
	5	25
	3	9
	5	25
	7	49
	5	25
	5	25
Totals	50	260

Then using the notation first introduced in Section 1-4, where Σ means "add up,"

$$\Sigma x = 50$$

and

$$\Sigma x^2 = 260$$

Then the variance is computed using the following formula

$$s^2 = \frac{\Sigma x^2 - \frac{(\Sigma x)^2}{n}}{n-1}$$

This formula is not as foreboding as it may first appear and is algebraically equivalent to the previous method. In the case of Set A:

$$s^2 = \frac{260 - \frac{(50)^2}{10}}{9} = \frac{10}{9}$$

Now, as usual, take the square root of this number to find the standard deviation.

$$s = \sqrt{\frac{10}{9}} \approx 1.05$$

Now we can quantitatively see that in fact Set A is less dispersed than either Set B or Set C since it has the smallest standard deviation of the three sets.

Note that another benefit of the formula presented here is that it does not require the calculation of the mean in order to determine the standard deviation. The formula can then be expressed as follows.

Alternate Formula for Variance

For a list of n elements $x_1, x_2, \ldots, x_n$, the variance can be found using the formula

$$s^2 = \frac{\Sigma x^2 - \frac{(\Sigma x)^2}{n}}{n-1}$$

EXAMPLE 2

Using the alternate formula find the standard deviation of the following set of 5 numbers.

16, 25, 29, 41, 55

Solution. The table below shows that $\Sigma x = 166$ and $\Sigma x^2 = 6428$. Since there are 5 elements in the list, $n = 5$.

	x	x^2
	16	256
	25	625
	29	841
	41	1681
	55	3025
Totals	166	6428

Therefore,

$$s^2 = \frac{6428 - \frac{(166)^2}{5}}{4} = \frac{916.8}{4} = 229.2$$

and

$$s = \sqrt{229.2} \approx 15.1$$

Formula for Frequency Distributions

Recall that a frequency distribution is just a shorthand method for writing a list with repeated entries. Therefore, the formula for variance for a frequency distribution should be very close to that of a list. In fact, the change in the formula is analogous to what happened to the formula for the mean when a frequency distribution was introduced, and the sum of the x values was replaced by the sum of the xf values. In the case of the variance, we will add another column that contains the products x^2f. In addition, the sample size is no longer given by n, but rather is given by the sum of the frequencies as in the case of the mean.

The formula for the variance when the data is presented in a frequency distribution is then given in the following display.

Variance Formula for Frequency Distributions

If a set has f_1 entries equal to x_1, f_2 entries equal to x_2, ..., and f_n entries equal to x_n, then the variance is given by

$$s^2 = \frac{\Sigma x^2 f - \dfrac{(\Sigma xf)^2}{\Sigma f}}{\Sigma f - 1}$$

EXAMPLE 3

An ecologist investigates the reproductive behavior of certain water fowl to determine whether DDT is causing infertile eggs. The species could become endangered if the eggs are infertile. A frequency distribution gives the number of nests which hold between zero and five eggs. Compute the mean and standard deviation for the number of eggs per nest.

eggs in nest	0	1	2	3	4	5
frequency	42	44	32	19	11	2

Solution. Create the following table

x	f	xf	x^2	x^2f
0	42	0	0	0
1	44	44	1	44
2	32	64	4	128
3	19	57	9	171
4	11	44	16	176
5	2	10	25	50
Totals	150	219		569

From the table, $\Sigma f = 150$, $\Sigma xf = 219$, and $\Sigma x^2 f = 569$. Therefore, the mean is

$$\bar{x} = \frac{219}{150} = 1.46$$

and the variance is

$$s^2 = \frac{569 - \dfrac{(219)^2}{150}}{149} = \frac{249.26}{149} = 1.672885906$$

Consequently,

$$s = \sqrt{s^2} \approx 1.29$$

Thus the mean number of bird eggs in a nest is 1.46 with a standard deviation of 1.29.

Probability Distributions

In a probability distribution where the theoretical probabilities are known the mean and the standard deviation take on a somewhat different form. First of all, the mean is nothing more than the expected value, E, that was first discussed in Section 5-4. When thought of as a mean, the expected value is denoted by μ, which is the "lower case" Greek letter "mu." Consequently,

$$\mu = x_1p_1 + x_2p_2 + \ldots + x_np_n = \sum Xp$$

The variance and standard deviation are denoted by σ^2 and σ, respectively. σ is the "lower case" Greek letter "sigma." The formula for variance is given by

$$\sigma^2 = x_1^2p + x_2^2p + \ldots + x_n^2p - \mu^2 = \sum X^2p - \mu^2$$

As usual, the standard deviation, σ, is found by taking the square root of the variance, σ^2. These formulas can then be summed up as follows.

The Mean and Variance for Probability Distributions

If the probabilities of the outcomes $x_1, x_2, \ldots, x_n$ are $p_1, p_2, \ldots, p_n$, respectively, then the mean of the probability distribution is given by

$$\mu = \sum Xp$$

and the variance is given by

$$\sigma^2 = \sum X^2p - \mu^2$$

EXAMPLE 4

If a coin is flipped three times and the number of heads is recorded, the following probability distribution results, where X represents the number of heads that occur.

X	p
0	1/8
1	3/8
2	3/8
3	1/8

Find the mean and standard deviation of the number of heads that occur.

Solution. Expand the table to include columns for Xp, X^2, and X^2p.

X	p	Xp	X^2	X^2p
0	.125	0	0	0
1	.375	.375	1	.375
2	.375	.75	4	1.5
3	.125	.375	9	1.125
Totals		1.5		3

The mean is given by the total of the *xp* column. Hence

$$\mu = 1.5$$

To find the standard deviation, first use the variance formula to find

$$\sigma^2 = \sum X^2 p - \mu^2 = 3 - 1.5^2 = 0.75$$

Then take the square root of the variance to find the standard deviation.

$$\sigma = \sqrt{0.75} \approx 0.87$$

Example 5 uses the expected value and standard deviation to compare two sets of data.

EXAMPLE 5

A farmer must determine which of two hybrid grains, A or B, has the highest yield and smallest variation. The following distributions are known, where the units for x are in thousands of bushels.

Hybrid A

X	p
1	.2
2	.1
3	.4
4	.2
5	.1

Hybrid B

X	p
2	.4
3	.4
4	.1
5	.1

Determine the expected value and standard deviation for each type of hybrid grain. Which type should the farmer plant?

Solution. Expand the tables for both types of hybrid grain.

HYBRID A

X	p	Xp	X^2	$X^2 p$
1	.2	.2	1	.2
2	.1	.2	4	.4
3	.4	1.2	9	3.6
4	.2	.8	16	3.2
5	.1	.5	25	2.5
Totals		2.9		9.9

HYBRID B

X	p	Xp	X^2	$X^2 p$
2	.4	.8	4	1.6
3	.4	1.2	9	3.6
4	.1	.4	16	1.6
5	.1	.5	25	2.5
Totals		2.9		9.3

Since the total of the Xp column is the same for each hybrid, the two hybrids have the same mean of 2.9 thousand bushels. Hence, they produce the same yield.
For hybrid A, the variance is

$$\sigma^2 = \sum X^2 p - \mu^2 = 9.9 - 2.9^2 = 1.49$$

and the standard deviation is

$$\sigma = \sqrt{1.49} \approx 1.22$$

For hybrid B the variance is

$$\sigma^2 = \sum X^2 p - \mu^2 = 9.3 - 2.9^2 = 0.89$$

and the standard deviation is

$$\sigma = \sqrt{0.89} \approx 0.94$$

Since the standard deviation for hybrid B is smaller than the standard deviation for hybrid A, there will be a more consistent yield with hybrid B, and consequently, the farmer should plant type B.

Chebyshev's Theorem

The expression "within k standard deviations of the mean" refers to those entries whose distance to the mean μ is less than or equal to $k\sigma$. We write "within $k\sigma$" for short. The entries can lie to the left or right of μ, as Figure 6–9 shows.

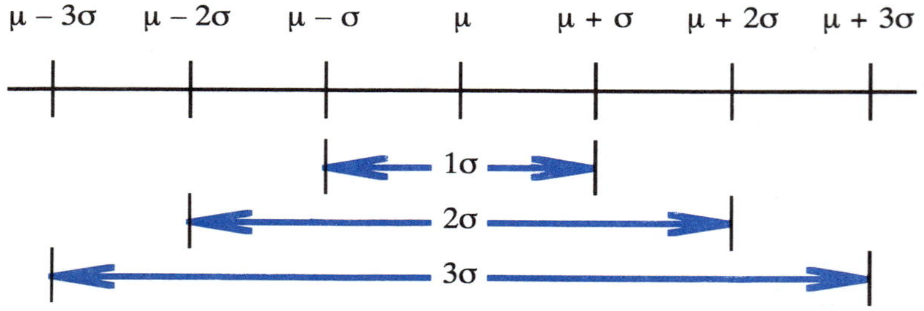

Figure 6–9

Recall set B which was given at the beginning of this section.

Set B: 6, 9, 7, 6, 5, 5, 2, 1, 4, 5

We showed that $\mu = 5$ and $\sigma \approx 2.2$. An entry lies within 2σ of the mean if its value is between $\mu - 2\sigma \approx 5 - 4.4 = 0.6$ and $\mu + 2\sigma \approx 5 + 4.4 = 9.4$. All 10 entries of B lie within this interval.

Since the standard deviation measures the spread of data, it seems reasonable that as k increases, the number of entries within $k\sigma$ of the mean will also increase. This fact, which was proved by the Russian mathematician Pafnuti Chebyshev (1821-1894), is called Chebyshev's Theorem.[14]

Chebyshev's Theorem

In any experiment, the probability that an entry lies within k standard deviations of the mean is at least $1 - (1/k)^2$.

For $k = 2$, Chebyshev's Theorem asserts that the proportion of entries within 2σ of the mean is at least $1 - (1/2)^2 = 3/4$, or 75%. Indeed, for set B, 100% of the entries lie in this interval. The importance of Chebyshev's result is that it refers to *any* distribution. Later we will improve the approximation for data with a normal distribution.

EXAMPLE 6

When a coin is tossed 100 times, the expected number of heads is $\mu = 50$. It will be shown in Section 6-5 that $\sigma = 5$. What is the lower bound for the probability of tossing between 35 and 65 heads?

Solution. We must first determine the number of standard deviations from the mean for $x = 35$ and $x = 65$. Set $x = 65$. Then $x - \mu = 65 - 50 = 15$. Since $\sigma = 5$ and $k\sigma = 15$, we get $k = 15/5 = 3$. Therefore $x = 65$ lies 3σ to the right of the mean. (See Figure 6–10.) Similarly, $x = 35$ lies 3σ to the left of the mean. By Chebyshev's Theorem, the probability of an entry lying within 3σ of the mean is at least

$$1 - (1/3)^2 = 1 - (1/9) = 8/9$$

Therefore, whenever a coin is tossed 100 times, we can expect that in the long run it will land heads between 35 and 65 times more than 8/9 (or approximately 89%) of the time.

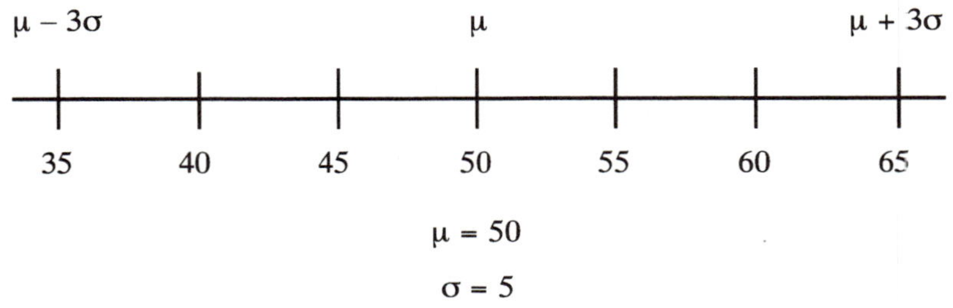

Figure 6–10

[14]Stephen M. Stigler has discovered that the French mathematician Bienaymé "stated and proved the Chebyshev inequality a decade before and in greater generality than Chebyshev's first work on the topic." See S. M. Stigler, "Stigler's Law of Eponymy," *Transactions of the New York Academy of Sciences*, April 1980, p. 148.

EXERCISE SET 6-3

In problems 1 to 10 calculate the variance and standard deviation.

1. 0, 1, 3, 1, 2, 0, 6, 2, 7, 3
2. –1, 0, –3, 1, 4, 0, –2, 5, 1, –3
3. 10, 15, 18, 11, 12, 15, 20, 10, 12, 18
4. 25, 20, 18, 15, 24, 22, 16, 20, 17, 23
5. 22, 18, 32, 40, 20, 28, 38, 42, 15, 45
6. 40, 48, 60, 55, 58, 65, 42, 45, 52, 35

7.

x	1	3	6	9	11
f	3	7	5	10	5

8.

x	2	6	10	14	18
f	15	18	20	28	20

9.

x	11	13	15	17	19	21
f	1	8	10	5	15	3

10.

x	11	13	15	17	19	21
f	1	8	10	5	15	3

In problems 11 to 14 find the variance and standard deviation of the probability distribution. The expected values were calculated in problems 9 to 12 in Section 6-2.

11.

x	2	3	5	7
p	1/8	1/2	1/4	1/8

12.

x	2	4	6
p	1/2	1/3	1/6

13.

x	2	3	4
p	0.2	0.3	0.5

14.

x	1	3	5	7
p	0.2	0.4	0.1	0.3

15. An automobile dealer has four types of cars selling for \$21,000, \$17,000, \$14,000, and \$9000. The probability that a shopper will buy each type of car is, respectively,

0.1, 0.2, 0.4, and 0.3. What is the average sales price? What is the standard deviation?

16. A student taking an economics exam has a probability of 0.75 of scoring 80%, 0.15 of scoring 85%, 0.05 of scoring 90%, and 0.05 of scoring 100%. What is the expected score? What is the standard deviation?

In problems 17 and 18 a coin is tossed 100 times.

17. What is the lower bound for the probability that it will land heads between 40 and 60 times?

18. What is the lower bound for the probability that it will land heads between 20 and 80 times?

In problems 19 to 24 calculate the variance and standard deviation.

19. 2.5, 2.0, 1.8, 1.5, 2.4, 2.2, 1.6, 2.0, 1.7, 2.3

20. 1.0, 1.5, 1.8, 1.1, 1.2, 1.5, 2.0, 1.0, 1.2, 1.8

21. 4.0, 4.8, 6.0, 5.5, 5.8, 6.5, 4.2, 4.5, 5.2, 3.5

22. 2.2, 1.8, 3.2, 4.0, 2.0, 2.8, 3.8, 4.2, 1.5, 4.5

23.

x	20	60	100	140	180
f	15	18	20	28	20

24.

x	110	130	150	170	190	210
f	1	8	10	5	15	3

In problems 25 to 28 find the variance and standard deviation of the probability distribution. The expected values were calculated in problems 23 to 26 in Section 6-2.

25.

x	−3	−1	0	1	3
p	0.2	0.2	0.1	0.4	0.1

26.

x	−2	−1	0	1	2
p	0.2	0.1	0.4	0.2	0.1

27.

x	−10	−5	0	5	10
p	0.2	0.4	0.1	0.2	0.1

28.

x	−5	−3	0	1	4
p	0.1	0.3	0.1	0.2	0.3

In problems 29 to 32 find the standard deviation from the given histogram. The mean or expected value was computed in Problems 13 to 16 of Section 6-2.

29.

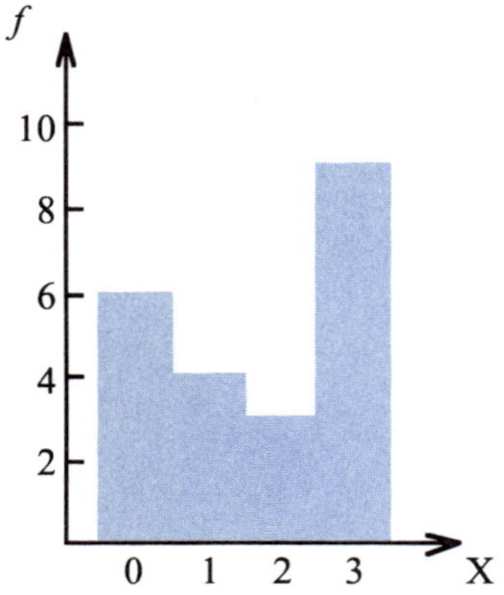

30.

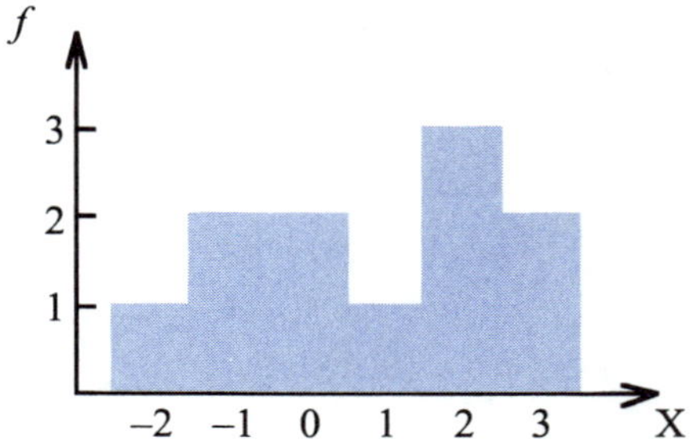

31.

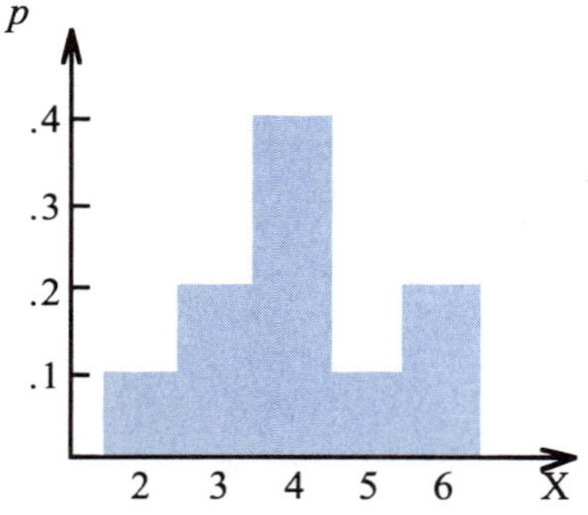

32.

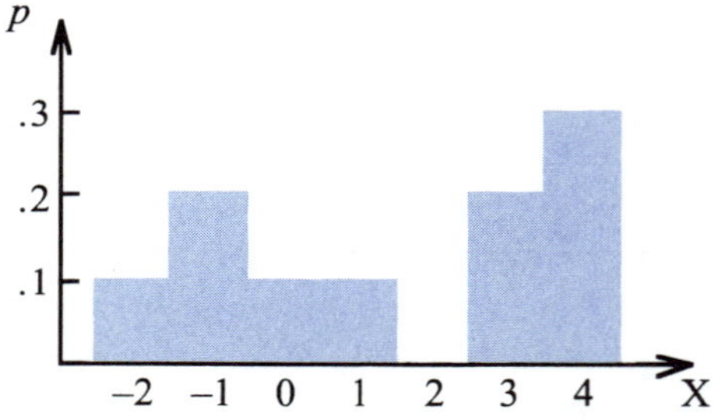

33. Toss a coin 3 times and let X be the number of heads. In Problem 17 of Section 6-2 it was found that $E(X) = 3/2$. Find the standard deviation.

34. Two cookies are selected without replacement from a jar with four vanilla cookies and three chocolate cookies. Find the expected number of vanilla cookies, and the standard deviation.

35. A biologist investigating airborne bacteria places a petri plate containing glucose under a ventilation duct. The number of bacteria colonies that appear per square milliliter is tabulated, as follows:

colonies per cell	0	1	2	3	4	5
cells	5	7	10	6	2	1

(a) What is the average number of colonies per cell?
(b) What is the standard deviation?

36. An economist determines trends in the stock market by comparing the average price per share of 10 selected stocks from month to month. The prices for the first month are \$15.75, \$17, \$6.25, \$7.25, \$47, \$32, \$21.50, \$19, \$40, and \$12.88. What is the average price per share, and how much variation is there?

37. (a) Find the standard deviation of the numbers 1, 2, 3, 4, 5, 6, 7.
    (b) Find the standard deviation of the numbers 2, 3, 4, 5, 6, 7, 8.
    (c) Find the standard deviation of another set of seven consecutive integers.
    (d) Based on parts (a), (b), and (c) what do you conclude about the standard deviation of seven consecutive integers?

38. (a) Which distribution in the accompanying figure has the largest variance?
    (b) Which distribution has the smallest variance?

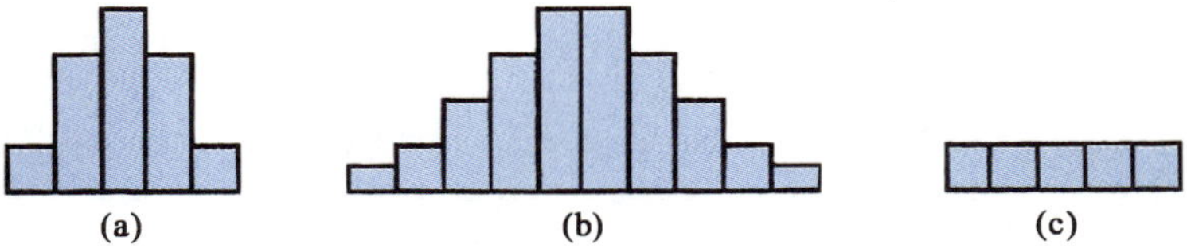

39. A fair coin is tossed four times.
    (a) What is the expected number of tails that appears?
    (b) What is the variance?

40. A fair coin is tossed five times.
    (a) What is the expected number of tails that appears?
    (b) What is the variance?

In problems 41 and 42 a coin is flipped 100 times. The expected number of heads is 50 and the standard deviation is 5.

41. What is a lower bound for the probability of tossing between 40 and 60 heads?

42. What is a lower bound for the probability of tossing between 30 and 70 heads?

43. If a coin is tossed n times, the expected number of heads is $n/2$. What is the standard deviation?

Referenced Exercise Set 6-3

1. The expected birth month for the Phillips study on strong wills prolonging life was found in Problem 3 of Referenced Exercise Set 6-2. Calculate the standard deviation.

2. Problem 2 in Referenced Exercise Set 6-2 computed the mean number of heads in J. E. Kerrich's coin tossing experiment from this data: 502, 511, 497, 529, 504, 476, 507, 528, 504, 529. Find the standard deviation.

3. The national speed limit was reduced to 55 miles per hour in 1974. Table 6–15 would seem to confirm the reasonableness of this policy, yet some critics object to it. Arnold Barnett, of the Sloan School of Management at MIT, used elementary statistics to rebut the critics.[15]
   (a) Compute the mean number of road fatalities and the standard deviation for the years 1970-1973.

---

[15] Arnold Barnett, "Misapplications Reviews: Speed Kills?," *Interfaces*, March-April 1986, pp. 63-68.

(b) Within how many standard deviations of the mean does the number of deaths in 1974 lie?

Year	Deaths
1970	54,800
1971	54,700
1972	56,600
1973	55,700
1974	46,600

Table 6–15

4. For assets which are not held in a diversified portfolio, it is commonplace in business to identify the standard deviation with the measure of risk.[16] Eastern Communications, Inc., a manufacturer and retailer of business and consumer telephones, projected its rate of return on two new lines, Standard and Designer, over a wide range of states of the economy. Table 6–16 summarizes the projections.
 (a) Compute (as a percent) for the Standard Phone line.
 (b) Compute (as a percent) for the Designer Phone line.
 (c) Which line had the smaller chance of significant loss from investing in that product line?

State of the Economy	Probability	Designer Line Rate of Return (%)	Standard Line Rate of Return (%)
Boom	0.3	100	20
Normal	0.4	15	15
Recession	0.3	B70	10

Table 6–16

5. Table 6–17 lists the expected EPS (earnings-per-share) and the standard deviation of the EPS for various D/A (debts/assets) ratios for Santa Clara Industries.[17]
 (a) For which D/A ratios is there a negative EPS within one standard deviation of the expected EPS?
 (b) For which D/A ratios is there a negative EPS within 1.5 standard deviations of the expected EPS?

D/A Ratio (%)	Expected EPS	Standard Deviation of EPS
0	$2.40	$1.52
10	$2.56	$1.69
20	$2.75	$1.90
30	$2.97	$2.17
40	$3.20	$2.53
50	$3.36	$3.04
60	$3.30	$3.7947

Table 6–17

[16] B. J. Campsey and Eugene F. Brigham, *Introduction to Financial Management*, The Dryden Press, Chicago, 1985, pp. 426-432.

[17] B. J. Campsey and Eugene F. Brigham, *Introduction to Financial Management*, The Dryden Press, Chicago, 1985, p. 539.

6. The well-known writer Stephen Jay Gould made use of statistics to support his claim that old-time baseball players, several of whom hit .400, were no better than the best players of today.[18] Gould wrote, "As [standard] variation shrinks around a constant mean batting average, .400 hitting disappears. It is, I think, as simple as that." Are you convinced by this argument?

7. Many studies use statistical analyses, based on the mean and standard deviation, to draw conclusions. One such study asked whether teaching geometry to general students in high school using computers (the Logo group) would improve student achievement over the standard pencil-and-paper approach (the traditional group).[19] This problem asks you to derive conclusions from Dr. Cook's findings.
   (a) An achievement test was administered to determine if the two groups of students were equivalent at the beginning of the study. The relevant results are tabulated in Table 6–18. Do you think that the groups were equivalent?

Group	Mean number of correct answers
Logo	29.28
Traditional	27.47

Table 6–18

   (b) The students took both a pretest and a posttest on achievement in geometry. The relevant data are tabulated in Table 6–19. Do you think that the use of the computer improved the scores over the standard approach?

	Pretest		Posttest	
Group	$\bar{x}$	σ	$\bar{x}$	σ
Logo	10.47	2.51	17.03	4.40
Traditional	9.57	3.33	16.27	3.38

Table 6–19

   (c) The students took both a pretest and a posttest on their attitude toward mathematics. The relevant data are tabulated in Table 6–20. Do you think that the use of a computer changed their attitudes significantly toward mathematics?

	Pretest		Posttest	
Group	$\bar{x}$	σ	$\bar{x}$	σ
Logo	10.47	2.51	17.03	4.40
Traditional	9.57	3.33	16.27	3.38

Table 6–20

8. S. Yeshurun wrote a provocative article on cognitive psychology that dealt with left-brain/right-brain differences and with information processing approaches to learning. The author revealed that the mean number of items that the brain's short-term memory can hold is 7, with standard deviation 2.[20] How many standard

---

[18] Stephen Jay Gould, "Entropic Homogeneity Isn't Why No One Hits .400 Any More," *Discover*, August 1986, pp. 60-66.

[19] Paula J. Cook, "The effects of an instructional unit utilizing logo and the computer on achievement in geometry and attitude toward mathematics of selected high school general mathematics students," Doctoral Dissertation, Temple University, 1987.

[20] S. Yeshurun, "A first look at psychomathematics and neuropsycho-

deviations from the mean is a brain whose short-term memory holds between 3 and 11 items?

Cumulative Exercise Set 6-3

1. Toss a coin four times and record the sequence of heads and tails. Let X be the number of heads recorded. Construct the probability distribution.
2. Draw the histogram for the cost of 1991 automobiles from the table below.

cost (in $1000)	5-11	11-17	17-23	23-29	29-35
frequency	8	8	14	6	5

3. Using the probability distribution below where x represents the number of heads when a penny is tossed five times, find the probability of tossing at least 4 heads.

x	0	1	2	3	4	5
p	1/32	5/32	10/32	10/32	5/32	1/32

4. Find the mean for the data: 5, 5, 6, 6, 6, 6, 9, 10, 10.
5. Consider the data in the table below.
(a) Find the mean.
(b) Find the median.

x	10	9	8	7	6	5
f	7	7	5	1	1	3

6. The number of new accounts opened by a new salesperson during her first ten days on the job are 5, 4, 4, 6, 3, 5, 5, 4, 7, 6. Find the mean number of accounts opened per day.
7. Consider the data in the table below.
(a) Find the mean.
(b) Find the median.

height (in inches)	56-58	58-60	60-62	62-64	64-66
frequency	400	800	1000	700	500

8. The department of transportation reported that the probability of no accident occurring per day on Route 2 near Boston is .28, the probability of one accident is .12, the probability of two accidents is .25, and the probability of three accidents is .35. What is the expected number of accidents per day?
9. A set of data has 5 entries. Each entry is either 6, 7, or 8. What is the set if the mean is 6.2 and the median is 6?
10. Find the expected value of the random variable X from the probability distribution table:

---

mathematics," *International Journal of Mathematical Education in Science and Technology*, March-April 1987, pp. 229-251.

x	-2	-1	2	4
p	0.1	0.4	0.2	0.3

11. Calculate the standard deviation for the data 3, 4, 5, 6, 7, 8, 9.

12. Calculate the standard deviation for the data in the table.

x	2	3	4	5
p	0.4	0.4	0.1	0.1

Multiple-Choice Exercise Set 6-3

1. Which of the following statements is true?
   (a) the standard deviation is always greater or equal to zero
   (b) the variance can be any number
   (c) the mean is always greater or equal to zero
   (d) the standard deviation can be any number
   (e) the mean is always bigger than the standard deviation

2. For a probability distribution the Greek letter μ (mu) stands for
   (a) $\bar{x}$ (b) E(X) (c) Σ (d) σ (e) none of these

3. The variance and the standard deviation measure
   (a) the average (b) the mean (c) the spread of the data from the mean
   (d) the median (e) none of these

4. The notation for the sample standard deviation is
   (a) $\bar{x}$ (b) μ (c) σ (d) s (e) none of these

5. The standard deviation is defined as
   (a) the mean of the squares
   (b) the sum of the distances to a line
   (c) the sum of the absolute values of the differences
   (d) the square root of the variance
   (e) none of these

6. If $\sigma^2 = 16$ then $\sigma =$
   (a) 256 (b) -16 (c) 2 (d) 8 (e) none of these

7. If $\sigma = 9$ then $\sigma^2 =$
   (a) 3 (b) -3 (c) 27 (d) 81 (e) none of these

8. The standard deviation of the set 7, 7, 7, 7, 7 is
   (a) 1 (b) 2 (c) 3 (d) 4 (e) none of these

9. The standard deviation of the set 1, 2, 3, 4, 5, 6, 7 is
   (a) $\sqrt{\frac{14}{3}}$ (b) 1 (c) $\sqrt{\frac{11}{3}}$ (d) $\sqrt{2}$ (e) none of these

10. The standard deviation of the set 3, 4, 5, 6, 7, 8, 9 is
(a) $\sqrt{\frac{14}{3}}$ (b) 1 (c) $\sqrt{\frac{11}{3}}$ (d) $\sqrt{2}$ (e) none of these

11. The standard deviation of the set 4, 3, 4, 4, 0 is
(a) 1 (b) $\sqrt{2}$ (c) $\sqrt{3}$ (d) 2 (e) none of these

12. The standard deviation of the set 6, 9, 7, 6, 5, 5, 2, 1, 4, 5 is approximately 2.3. What is the approximate standard deviation of the set 8, 6, 5, 4, 4, 1, 0, 3, 4?
(a) 0.3 (b) 1.3 (c) 2.3 (d) 3.3 (e) none of these

13. The standard deviation of the set 6, 9, 7, 6, 5, 5, 2, 1, 4, 5 is approximately 2.3. What is the approximate standard deviation of the set 19, 17, 16, 15, 15, 12, 11, 14, 15?
(a) 10.3 (b) 11.3 (c) 12.3 (d) 13.3 (e) none of these

14. If s = 4 for a set with 5 entries, then for a set with 6 entries, s
(a) = 4 (b) = 5 (c) = 6 (d) = 2 (e) cannot be determined

15. If s = 4 for a set with 5 entries, then for the set consisting of those 5 entries and the element 4, s
(a) = 4 (b) = 5 (c) = 6 (d) = 2 (e) cannot be determined

16. Find the variance of the data from the frequency distribution below is approximately
(a) 0.54 (b) 0.49 (c) 17.3 (d) 4.1 (e) none of these

x	3	4	5
f	2	5	3

17. Find the variance of the data from the given probability distribution.
(a) 0.7 (b) 4.9 (c) 41 (d) 7 (e) none of these

x	3	4	5
p	0.2	0.5	0.3

18. Two dimes are flipped and the number of heads is recorded. The expected value is 1. What is the standard deviation?
(a) 3/2 (b) 2 (c) $\sqrt{\frac{3}{2}}$ (d) $\sqrt{\frac{1}{2}}$ (e) none of these

19. In any experiment, the probability that an entry lies within two standard deviations of the mean is at least
(a) 8/9 (b) 2 (c) 1/4 (d) 3/4 (e) none of these

20. A coin is tossed 100 times. What is the lower bound for the probability of tossing between 40 and 50 heads?
(a) 3/4 (b) 2 (c) 1/4 (d) 10 (e) none of these

6-4 THE NORMAL CURVE

Suppose an expert archer and a complete novice shoot at a target on an archery range. Although the expert's shots will be more accurate, the novice's shots will land in a surprisingly similar manner. This does not mean that the outcome of each single shot can be predicted. Rather, if each archer shoots 1000 arrows and the distance between the hits and the target is measured, then each histogram of the distances will assume a definite shape that does not depend upon their relative abilities.

The shape of each histogram will approximate an ideal distribution known as the normal curve. The more shots that are made the closer the histogram will be to this curve. Because the normal curve resembles a bell, it is sometimes called a "bell-shaped curve."

The normal curve describes many mental and physical characteristics of human beings. This is fortunate because many important facts are known about the normal curve, including its mean and standard deviation. This section examines these facts and shows how to use tables of values to apply them to questions involving height and IQ (intelligence quotient.)

Fundamental Properties

Consider a probability experiment of tossing six coins and recording the number of heads. Its histogram is shown in Figure 6–11. We have superimposed upon it a curve that passes through the midpoints at the top of the rectangles. It resembles a curve known as the normal curve.

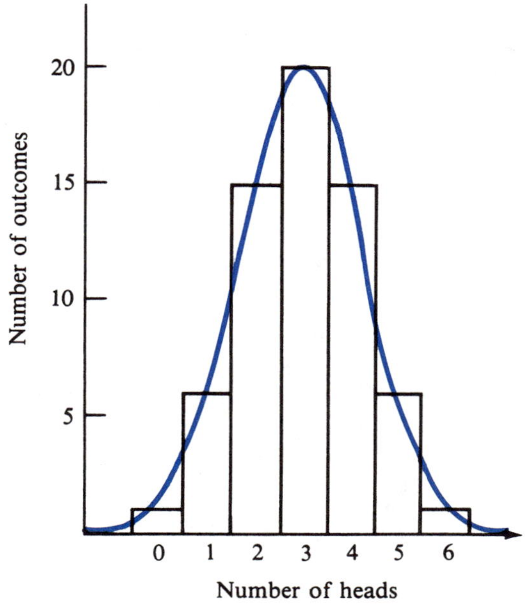

The normal curve has an unwieldy-looking equation.

$$y = \frac{1}{\sqrt{2\pi}} e^{-z^2/2}$$

where $\pi \approx 3.142$ and $e \approx 2.718$. The graph of the normal curve is given in Figure 6–12. The values on the z-axis are standard deviations. The units on the y-axis correspond to the relative frequencies on a histogram, and therefore a normal curve can be used to approximate the histogram.

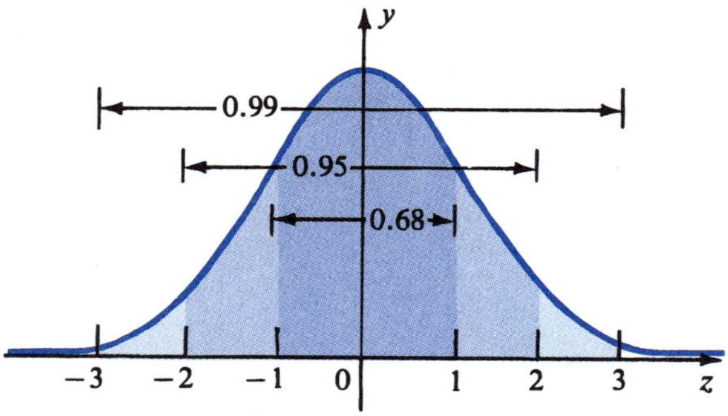

Figure 6–12

There are four basic properties of the normal curve:

1. The mean corresponds to the value $z = 0$.
2. The curve is symmetric about the mean.
3. The curve always lies above the z-axis.
4. The area under the curve and above the z-axis is exactly 1.

We make use of these properties to find areas that lie under the curve and above the z-axis. The reason for the emphasis on areas is *that the area under a normal curve corresponds to the probability of an event.* There are three fundamental areas with which we are concerned:

The area between $z = -1$ and $z = 1$ is about .68.
The area between $z = -2$ and $z = 2$ is about .95.
The area between $z = -3$ and $z = 3$ is about .99.

These areas refer to the proportion of entries that lie within 1σ, 2σ, and 3σ of the mean. Therefore for a normal curve, about two-thirds of the entries will lie within 1σ of the mean, about 95% will lie within 2σ, and virtually all will lie within 3σ. These properties are shown in Figure 6–12.

According to Chebyshev's Theorem for $k = 2$, at least three-fourths of the entries in any distribution lie within 2σ of the mean. For data with a normal distribution, this approximation can be improved to 95%. For 3σ, Chebyshev's Theorem gives a figure of at least $8/9 = .89$, while the corresponding figure for the normal curve is known to be .99.

Examples 1 and 2 combine the four basic properties of the normal curve with the three designated areas to derive other areas under the normal curve.

EXAMPLE 1

What is the area under the normal curve between $z = 0$ and $z = 1$?

Solution. The area under the normal curve between $z = -1$ and $z = 1$ is about .68. Since the normal curve is symmetric about the mean, the mean divides this area into two equal parts. (See Figure 6–13.) The area between $z = 0$ and $z = 1$ is one-half of the area between $z = -1$ and $z = 1$, so its area is about (.5)(.68) = .34.

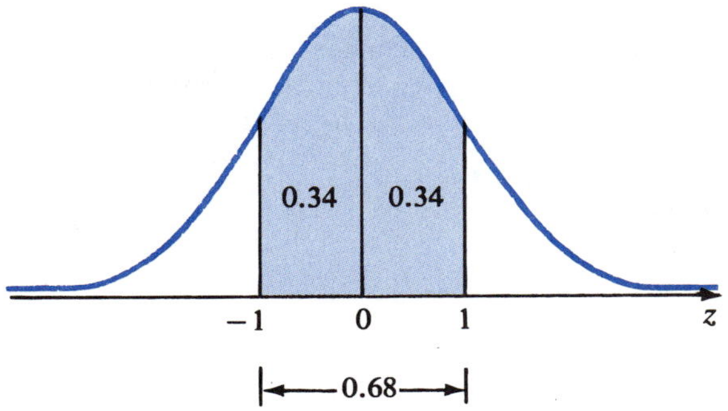

Figure 6–13

EXAMPLE 2

What is the area under the normal curve to the right of $z = 2$?

Solution. The area under the normal curve to the right of $z = 2$ is derived from the fact that the area between $z = -2$ and $z = 2$ is about .95. There is about .05 remaining. There are two regions under the curve that comprise this area, the region to the right of $z = 2$ and the region to the left of $z = -2$. These regions have equal areas because of the symmetry of the normal curve. Hence the area under the normal curve to the right of $z = 2$ is one-half of .05, or .025. (See Figure 6–14.)

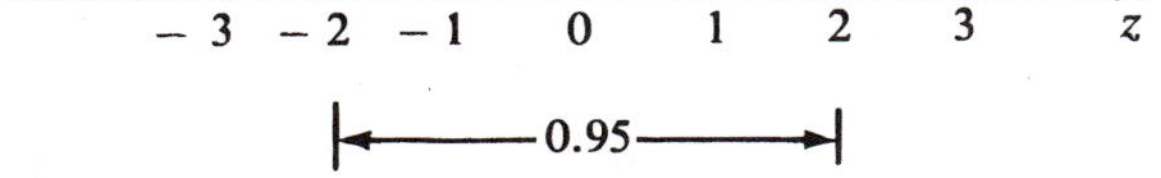

Figure 6–14

The area under the normal curve to the left of $z = 2$ can be obtained from Example 2 by using Property 4. Since the total area is 1 and the area to the right is about .025, the area to the left is approximately $1 - .025 = .975$. An alternate way to obtain this area is to look it up in a table.

Using a Table

The values we have supplied for areas under the normal curve are very limited; they are rough approximations that apply only to integral multiples of σ. Table 6–21 contains several more values that are approximated to four decimal places.

z	A(z)
2.00	.9773
1.50	.9332
1.00	.8413
0.50	.6915
0.00	.5000
–0.50	.3085
–1.00	.1587
–1.50	.0668
–2.00	.0227

Table 6–21

A more complete table is printed in the Appendix. For each value of z, called the **z-score**, A(z) gives the area under the normal curve to the left of z. The area A(z) is shaded in Figure 6–15.

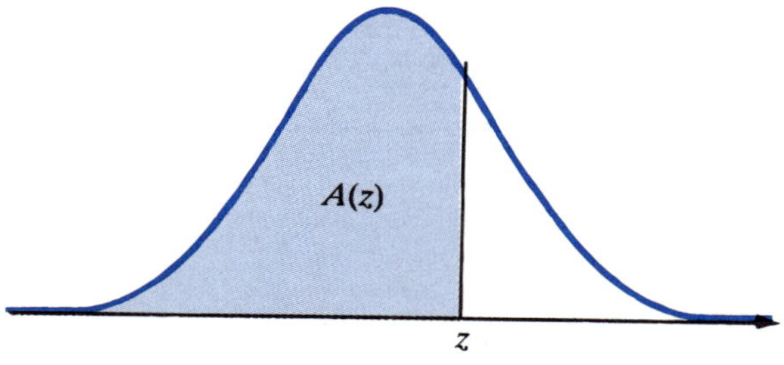

Figure 6–15

There are three kinds of areas under the normal curve:

L: the area to the left of z
R: the area to the right of z
B: the area between two values of z

Table 6–21 accounts for L-areas. For instance, the area to the left of $z = 2$ is $A(2) = .9773$. This value is a better approximation than the value .975 obtained after Example 2. Example 3 shows how to compute R-areas.

EXAMPLE 3

Find the area under the normal curve to the right of $z = 1$.

Solution. From Table 6–21, the area to the left of $z = 1$ is $A(1) = .8413$. Since the total area under the normal curve is 1, the area to the right of $z = 1$ is $1 - .8413 = .1587$. (See Figure 6–16.)

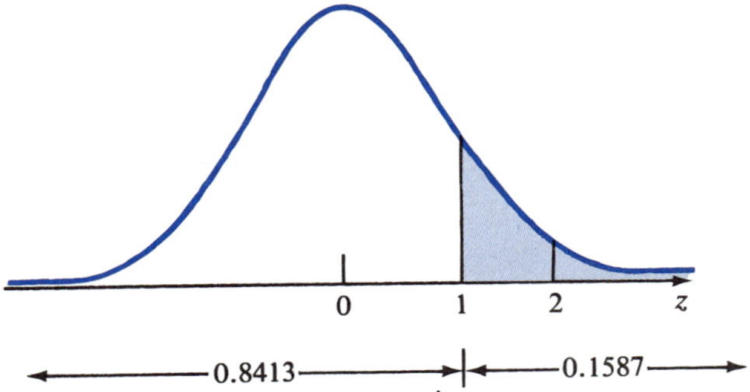

Figure 6–16

The following rule emerges from Example 3.

> **Finding the Area to the Right of a *z*-score**
>
> The area under the normal curve to the right of z is $1 - A(z)$.

EXAMPLE 4

Find the area under the normal curve between $z = -0.5$ and $z = 0.5$.

Solution. First set $z = 0.5$, then $A(0.5) = .6915$ is the area to the left of $z = 0.5$. Similarly $A(-0.5) = .3085$ is the area to the left of $z = -0.5$. The desired area is the difference between these two areas. Figure 6–17 shows why this is true. Thus the area between $z = -0.5$ and $z = 0.5$ is $A(0.5) - A(-.5) = .6915 - .3085 = .3830$.

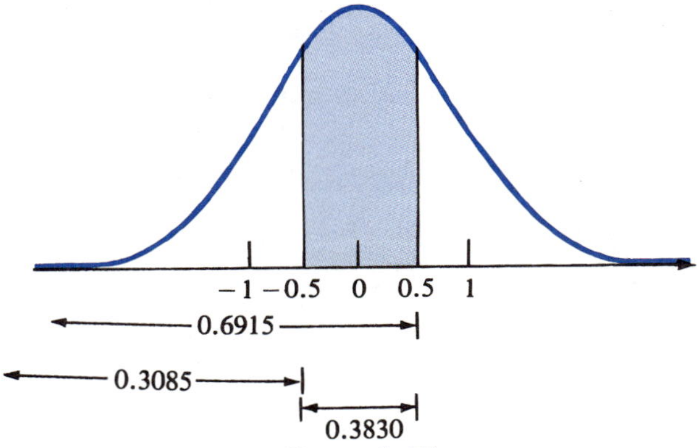

Figure 6–17

The following rule emerges from Example 4.

Finding the Area Between Two *z*-scores

The area under the normal curve between z_1 and z_2, where $z_1 < z_2$, is $A(z_2) - A(z_1)$.

The normal curve is a mathematical abstraction that is not derived from any specific real-world phenomena. However, it is very useful for describing any set whose histogram resembles the normal curve. The main difference between the normal curve and such a set is that the normal curve is *continuous* while data sets are necessarily *discrete*. In spite of this difference the normal curve's utility is vast. Let us turn to its applications.

Applications

So far we have referred to *the* normal curve. Actually, it is called the **standard normal curve**. Any list of data that is assumed to be normally distributed will have a histogram, and hence a normal curve, that resembles the standard normal curve. The **normal approximation** is a procedure for finding the area under any normal curve by adjusting the units of the horizontal and vertical axes, so that the resulting curve resembles the standard normal curve. Let us explain this procedure.

In addition, for the most part, we have worked with *sample* statistics – the *sample* mean, the *sample* variance, and the *sample* standard deviation. These statistics refer to the mean, variance and standard deviation of only a portion of the entire group with which we are concerned. For example, if we were concerned with questions concerning the starting salary for accountants, we would be unable to survey *all* new accountants, but rather we would take a sample of a few, a few hundred or a few thousand such accountants. Computing the mean, variance, or standard deviation for this sample would require the use of the formula for $\bar{x}$, s^2, or s, respectively. From these calculated sample statistics, we would try to draw conclusions about *all* new accountants, called the **population**.

Much of statistics is built upon estimating the population statistics based on the sample statistics. The **population mean** is denoted by μ, the **population variance** by σ^2, and the **population standard deviation** by σ. (Note that these are the same symbols used for a probability distribution.) For the remaining part of this chapter, we will assume that the population statistics are known for a variety of situations such as IQ scores, test scores and as in the following discussion, heights.

The Health Examination Survey, conducted by the Department of Health and Human Services, found that the average height of women in the United States forms a normal curve with $\mu = 64$ inches with standard deviation $\sigma = 2.5$ inches.

What percentage of women are less than 67.75 inches tall?

To answer this question, draw an *x*-axis and mark the heights at μ, $\mu \pm 1\sigma$ (66.5 inches and 61.5 inches), and $\mu \pm 2\sigma$ (69 inches and 59 inches). Draw a *z*-axis below it, marking the corresponding number of standard deviations. (See Figure 6–18.)

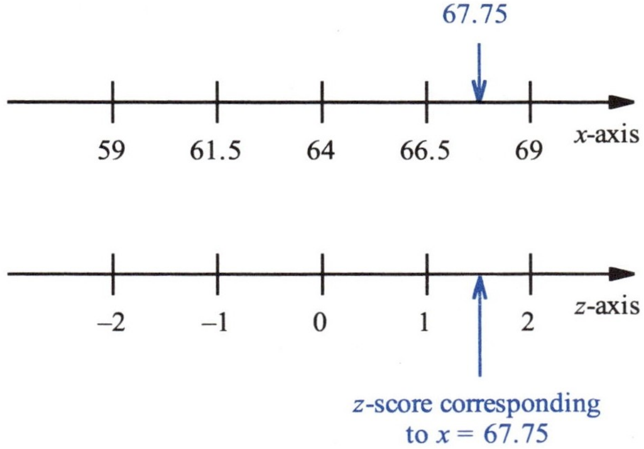

Figure 6–18

The value $x = 67.75$ on the upper line lies between 66.5 and 69, so the corresponding z-score on the lower line is between 1 and 2. Since $\mu = 64$ inches and $\sigma = 2.5$, the precise z-score is obtained from the following formula.

$$z = \frac{\text{distance between } x \text{ and } \mu}{\text{standard deviation}} = \frac{67.75 - 64}{2.5} = \frac{3.75}{2.5} = 1.5$$

Hence $x = 67.5$ inches is $z = 1.5$ standard deviations from the mean $\mu = 64$ inches. The formula for the z-score is described in the display.

Computing *z*-scores

If a set of data has mean μ and standard deviation σ, the z-score corresponding to a value of x is

$$z = \frac{x - \mu}{\sigma}$$

The question about the percentage of women whose height is less than $x = 67.75$ inches translates into z-scores that are less than $z = 1.5$. This problem is an L-area. From Table 6–21 the area to the left of $z = 1.5$ is $A(1.5) = .9332$. Therefore about 93.32% of American women are less than 67.75 inches tall.

In summary, the normal approximation is a three-part procedure. First the x-scores are converted to z-scores. Then the problem is translated into one of the fundamental areas: L, R, or B. Finally, a table of z-scores for the standard normal curve provides the area, which is interpreted as either a probability or a percentage. Example 5 illustrates this procedure for a B-area.

EXAMPLE 5

Assume that the heights of women in the United States are normally distributed with $\mu = 64$ inches and $\sigma = 2.5$ inches. What percentage of heights are between 5' and 5' 2"?

Solution. Since 5' = 60", set $x_1 = 60$. The corresponding z-score is

$$z_1 = \frac{60-64}{2.5} = -1.6$$

From the Appendix, A(–1.6) = .0548. Next, set $x_2 = 62$. Then

$$z_2 = \frac{62-64}{2.5} = -0.8$$

and from the Appendix, A(–0.8) = .2119.

The percentage of heights between $x_1 = 60$ inches and $x_2 = 62$ inches translates to the area between $z_1 = -1.6$ and $z_2 = -0.8$. This area is given by

$$A(z_2) - A(z_1) = A(-0.8) - A(-1.6) = .2119 - .0548 = .1571.$$

Consequently 15.71% of American women are between 5' and 5' 2". This percentage is roughly 1 in 6.

Example 6 applies the normal approximation procedure to IQ (intelligence quotient) tests under the usual assumption that IQ scores are normally distributed. It illustrates the method for R-areas.

EXAMPLE 6

Many IQ tests are designed so that the mean score is 100 and the standard deviation is 15. If a person is selected at random, what is the probability that the IQ score is greater than 130?

Solution. The z-score corresponding to $x = 130$ is

$$z = \frac{130-100}{15} = 2$$

From Table 6–21, A(2) = .9773. Scores above 130 translate to the area to the right of $z = 2$. This region is an R-area, which is equal to 1 – A(2) = 1 – .9773 = .0227. Thus the probability is .0227.

In many public schools the cut-off point for a student to be classified as "gifted" is an IQ score of 130. Example 6 shows that about 2.27% of the population is gifted under the assumption that $\mu = 100$ and $\sigma = 15$.

EXERCISE SET 6-4

In problems 1 to 8 find the area under the normal curve within the given limits, using the four basic properties and three fundamental areas of the normal curve.

1. between $z = 0$ and $z = 2$.
2. between $z = 0$ and $z = 3$.
3. $z = -1$ and $z = 0$.
4. $z = -2$ and $z = 0$.
5. to the right of $z = 1$.
6. to the right of $z = 3$.
7. to the left of $z = -2$.
8. to the left of $z = -1$.

In problems 9 to 12 use Table 6–21 to find the area under the normal curve within the given limits.

9. to the right of $z = 1.5$.
10. to the right of $z = -0.5$.
11. between $z = -1$ and $z = 1$.
12. between $z = -0.5$ and $z = 1.5$.

In problems 13 to 16 assume that the heights of women in the U.S. are normally distributed with $\mu = 64$ inches and $\sigma = 2.5$ inches. Determine the percentage of heights that are in the range.

13. between 5' 3" and 5' 6"
14. between 5' 1" and 5' 5"
15. taller than 5' 7"
16. shorter than 5'

In problems 17 to 20 the mean IQ is 100 and the standard deviation is 15. Determine the percentage of IQ's that are in the given range.

17. between 110 and 130
18. between 90 and 115
19. greater than 95
20. less than 110

In problems 21 to 26 find the area under the normal curve within the given limits, using the four basic properties and three basic areas of the normal curve.

21. $z = 1$ and $z = 2$.
22. $z = 2$ and $z = 3$.
23. $z = -1$ and $z = 2$.
24. $z = -2$ and $z = 1$.
25. to the right of $z = -1$
26. to the left of $z = 3$

In problems 27 to 30 use the table in the Appendix to find the area under the normal curve within the given limits.

27. to the left of $z = 0.38$.
28. to the left of $z = -1.05$.
29. between $z = -1.74$ and $z = 0.51$.
30. between $z = -0.56$ and $z = 2.15$.

31. If the mean IQ is 100 with standard deviation 15, what percentage of IQ scores is between 102 and 118?
32. If the mean on an exam is 475 with standard deviation 100, what percentage of scores is between 465 and 527?
33. If the mean of a set of data that is normally distributed is 4.1 with standard deviation 0.1, what percentage of data is greater than 4.21?

34. If the mean of a set of data that is normally distributed is 1.1 with standard deviation 0.5, what percentage of data is less than 1.2?

Referenced Exercise Set 6-4

The mean and median scores on the SAT exams vary slightly from year to year.[21] They were re-centered in 1995. In problems 1 to 4 assume that the SAT scores form a normal distribution and that, for the mathematics section, $\mu = 470$ and $\sigma = 100$.

1. What percentage of students scored above 500?
2. What percentage of students scored between 400 and 600?
3. If a student is selected at random, what is the probability that the student scored between 430 and 530?
4. If a student is selected at random, what is the probability that the student scored above 440?

Problems 5 and 6 refer to Table 6–22, which lists the data from a computer simulation of 20,000 games of blackjack in Atlantic City and Las Vegas.[22] (The actual playing of 20,000 hands would consume some 300 to 400 human hours.)

	Proportion of wins	Mean Win (chips)	σ (chips)	Mean Player Edge (%)	Largest Player Deficit (%)
Las Vegas	0.87	2084	1786	1.35	4961
Atlantic City	0.79	1108	1471	0.91	3621

Table 6–22

5. (a) What is the probability of a player winning at least 300 chips in Las Vegas?
   (b) What is the probability of a player winning at least 300 chips in Atlantic City?
6. (a) What is the probability of a player winning fewer than 100 chips in Las Vegas?
   (b) What is the probability of a player winning fewer than 100 chips in Atlantic City?

Multiple-Choice Exercise Set 6-4

1. If an archer shoots 1000 arrows and the distance between the hit and the target is measured, then each histogram of the distances will assume the shape of a(n)
   (a) line (b) parabola (c) exponential curve
   (d) normal curve (e) none of these

---

[21] William H. Angoff, "How We Calibrate College Board Scores," *The College Board Review*, No. 68, Summer 1968.

[22] *Source*: Martin H. Millman, "A Statistical Analysis of Casino Blackjack," *American Mathematical Monthly*, Aug-Sept 1983, pp. 431-436.

2. A normal curve is also known as a
(a) histogram (b) standard deviation (c) median
(d) mean (e) none of these

3. The values on the z-axis of a normal curve are
(a) standard deviations (b) means (c) medians
(d) relative frequencies (e) none of these

4. The units on the y-axis of a normal curve correspond to
(a) standard deviations (b) means (c) medians
(d) relative frequencies (e) none of these

5. The mean of a normal curve corresponds to the value $z =$
(a) 1 (b) 0 (c) $\frac{1}{\sqrt{2\pi}}$ (d) 2 (e) none of these

6. The normal curve is symmetric about the
(a) mean (b) x-axis (c) z-axis
(d) standard deviation (e) none of these

7. The normal curve always lies above the
(a) mean (b) z-axis (c) y-axis
(d) standard deviation (e) none of these

8. The area under the normal curve and above the z-axis is
(a) exactly $\frac{1}{\sqrt{2\pi}}$ (b) approximately $\frac{1}{\sqrt{2\pi}}$ (c) exactly 1
(d) approximately 1 (e) none of these

9. For a normal curve, the area within 1σ of the mean is about
(a) 1 (b) 0.95 (c) 0.68 (d) 0.99 (e) none of these

10. For a normal curve, the area within 2σ of the mean is about
(a) 1 (b) 0.95 (c) 0.68 (d) 0.99 (e) none of these

11. For a normal curve, the fraction of the entries that will lie within 1σ of the mean is
(a) 1/2 (b) 1/4 (c) 3/4 (d) 2/3 (e) none of these

12. The area under the normal curve between $z = 0$ and $z = 2$ is
(a) 0.0227 (b) 0.68 (c) 0.4773 (d) 0.9773 (e) none of these

13. The area under the normal curve between $z = 0$ and $z = 1.5$ is
(a) 0.0668 (b) 0.4332 (c) 0.68 (d) 0.9332 (e) none of these

14. The area under the normal curve between $z = 0$ and $z = 0.7$ is
(a) 0.0279 (b) 0.68 (c) 0.758 (d) 0.242 (e) none of these

15. The area under the normal curve to the right of $z = 2$ is
(a) 0.0227 (b) 0.68 (c) 0.9773 (d) 0.9887 (e) none of these

16. The area under the normal curve to the right of $z = 1.5$ is
(a) 0.0668 (b) 0.4332 (c) 0.68 (d) 0.9332 (e) none of these

17. The area under the normal curve to the right of $z = 0.7$ is
(a) 0.34 (b) 0.68 (c) 0.758 (d) 0.242 (e) none of these

18. The area under the normal curve between $z = 1$ and $z = 2$ is
(a) 0.8413 (b) 0.34 (c) 0.136 (d) 0.1587 (e) none of these

19. The area under the normal curve between $z = 1.5$ and $z = 2$ is
(a) 0.441 (b) 0.0441 (c) 0.0668 (d) 0.559 (e) none of these

20. The area under the normal curve between $z = 0.7$ and $z = 2$ is
(a) 0.2193 (b) 1.393 (c) 0.7807 (d) 0.758 (e) none of these

21. The area under the normal curve between $z = -1$ and $z = 0$ is
(a) 0.3413 (b) 0.68 (c) 0.9773 (d) 0.9887 (e) none of these

22. The area under the normal curve between $z = -1.5$ and $z = 0$ is
(a) 0.1587 (b) 0.8413 (c) 0.4332 (d) 0.9332 (e) none of these

23. The area under the normal curve between $z = -0.7$ and $z = 0$ is
(a) 0.1392 (b) 0.8608 (c) 0.758 (d) 0.242 (e) none of these

24. The area under the normal curve to the left of $z = -2$ is
(a) 0.0227 (b) 0.68 (c) 0.9773 (d) 0.9887 (e) none of these

25. The area under the normal curve to the left of $z = -1.5$ is
(a) 0.5668 (b) 0.4332 (c) 0.68 (d) 0.9332 (e) none of these

26. The area under the normal curve to the left of $z = -0.7$ is
(a) 0.34 (b) 0.68 (c) 0.758 (d) 0.242 (e) none of these

27. The area under the normal curve between $z = 1.75$ and $z = 2.1$ is
(a) 0.35 (b) 0.222 (c) 0.4778 (d) 0.0222 (e) none of these

28. The area under the normal curve between $z = -2.1$ and $z = -1.75$ is
(a) –0.35 (b) –0.222 (c) –0.4778 (d) –0.0222 (e) none of these

29. The area under the normal curve between $z = -1.7$ and $z = -0.23$ is
(a) 0.3644 (b) 0.6356 (c) –0.3644 (d) –0.6356 (e) none of these

30. If a set of data is normally distributed with $\mu = 4$ and $\sigma = 1$, the z-score corresponding to a value of $x = 6$ is
(a) 1.5 (b) –2 (c) 2 (d) 0 (e) none of these

31. If a set of data is normally distributed with $\mu = 4$ and $\sigma = 1$, the z-score corresponding to a value of $x = 2$ is
(a) 1.5 (b) –2 (c) 2 (d) 0 (e) none of these

32. If $\mu = 100$ and $\sigma = 10$, about what percentage of entries are less than 110?
(a) 34% (b) 84% (c) 68% (d) 16% (e) none of these

33. If $\mu = 100$ and $\sigma = 10$, about what percentage of entries are greater than 110?
(a) 34% (b) 84% (c) 68% (d) 16% (e) none of these

34. If $\mu = 100$ and $\sigma = 10$, about what percentage of entries are between and 110?
(a) 34% (b) 84% (c) 68% (d) 16% (e) none of these

35. If $\mu = 100$ and $\sigma = 10$, about what percentage of entries are between and 120?
(a) 34% (b) 84% (c) 68% (d) 16% (e) none of these

36. If $\mu = 500$ and $\sigma = 100$, about what percentage of entries are less than 575?
(a) 0.2266 (b) 0.7266 (c) 0.7734 (d) 0.2734 (e) none of these

37. If $\mu = 500$ and $\sigma = 100$, about what percentage of entries are greater than 475?
(a) 0.4013 (b) 0.4987 (c) 0.9013 (d) 0.5987 (e) none of these

38. If $\mu = 500$ and $\sigma = 100$, about what percentage of entries are between 380 and 620?
(a) 0.7698 (b) 0.2302 (c) 0.0956 (d) 0.9044 (e) none of these

39. If $\mu = 500$ and $\sigma = 100$, about what percentage of entries are between 485 and 515?
(a) 0.1192 (b) 0.8664 (c) 0.1336 (d) 0.8808 (e) none of these

6-5 THE NORMAL APPROXIMATION TO BINOMIAL EXPERIMENTS

When you apply to a college the Admissions Office either accepts or rejects your application. Each decision from a college represents a Bernoulli trial because it consists of two possible outcomes. Some other examples of Bernoulli trials are the effectiveness of a medication (the patient is or is not cured), the movement of the Dow-Jones average (up or down), and the firing of a missile (hit or miss).

A binomial experiment consists of n Bernoulli trials. In Section 5-3 we saw that if the probability of success is p and the probability of failure is $q = 1 - p$, then the probability of obtaining exactly x successes in n trials is

$$\Pr(X = x) = C(n, x)p^x q^{n-x}$$

This formula is useful when the values of n are relatively small, but the computations become tedious when n is large. Fortunately the normal curve provides a very close approximation to a binomial experiment in such a case. This section shows how to carry out and apply the normal approximation procedure.

Coin Tossing

Coin tossing serves as a convenient model for binomial experiments. As an example, toss a coin 6 times and let the random variable X represent the number of heads that appear. Then $n = 6$ and $p = 1/2$. The probability of tossing exactly four heads is

$$\Pr(X = 4) = C(6, 4)\left(\frac{1}{2}\right)^4\left(\frac{1}{2}\right)^2 = \frac{15}{64} \approx .234$$

The entire probability distribution, including the particular case X = 4, is given in Table 6–23. Notice the values in the row of totals. The column of probabilities p adds up to 1, which serves as a check on the individual probabilities. The sum of the xp column is the expected value $\mu = 192/64 = 3$. Common sense confirms that, on average, six tosses of a coin should result in three heads. Recall from Section 5-4 that another way of obtaining this value is from the formula $\mu = np$. Computing in this manner, $\mu = (6)(1/2) = 3$.

x	p	xp	x^2p
0	1/64	0/64	0/64
1	6/64	6/64	6/64
2	15/64	30/64	60/64
3	20/64	60/64	180/64
4	15/64	60/64	240/64
5	6/64	30/64	150/64
6	1/64	6/64	36/64
Totals	1	192/64	672/64

Table 6–23

Recall that the variance σ^2 is equal to the total of the $x^2 p$ column of Table 6–23 minus the square of μ.

$$\sigma^2 = \frac{672}{64} - 3^2 = 1.5$$

Therefore, the standard deviation can be found by taking the square root of this value.

$$\sigma = \sqrt{1.5} \approx 1.225$$

For a binomial experiment it is possible to compute this value from the formula $\sigma^2 = npq$. In this example, $\sigma^2 = 6\ (1/2)\ (1/2) = 3/2$, which confirms the value derived from Table 6–23. The appropriate formulas are given in the display.

μ and σ for Binomial Experiments

If a binomial experiment consists of n Bernoulli trials, each with probability of success p and probability of failure q, then

- The mean is $\mu = np$.
- The variance is $\sigma^2 = npq$.
- The standard deviation is $\sigma = \sqrt{npq}$

EXAMPLE 1

Toss a coin 100 times and record the number of heads that appear. What are the values of
(a) the mean, and
(b) the standard deviation?

Solution. In this experiment, we have $n = 100$ since the coin is being tossed 100 times, and $p = 1/2$ since the probability of heads *on one toss* is 1/2.
(a) Using the formula for mean, $\mu = np = 50$.
(b) Since $p = 1/2$ it follows that $q = 1 - p = 1/2$, so using the formula for standard deviation,

$$\sigma = \sqrt{(100)\left(\frac{1}{2}\right)\left(\frac{1}{2}\right)} = \sqrt{25} = 5$$

Figure 6–19 shows the histogram for the probability distribution of tossing six coins and recording the number of heads. The histogram can be approximated by the normal curve which is superimposed upon it. From the values calculated prior to Example 1, the distribution has $\mu = 3$ and $\sigma \approx 1.225$. Let us see how to use this normal curve to approximate probabilities.

Consider the value of $X = 4$ heads. Its probability is equal to the area of the shaded bar above 4 in the histogram. This area can be approximated by the area under the normal curve between $x = 3.5$ and $x = 4.5$. The procedure for finding this area, as explained in Section 6-4, uses the formula $z = (x - \mu) / \sigma$.

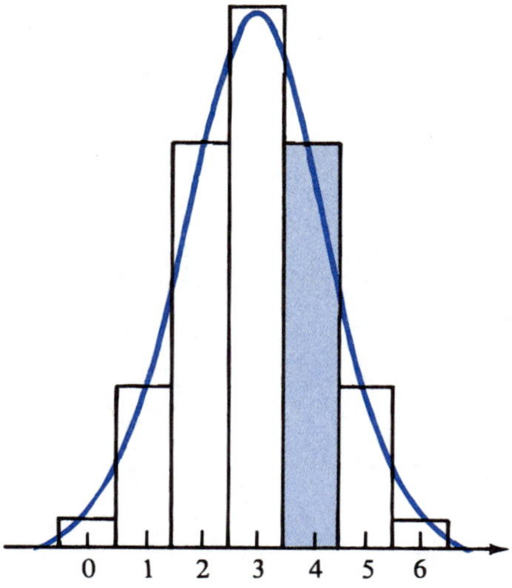

Figure 6–19

for $x_1 = 3.5$, $z_1 = \dfrac{3.5-3}{1.225} \approx 0.41$

for $x_2 = 4.5$, $z_1 = \dfrac{4.5-3}{1.225} \approx 1.22$

From the table in the Appendix, $A(z_1) = .6591$ and $A(z_2) = .8888$. Therefore, the area under the normal curve between $x = 3.5$ and $x = 4.5$ is approximately equal to

$$A(z_2) - A(z_1) = .8888 - .6591 = .2297$$

This value compares favorably with the theoretical probability of $15/64 \approx .2344$ which was found prior to Example 1 and is listed in Table 6–23.

The probability of tossing at least four heads in six tosses is given by

$$\Pr(4) + \Pr(5) + \Pr(6) = 22/64 \approx .3438.$$

This value can be approximated by the area under the normal curve to the right of $x = 3.5$ since the area of the histogram in Figure 6–19 corresponds to this probability. For $x = 3.5$ we found that $z = 0.41$ and $A(0.41) = .6591$. The desired area, which lies to the right of $z = 0.39$, is $1 - A(z) = 1 - .6591 = .3409$. Again, this value compares favorably with the theoretical probability of .3438.

Example 2 applies a similar approximation to the question of whether a coin is fair. Do you think a coin which lands heads 60 times in 100 tosses is fair?

EXAMPLE 2

What is the probability of a coin landing heads 60 or more times in 100 tosses?

Solution. From Example 1, $\mu = 50$ and $\sigma = 5$. The problem is to find $\Pr(X \geq 60)$. The normal curve is shown in Figure 6–20, where the desired area is shaded. The value $x = 59.5$ is used for $X = 60$ because the bar in the histogram corresponding to 60 runs from 59.5 to 60.5. Then $z = (59.5 - 50)/5 = 1.9$ From the table in the Appendix, the area to the left of $z = 1.9$ is $A(1.9) = .9713$. Therefore, the area to the right is $1 - .9713 = .0287$. Consequently, $\Pr(X \geq 60) = .0287$.

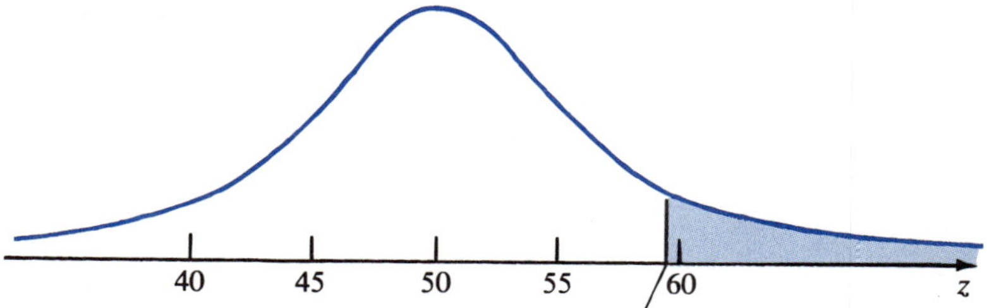

Figure 6–20

It follows from Example 2 that if a fair coin is tossed 100 times, it will land heads 60 or more times less than 3% of the time. It is reasonable to conclude that if a coin produces 60 or more heads in 100 tries, it is probably not fair.

The display lists the steps for applying the normal approximation procedure to a binomial experiment.

Procedure for Approximating a Binomial Experiment by the Normal Curve

1. Compute $\mu = np$ and $\sigma = \sqrt{npq}$
2. Sketch the normal curve and shade the desired area.
3. Convert X to x, then x to z using $z = (x - \mu)/\sigma$.
4. Use a table to determine the area.

To appreciate its power, notice that the alternative to Example 2 is to carry out 41 computations since

$$\Pr(X \geq 60) = \Pr(X = 60) + \Pr(X = 61) + \text{L} + \Pr(X = 99) + \Pr(X \geq 100)$$

EXAMPLE 3

If a pair of dice is tossed 2880 times, what is the probability of obtaining between 450 and 500 doubles, inclusive? (In this section "between" means to include the first and last numbers.)

Solution. We have $n = 2880$. The probability of tossing doubles on one toss is 6/36, so $p = 1/6$ and $q = 5/6$. Then $\mu = np = 480$and $\sigma = \sqrt{npq} = 20$. The problem is to evaluate Pr $(450 \leq X \leq 500)$.

Set $x_1 = 449.5$ and $x_2 = 500.5$. Then

$$z_2 = \frac{449.5 - 480}{20} \approx -1.53 \text{ and } z_2 = \frac{500.5 - 480}{20} \approx 1.03$$

The area between z_1 and z_2 is $A(z_2) - A(z_1) = .8485 - .0630 = .7855$. Thus the probability of obtaining between 450 and 500 doubles is .7855.

Many situations involve either one outcome or another. We label each outcome a "success" or "failure." A "success" is merely the occurrence of the event in which we are interested, and a "failure" is the nonoccurrence of that event. A "success" does not have to be something "good," nor does a "failure" have to be "bad."

Consider the cause of accidents. Fatigue causes a certain percent p of accidents depending upon the particular occupation. An accident is either due to fatigue or is not, so it represents a Bernoulli trial. If an accident is due to fatigue we label it a "success" and assign it the probability p, even though in the human arena an accident caused by fatigue could hardly be called a success.

Example 4 considers industrial accidents.

EXAMPLE 4

Fatigue is the cause of 20% of all industrial accidents. Last year there were 2500 accidents in the steel industry. What is the probability that at least 525 of the accidents were caused by fatigue?

Solution. Let X represent the number of accidents caused by fatigue, then p, the probability that an accident was caused by fatigue, is 0.2, and n the total number of accidents, is 2500. Therefore,

$$\mu = (2500)(0.2) = 500 \quad \text{and} \quad \sigma = \sqrt{(2500)(0.2)(0.8)} = \sqrt{400} = 20$$

The problem is to find Pr (X ≥ 525), and the desired region is sketched in Figure 6–21. Set $x = 524.5$. Then

$$z = \frac{524.5 - 500}{20} \approx 1.23$$

From the table in the Appendix, A(1.23) = .8907. The desired area lies to the right of $z = 1.23$, so

$$\Pr(X \geq 525) = 1 - 0.8907 = .1093$$

Therefore the probability of at least 525 of the accidents being caused by fatigue is about 11%.

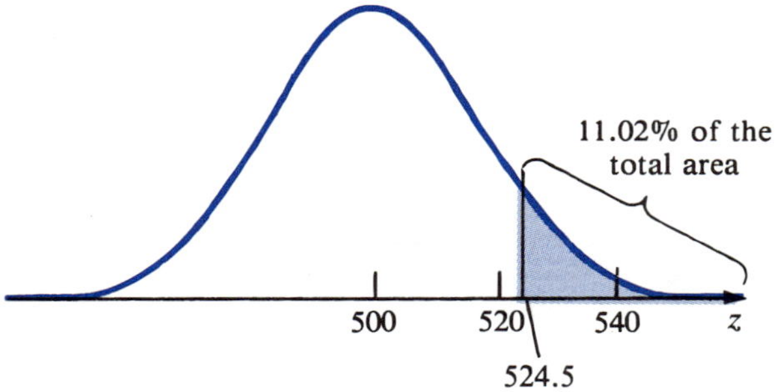

Figure 6–21

Example 5 shows how statistical studies allow some professions to prepare for certain events beforehand.

EXAMPLE 5

Next year 62,500 people will use a drug which has caused severe side effects (loss of weight, rash, depression) in 10% of all people who have already taken it. What is the probability that at least 6200 people will suffer side effects?

Solution. Since 10% of all people suffer side effects, $p = 0.1$, and since 62,500 people will use the drug, $n = 62{,}500$. Consequently, $q = 0.9$. Applying the formulas,

$$\mu = 6250 \text{ and } \sigma = 75$$

The problem is to compute Pr (X ≥ 6200). The desired region is drawn in Figure 6–22. Set $x = 6199.5$, and compute the corresponding z-score.

$$z = \frac{6199.5 - 6250}{75} = \frac{-50.5}{75} \approx -0.67$$

The table in the Appendix gives A(–0.67) = .2514. Therefore,

$$\Pr(X \geq 6200) = 1 - .2514 = .7486$$

Therefore there is approximately a 75% chance that at least 6200 people will suffer side effects.

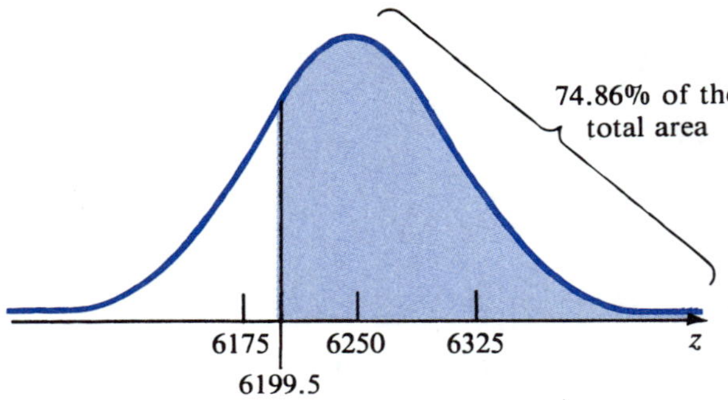

Figure 6–22

EXERCISE SET 6-5

In problems 1 to 4 toss n coins and record the number of heads. Find μ and σ.

1. $n = 144$
2. $n = 36$
3. $n = 81$
4. $n = 121$

In problems 5 to 8 a coin is tossed 100 times. Find the probability of each event.(a)

5. At least 55 heads.
6. At least 45 heads.
7. Between 45 and 50 heads.
8. At most 60 heads.

In problems 9 to 12 a coin is tossed 36 times. Find the probability of each event.

9. Less than 15 heads
10. At least 20 heads
11. Between 15 and 20 heads
12. More than 20 heads.

In problems 13 to 16 a pair of dice is tossed 2880 times. Find the probability of each event.

13. At least 450 doubles.
14. At least 510 doubles.
15. Between 500 and 540 doubles.
16. At most 490 doubles.

In problems 17 to 20 a pair of dice is tossed 180 times. Find the probability of each event.

17. At most 25 doubles.
18. At most 35 doubles.
19. Between 25 and 35 doubles.
20. At least 20 doubles.

Problems 21 and 22 refer to an industry where 20% of the accidents are caused by fatigue. Suppose that 900 accidents occur. Compute the probability of each event.

21. At least 150 accidents were caused by fatigue.
22. Between 162 and 192 accidents were caused by fatigue.

Problems 23 and 24 refer to a drug which causes severe side effects in 10% of the people who take it. Suppose that 100 people take the drug. Compute the probability of each event.

23. Between 5 and 15 people suffer severe side effects.
24. At least 12 people suffer severe side effects.

In problems 25 to 28 a coin is tossed 400 times. Find the probability of each event.

25. At most 205 heads.
26. At most 195 heads.
27. Between 195 and 220 heads.
28. At least 210 heads.

In problems 29 to 34 a survey of 1200 families with exactly two children is conducted. Find the probability of each event.

29. At most 315 of the families have two boys.
30. At most 290 of the families have two girls.
31. At least 310 of the families have two girls.
32. At least 285 of the families have two boys.
33. Between 270 and 330 of the families have two girls.
34. Between 300 and 345 of the families have two girls.
35. About 2% of a dentist's patients cancel their appointment. What is the probability that out of 10,000 patients, at least 180 appointments will be canceled?
36. A person who claims to be clairvoyant draws a card from a an ordinary deck of 52 cards, announces the suit without looking, and replaces the card before drawing another. In all, 48 cards are chosen. What is the probability that the "clairvoyant" will name the correct suit at least 18 times?
37. A jar contains a total of 12 jelly beans -3 red jelly beans and 9 black jelly beans. If 75 jelly beans are selected with replacement. What is the probability that between 15 and 20 are red?
38. A fluoride treatment is rated 98% effective. If it is applied in 100 cases, what is the probability that between 90 and 98 are effective?

39. A student known as a "partier" must guess at every answer on a 50-question multiple-choice test, which has five possible answers to every question. What is the probability that the student will obtain a grade of 40% or better?

40. A manufacturer of disks has found that only 1% of the disks are defective. What is the probability that a shipment of 1100 disks will contain fewer than 12 defective disks?

41. The Dow-Jones average will rise on a given day about 55% of the time. What is the probability that it will rise on at least 210 days during a regular year?

42. Of 1000 two-children families, what is the probability that 600 or more of them will have one boy and one girl?

43. A pair of dice is tossed 180 times and the number of times that doubles occur is recorded. What is the probability that doubles will land fewer than 27 times?

44. Toss one die 180 times and record the number of times that the number turns up. What is the probability that a 3 will land between 24 and 32 times?

45. If 100 people select a Ping-Pong ball from 100 containers of Ping-Pong balls, and each container holds 10 Ping-Pong balls numbered 0 through 9, what is the probability that the number of people who select the number 7 is between 9 and 11?

46. If 100 people select a Ping-Pong ball from 100 containers of Ping-Pong balls, and each container holds 10 Ping-Pong balls numbered 0 through 9, what is the probability that at least 12 of them will select the number 7?

Referenced Exercise Set 6-5

1. USAir knows from experience that 15% of the reservations will not be honored.[23]
   (a) The airline sells 160 tickets for a flight that has 148 seats. What is the probability that more than 148 people will show up?
   (b) What is the maximum number of tickets the airline should sell so that there is a 95% chance that everyone who shows up will get a seat?

2. In Dade County, Florida, the number of malpractice lawsuits against doctors doubled from the year 1975 to the year 1985. At present, 25% of all physicians in the state of Florida are sued annually.[24] If a hospital has 48 physicians on staff, what is the probability that at least 15 of them will be sued during a year?

3. When Hewlett-Packard Company (H-P) entered into a joint venture with Yokogawa Corporation in 1983 to produce circuit boards for computers, it found that boards produced at the Tokyo plant contained eight defects per million connections compared with the rate of six defects per thousand connections at the H-P plant in Cupertino, California.[25]

---

[23]We are grateful to Dennis Dombroski of USAir for supplying us with these figures.

[24]William F. Buckley Jr., "Malpractice Brings a Truce, of Sorts," Nationally Syndicated Column, April 4, 1988.

[25]John Schneidawind, "Computer Company Capitalizes on Japanese Methods," Knight-Ridder News Service, May 13, 1988.

(a) If a batch of 50 million connections is produced at the Tokyo plant, what is the probability that fewer than 460 connections will be defective?
(b) If a batch of 50 million connections is produced at the Cupertino plant, what is the probability that less than 298,360 connections will be defective?

4. In American jury trials, prospective jurors can be "peremptorily" challenged by the counsel for either side. Although the state of Florida ordinarily allows six peremptory challenges for each side, the judge in the Theodore Bundy trial permitted 20 because of the national publicity surrounding Bundy's brutal murder of two sorority sisters. A pretrial survey conducted by the Public Defender's office revealed that 40 percent of the eligible jurors felt that Bundy was guilty.[26] If 30 people are questioned, what is the probability that at least 12 of them did not already feel that Bundy was guilty?

Cumulative Exercise Set 6-5

1. Toss a coin three times and record the sequence of heads and tails. Let X be the number of heads recorded.
   (a) Construct the probability distribution.
   (b) Draw the histogram.

2. A set of data has five entries. Each entry is either 6, 7, or 8. What is the set if the mean is 7.8 and the median is 8?

3. A player rolls two dice and wins \$5 if doubles occur and loses \$1 if not. What is the expected value?

4. Calculate the standard deviation for the data in the table.

x	1	2	3	4	5
p	.2	.1	.4	.2	.1

5. A fair coin is tossed three times. The random variable X is the number of heads that appears.
   (a) What is the expected number of heads?
   (b) What is the variance?

6. Approximate the area under the normal curve between $z = 0$ and $z = 2$ without using a table.

7. What is the area under the normal curve between $z = -0.5$ and $z = 1.5$?

8. Suppose scores on an exam form a normal distribution with a mean of 920 and a standard deviation of 50. What percentage of scores are at least 820?

9. Roll a pair of dice 180 times and record the number of times they land doubles.
   (a) What is the expected value for this experiment?
   (b) What is the standard deviation for this experiment?

---

[26]LeRoy A. Franklin, "Bayes' Theorem, Binomial Probabilities, and Fair Numbers of Peremptory Challenges in Jury Trials," *College Mathematics Journal*, September 1987, pp. 291-299.

10. If a quarter is flipped 100 times, what is the probability that it will land heads at most 52 times?

11. If a pair of dice is tossed 180 times, what is the probability that the sum of 7 will occur at least 35 times?

12. A shelf holds 25 cookie jars, each of which contains 8 chocolate cookies and 2 vanilla cookies. If one cookie is selected from each jar, what is the probability that the number of vanilla cookies selected is between 4 and 6?

Multiple-Choice Exercise Set 6-5

1. How many possible outcomes does a Bernoulli trial consist of?
(a) 0 (b) 1 (c) 2 (d) 3 (e) any positive integer

2. How many Bernoulli trials does a binomial experiment consist of?
(a) 0 (b) 1 (c) 2 (d) 3 (e) any positive integer

3. If a binomial experiment consists of n Bernoulli trials, each with probability of success p, then the formula for the mean μ is
(a) np (b) $np(1-p)$ (c) $np(p-1)$ (d) $\sqrt{npq}$ (e) none of these

4. If a binomial experiment consists of n Bernoulli trials, each with probability of success p, then the formula for the standard deviation σ is
(a) np (b) $np(1-p)$ (c) $np(p-1)$ (d) $\sqrt{npq}$ (e) none of these

5. If a binomial experiment consists of 100 Bernoulli trials, each with probability of success 3, then the mean μ is
(a) 75 (b) 25 (c) $\sqrt{18.75}$ (d) 5/2 (e) none of these

6. If a binomial experiment consists of 100 Bernoulli trials, each with probability of success 3, then the standard deviation σ is
(a) 75 (b) 25 (c) $\sqrt{18.75}$ (d) 5/2 (e) none of these

7. If a coin is tossed 144 times, what is the mean number of heads that will land?
(a) 36 (b) 6 (c) 12 (d) 72 (e) none of these

8. If a coin is tossed 144 times and the number of heads recorded, what is the standard deviation?
(a) 36 (b) 6 (c) 12 (d) 72 (e) none of these

In problems 9 to 11 a normal curve has mean $\mu = 100$ and standard deviation $\sigma = 10$.

9. What value of x is used to find the probability that at least 95 will occur?
(a) 94 (b) 94.5 (c) 95 (d) 95.5 (e) none of these

10. What value of x is used to find the probability that at most 97 will occur?
(a) 96 (b) 96.5 (c) 97 (d) 97.5 (e) none of these

11. What values of x are used to find the probability that between 98 and 103 will occur?
(a) 97.5 and 103.5 (b) 97.5 and 102.5 (c) 98.5 and 102.5
(d) 98.5 and 103.5 (e) none of these

12. What is the probability that a coin lands heads less than 40 times in 100 tosses?
(a) 0.0227 (b) 0.0179 (c) 0.0256 (d) 0.0287 (e) none of these

13. What is the probability that a coin lands heads more than 55 times in 100 tosses?
(a) 0.1587 (b) 0.1401 (c) 0.8159 (d) 0.8599 (e) none of these

In problems 14 to 19 a pair of dice is tossed 2880 times.

14. The probability of tossing doubles is
(a) 1/36 (b) 5/36 (c) 7/36 (d) 35/36 (e) none of these

15. The mean number of doubles is
(a) 475 (b) 425 (c) 375 (d) 325 (e) none of these

16. The standard deviation for the number of doubles is
(a) 480 (b) 360 (c) 25 (d) 20 (e) none of these

17. The probability of tossing at least 460 doubles is
(a) 0.1527 (b) 0.1587 (c) 0.8473 (d) 0.8413 (e) none of these

18. The probability of tossing fewer than 500 doubles is
(a) 0.1648 (b) 0.1527 (c) 0.8352 (d) 0.8473 (e) none of these

19. The probability of tossing between 500 and 520 doubles is
(a) 0.1433 (b) 0.1407 (c) 0.1286 (d) 0.1312 (e) none of these

20. A shelf in a store holds 48 candy jars, each with 3 red jelly beans and 9 black jelly beans. A jelly bean is selected at random from each jar. What is the probability that at least 15 red jelly beans are selected?
(a) 0.1587 (b) 0.2023 (c) 0.7977 (d) 0.8413 (e) none of these

21. A card is selected from a deck 1200 times, its suit is recorded each time. What is the probability that no more than 285 hearts are selected?
(a) 0.1507 (b)0.1587 (c) 0.8493 (d) 0.1668 (e) none of these

CASE STUDY
THE FEDERALIST PAPERS

Let us recall some events from the American Revolution. In the 1760s the British government adopted the policy of taxing the American colonies. The most famous and most important of these measures was the Stamp Act, which was later repealed and replaced by the Townshend Acts, both of which provided taxes to maintain the British army in America. This taxation led to the Boston Tea Party, and the decision by King George III that the colonies should be subdued.

Revolt followed.

Hostilities broke out in April 1775, and on July 4 of the following year the delegates to the second Continental Congress approved the declaration that "These United Colonies are, and of right ought to be, free and independent states." The war came to an official end in 1783 with the Treaty of Paris, in which Britain recognized the independence of the United States.

That sequence of events answered one of the two great questions from the revolutionary era: how should power be distributed between Britain and America? But another question remained unresolved: how should power be divided among the Americans? It is instructive to think not of *the* revolution, but of two revolutions, one external and the other internal. We just outlined the external revolution; now let us describe the internal revolution.

The Continental Congress functioned as a national government until 1781, when the Articles of Confederation, the first national constitution, were adopted. However, the Articles contained one major defect: they did not provide for a strong central government. This is understandable. American heads of state had no intention of replacing one form of despotism by another. Consequently the central government's lack of power to regulate commerce caused grave commercial problems, and its lack of power to tax led to a financial disaster. The situation was so serious that historians have come to call the period from 1783 to 1787 not *a* critical period but *the* critical period in American history.

A constitutional convention was convened in hot, steamy Philadelphia from May to September of 1787. There the delegates hammered out a document that would prove to be the cornerstone of our government B the Constitution. The 13 colonies were much less united than the 50 states are today, yet politicians from all ideologies recognized the importance of the passage of the Constitution. *The Federalist* papers were published by Alexander Hamilton, John Jay, and James Madison to persuade the citizens of New York State to vote for ratification. They have earned an important niche in political philosophy.

A problem arose afterwards because *The Federalist* papers had been published anonymously over the signature "Publius." Of the 85 papers there was general agreement that Hamilton wrote 51, Madison wrote 14, and Jay wrote 5. Three were of joint authorship. The remaining 12 were written by either Madison or Hamilton, but it was not clear which of these distinguished statesmen wrote them.

Madison or Hamilton B who was the author of the disputed *Federalist* papers? Even though they held opposing political views, neither would claim the honor in deference to the greater good of convincing the New York electorate to vote for ratification of the new Constitution. The usual means for determining authorship by political content and writing style failed to solve the problem. Besides, Madison and Hamilton consciously adopted and perfected a common legal writing style. Thus the usual linguistic methods of studying sentence length and word usage shed no further light on the problem. The papers became known as the "disputed papers."

The first conclusive evidence was provided nearly two centuries later in a historic study by Mosteller and Wallace[27]. It was based on certain words, called *marker words*, which are exceptionally common to one author's style and not the other's. However, the use of marker words presents two big difficulties: finding them, and proving that the difference in their usage is significant. The nagging uncertainty never disappears, as perfect markers can never be found, but sometimes a nearly conclusive case can be made.

The question of the authorship of *The Federalist* papers forged an important link between statistics and history. The standard deviation was the key element. We present the data gathered by Mosteller and Wallace for the marker word "upon" from 50 essays written by Madison, 48 essays written by Hamilton, and the 12 disputed papers.

$$\bar{x} = \frac{\Sigma xf}{\Sigma f} = \frac{160}{48} \approx 3.3 \qquad s = \sqrt{\frac{\Sigma x^2 - (\Sigma xf)^2 / \Sigma f}{\Sigma f - 1}} = \sqrt{\frac{634 - (160)^2 / 48}{47}} \approx 1.46$$

Table 6–24 and Table 6–25 present the distribution for the essays by Hamilton and Madison, respectively, while Table 6–26 gives the distribution for the 12 disputed papers.

Rate per 1000 Words Interval	Midpoint, x_i	No. of Papers, f_i	$f_i x_i$	x_i^2	$f x_i^2$
0.5-1.5	1	5	5	1	5
1.5-2.5	2	10	20	4	40
2.5-3.5	3	12	36	9	108
3.5-4.5	4	10	40	16	160
4.5-5.5	5	8	40	25	200
5.5-6.5	6	2	12	36	72
6.5-7.5	7	1	7	49	49
	Totals	48	160	140	634

$$\bar{x} = \frac{\Sigma xf}{\Sigma f} = \frac{160}{48} \approx 3.3$$

$$s = \sqrt{\frac{\Sigma x^2 - (\Sigma xf)^2 / \Sigma f}{\Sigma f - 1}} = \sqrt{\frac{634 - (160)^2 / 48}{47}} \approx 1.46$$

Table 6–24 Frequency Distribution of Rate per 1000 Words in 48 Hamilton Papers for the Word "upon"

[27]Frederick Mosteller and David L. Wallace, *Inference and Disputed Authorship: The Federalist Papers*, Reading, MA, Addison-Wesley, 1964. This study was updated in a second edition of the book under the title *Applied Bayesian and Classical Inference: The Case of The Federalist Papers*, New York, Spring-Verlag, 1984.

Rate per 1000 Words Interval	Midpoint, x_i	No. of Papers, f_i	$f_i x_i$	x_i^2	$f_i x_i^2$
0.0-0.4	0.2	41	8.2	0.04	1.64
0.4-0.8	0.6	2	1.2	0.36	0.72
0.8-1.2	1.0	4	4.0	1.00	4.00
1.2-1.6	1.4	1	1.4	1.96	1.96
1.6-2.0	1.8	2	3.6	3.24	6.48
	Totals	50	18.4	6.60	14.80

$$\bar{x} = \frac{\Sigma xf}{\Sigma f} = \frac{18.4}{50} = 0.368$$

$$s = \sqrt{\frac{\Sigma x^2 - (\Sigma xf)^2 / \Sigma f}{\Sigma f - 1}} = \sqrt{\frac{14.8 - (18.4)^2 / 50}{49}} \approx 0.40$$

Table 6–25 Frequency Distribution of Rate per 1000 Words in 50 Madison Papers for the Word "upon"

Rate per 1000 Words Interval	Midpoint, x_i	No. of Papers, f_i	$f_i x_i$	x_i^2	$f_i x_i^2$
0.0-0.4	0.2	11	2.2	0.04	0.44
0.4-0.8	0.6	0	0.0	0.36	0.00
0.8-1.2	1.0	0	0.0	1.00	0.00
1.2-1.6	1.4	1	1.4	1.96	1.96
		12	3.6	3.36	2.40

$$\bar{x} = \frac{\Sigma xf}{\Sigma f} = \frac{3.6}{12} = 0.3$$

$$s = \sqrt{\frac{\Sigma x^2 - (\Sigma xf)^2 / \Sigma f}{\Sigma f - 1}} = \sqrt{\frac{2.40 - (3.6)^2 / 12}{11}} \approx 0.35$$

Table 6–26 Frequency Distribution of Rate per 1000 Words in 12 Disputed Papers for the Word "upon"

The means and standard deviations are summarized in Table 6–27. They show that the usage of the marker word "upon" in the disputed papers fits the usage in the Madison essays much more closely than it fits the Hamilton essays. The same conclusion was obtained for many other marker words. The inference from the Mosteller and Wallace study is clear: Madison is the author of the disputed *Federalist* papers.

	Hamilton	Madison	Disputed
Mean	3.33	0.37	0.30
Standard Deviation	1.45	0.40	0.33

Table 6–27 Mean and Standard Deviation for 48 Hamilton, 50 Madison, and 12 Disputed Papers for the Word "upon"

CASE STUDY EXERCISE SET

Problems 1 to 10 provide the data for ten marker words in *The Federalist* Papers. The problems are adapted from Table 4.8-3 of the 1984 edition of the Mosteller and Wallace book, pp. 167-172. H stands for Hamilton, M for Madison, and D for disputed. The intervals refer to the rate per 1000 words, and the numbers in the other columns give the number of papers having the stated rate. (There were 48 papers of Hamilton, 50 of Madison, and 12 disputed.) For each marker word, calculate the mean and standard deviation for of the three authors.

1. marker word: *also*

Interval	H	M	D
0 - 0.4	28	18	0
0.4 - 0.8	13	11	8
0.8 - 1.2	5	16	3
1.2 - 1.6	2	1	0
1.6 - 2.0	0	3	1
3.2 - 3.6	0	1	0

2. marker word: *an*

Interval	H	M	D
0 - 2	1	1	2
2 - 4	7	20	5
4 - 6	20	19	4
6 - 8	15	8	1
8 - 10	4	2	0
10 - 12	0	0	0
12 - 14	1	0	

3. marker word: *by*

Interval	H	M	D
0 - 2	1	0	0
2 - 4	3	0	8
4 - 6	14	4	3
6 - 8	12	2	0
8 - 10	12	11	1
10 - 12	6	11	0
12 - 14	0	12	0
14 - 16	0	4	0
16 - 18	0	4	0
18 - 20	0	2	0

4. marker word: *of*

Interval	H	M	D
32 - 36	0	1	0
44 - 48	0	1	0
48 - 52	2	4	0
52 - 56	2	10	3
56 - 60	13	16	7
60 - 64	10	8	0
64 - 68	8	7	1
68 - 72	3	3	1
72 - 76	6	0	0
76 - 80	4	0	0

5. marker word: *on*

Interval	H	M	D
0 - 2	9	0	0
2 - 4	23	2	1
4 - 6	13	14	1
6 - 8	2	14	2
8 - 10	1	9	4
10 - 12	0	8	4
12 - 14	0	2	0
14 - 16	0	1	0

6. marker word: *there*

Interval	H	M	D
0 - 1	2	24	6
1 - 2	8	16	3
2 - 3	12	5	3
3 - 4	10	4	0
4 - 5	10	1	0
5 - 6	4	0	0
6 - 7	1	0	0
9 - 10	1	0	0

7. marker word: *this*

Interval	H	M	D
0 - 2	1	2	0
2 - 4	1	8	1
4 - 6	8	13	3
6 - 8	14	18	5
8 - 10	19	9	2
10 - 12	5	0	1

8. marker word: *to*

Interval	H	M	D
23 - 26	0	3	3
26 - 29	2	2	2
29 - 32	2	11	1
32 - 35	4	11	3
35 - 38	6	7	2
38 - 41	10	7	1
41 - 44	8	6	0
44 - 47	10	1	0
47 - 50	3	2	0
50 - 53	1	0	0
53 - 56	1	0	0
56 - 59	1	0	0

9. marker word: *both*

Interval	H	M	D
0 - 0.4	23	10	3
0.4 - 0.8	10	12	4
0.8 - 1.2	11	10	2
1.2 - 1.6	3	6	3
1.6 - 2.0	1	6	0
2.0 - 2.4	0	1	0
2.4 - 2.8	0	2	0
2.8 - 3.2	0	1	0
3.2 - 3.6	0	2	0

10. marker word: *though*

Interval	H	M	D
0 - 0.4	9	21	8
0.4 - 0.8	13	22	3
0.8 - 1.2	12	5	1
1.2 - 1.6	6	1	0
1.6 - 2.0	5	0	0
2.0 - 2.4	1	0	0
2.4 - 2.8	2	0	0
2.8 - 3.2	0	0	0
3.2 - 3.6	0	1	0

CHAPTER REVIEW

Key Terms

6-1 Frequency & Probability Distributions

Frequency Distribution
Frequency
Probability Distribution
Histogram
Random Variable
Relative Frequency
Uniform Distribution
Normal Distribution

6-2 Measures of Central Tendency

Mean
Median
Expected Value

6-3 Standard Deviation

Variance
Standard Variation
Chebyshev's Theorem

6-4 The Normal Curve

Normal Curve
z-score
Standard Normal Curve
Normal Approximation

6-5 The Normal Approximation to Binomial Experiments

Bernoulli Trial
Binomial Experiment

REVIEW EXERCISE SET

1. If four coins are tossed, what is the probability that at most 1 head is observed?

Problems 2 to 4 refer to the probability experiment of tossing two dice and letting X be the sum of the top faces.

2. Draw the histogram.
3. Construct the probability distribution.
4. Find Pr $(3 \le X \le 7)$.

Problems 5 to 7 refer to the probability experiment of choosing three cards from an ordinary deck without replacement and letting X be the number of clubs selected.

5. Construct the probability distribution.
6. Draw the histogram.
7. Find $\Pr(0 \le X \le 2)$.

8. Find (a) the mean and (b) the median using the midpoints of the intervals as the data given in the following table.

interval	0-2	2-4	4-6	6-8	8-10
f	3	1	3	2	1

Problems 9 and 10 refer to the frequency distribution in the table.

x	1	2	3	4	5
f	6	7	2	3	2

9. Find the (a) mean and (b) median.

10. Find the (a) variance and (b) standard deviation.

Problems 11 and 12 refer to the probability distribution in the table.

x	–2	–1	1	2
p	1/8	1/2	1/4	1/8

11. Find the expected value.

12. Find (a) the variance and (b) the standard deviation.

13. Ten employees in a company have a mean salary of $15,000.
    (a) If another person is hired at a salary of $15,000, what is the mean salary of the 11 employees?
    (b) If a twelfth person is hired at a salary of $50,000, what is the mean salary of the 12 employees?

14. A leap year has 4 months with 30 days, 7 months with 31 days, and 1 month (February) with 29 days.
    (a) Find the mean number of days in a month during a leap year.
    (b) Find the standard deviation of the number of days in a month during a leap year.

15. A college wishes to estimate the average cost of books per student during the year. Compute the average cost and the standard deviation from the table.

	Class			
	freshman	sophomore	junior	senior
percent of all books	35%	25%	21%	19%
cost of books	$100	$160	$180	$190

In problems 16 and 17 find the area under the normal curve within the given limits.

16. (a) Between $z = -2$ and $z = 2$.
    (b) To the right of $z = 2$.

17. (a) Between $z = -0.73$ and $z = 1.44$.
    (b) To the right of $z = -1.05$.

In problems 18 and 19 suppose that the scores on a particular test form a normal distribution with $\mu = 470$ and $\sigma = 20$.

18. What percentage of scores are between 430 and 480?

19. If a student is selected at random, what is the probability that his or her score is at least 500?

20. If a pair of dice is tossed 180 times, what is the probability of obtaining a sum of 7 at least 25 times?

Problems 21 to 24 refer to a binomial experiment consisting of 18 trials, each with a probability of success equal to 2/3.

21. Compute the probability of obtaining exactly 14 successes.

22. What is the mean number of successes?

23. What is the standard deviation?

24. Compute the probability of obtaining at least 14 successes.

Problems 25 to 27 refer to the probability experiment of drawing one card from each of 4800 decks of ordinary playing cards and recording the number of hearts drawn.

25. What is the expected value?

26. What is the standard deviation?

27. What is the probability that more than 1150 hearts will be drawn?

APPENDIX: Areas Under the Standard Normal Curve

z	A(z)	z	A(z)	z	A(z)
-3.00	0.0013	-2.46	0.0069	-1.92	0.0274
-2.99	0.0014	-2.45	0.0071	-1.91	0.0281
-2.98	0.0014	-2.44	0.0073	-1.90	0.0287
-2.97	0.0015	-2.43	0.0075	-1.89	0.0294
-2.96	0.0015	-2.42	0.0078	-1.88	0.0301
-2.95	0.0016	-2.41	0.0080	-1.87	0.0307
-2.94	0.0016	-2.40	0.0082	-1.86	0.0314
-2.93	0.0017	-2.39	0.0084	-1.85	0.0322
-2.92	0.0018	-2.38	0.0087	-1.84	0.0329
-2.91	0.0018	-2.37	0.0089	-1.83	0.0336
-2.90	0.0019	-2.36	0.0091	-1.82	0.0344
-2.89	0.0019	-2.35	0.0094	-1.81	0.0351
-2.88	0.0020	-2.34	0.0096	-1.80	0.0359
-2.87	0.0021	-2.33	0.0099	-1.79	0.0367
-2.86	0.0021	-2.32	0.0102	-1.78	0.0375
-2.85	0.0022	-2.31	0.0104	-1.77	0.0384
-2.84	0.0023	-2.30	0.0107	-1.76	0.0392
-2.83	0.0023	-2.29	0.0110	-1.75	0.0401
-2.82	0.0024	-2.28	0.0113	-1.74	0.0409
-2.81	0.0025	-2.27	0.0116	-1.73	0.0418
-2.80	0.0026	-2.26	0.0119	-1.72	0.0427
-2.79	0.0026	-2.25	0.0122	-1.71	0.0436
-2.78	0.0027	-2.24	0.0125	-1.70	0.0446
-2.77	0.0028	-2.23	0.0129	-1.69	0.0455
-2.76	0.0029	-2.22	0.0132	-1.68	0.0465
-2.75	0.0030	-2.21	0.0136	-1.67	0.0475
-2.74	0.0031	-2.20	0.0139	-1.66	0.0485
-2.73	0.0032	-2.19	0.0143	-1.65	0.0495
-2.72	0.0033	-2.18	0.0146	-1.64	0.0505
-2.71	0.0034	-2.17	0.0150	-1.63	0.0516
-2.70	0.0035	-2.16	0.0154	-1.62	0.0526
-2.69	0.0036	-2.15	0.0158	-1.61	0.0537
-2.68	0.0037	-2.14	0.0162	-1.60	0.0548
-2.67	0.0038	-2.13	0.0166	-1.59	0.0559
-2.66	0.0039	-2.12	0.0170	-1.58	0.0571
-2.65	0.0040	-2.11	0.0174	-1.57	0.0582
-2.64	0.0041	-2.10	0.0179	-1.56	0.0594
-2.63	0.0043	-2.09	0.0183	-1.55	0.0606
-2.62	0.0044	-2.08	0.0188	-1.54	0.0618
-2.61	0.0045	-2.07	0.0192	-1.53	0.0630
-2.60	0.0047	-2.06	0.0197	-1.52	0.0643
-2.59	0.0048	-2.05	0.0202	-1.51	0.0655
-2.58	0.0049	-2.04	0.0207	-1.50	0.0668
-2.57	0.0051	-2.03	0.0212	-1.49	0.0681
-2.56	0.0052	-2.02	0.0217	-1.48	0.0694
-2.55	0.0054	-2.01	0.0222	-1.47	0.0708
-2.54	0.0055	-2.00	0.0228	-1.46	0.0721
-2.53	0.0057	-1.99	0.0233	-1.45	0.0735
-2.52	0.0059	-1.98	0.0239	-1.44	0.0749
-2.51	0.0060	-1.97	0.0244	-1.43	0.0764
-2.50	0.0062	-1.96	0.0250	-1.42	0.0778
-2.49	0.0064	-1.95	0.0256	-1.41	0.0793
-2.48	0.0066	-1.94	0.0262	-1.40	0.0808
-2.47	0.0068	-1.93	0.0268	-1.39	0.0823

z	A(z)	z	A(z)	z	A(z)
-1.38	0.0838	-0.82	0.2061	-0.26	0.3974
-1.37	0.0853	-0.81	0.2090	-0.25	0.4013
-1.36	0.0869	-0.80	0.2119	-0.24	0.4052
-1.35	0.0885	-0.79	0.2148	-0.23	0.4090
-1.34	0.0901	-0.78	0.2177	-0.22	0.4129
-1.33	0.0918	-0.77	0.2206	-0.21	0.4168
-1.32	0.0934	-0.76	0.2236	-0.20	0.4207
-1.31	0.0951	-0.75	0.2266	-0.19	0.4247
-1.30	0.0968	-0.74	0.2296	-0.18	0.4286
-1.29	0.0985	-0.73	0.2327	-0.17	0.4325
-1.28	0.1003	-0.72	0.2358	-0.16	0.4364
-1.27	0.1020	-0.71	0.2389	-0.15	0.4404
-1.26	0.1038	-0.70	0.2420	-0.14	0.4443
-1.25	0.1056	-0.69	0.2451	-0.13	0.4483
-1.24	0.1075	-0.68	0.2483	-0.12	0.4522
-1.23	0.1093	-0.67	0.2514	-0.11	0.4562
-1.22	0.1112	-0.66	0.2546	-0.10	0.4602
-1.21	0.1131	-0.65	0.2578	-0.09	0.4641
-1.20	0.1151	-0.64	0.2611	-0.08	0.4681
-1.19	0.1170	-0.63	0.2643	-0.07	0.4721
-1.18	0.1190	-0.62	0.2676	-0.06	0.4761
-1.17	0.1210	-0.61	0.2709	-0.05	0.4801
-1.16	0.1230	-0.60	0.2743	-0.04	0.4840
-1.15	0.1251	-0.59	0.2776	-0.03	0.4880
-1.14	0.1271	-0.58	0.2810	-0.02	0.4920
-1.13	0.1292	-0.57	0.2843	-0.01	0.4960
-1.12	0.1314	-0.56	0.2877	0.00	0.5000
-1.11	0.1335	-0.55	0.2912	0.01	0.5040
-1.10	0.1357	-0.54	0.2946	0.02	0.5080
-1.09	0.1379	-0.53	0.2981	0.03	0.5120
-1.08	0.1401	-0.52	0.3015	0.04	0.5160
-1.07	0.1423	-0.51	0.3050	0.05	0.5199
-1.06	0.1446	-0.50	0.3085	0.06	0.5239
-1.05	0.1469	-0.49	0.3121	0.07	0.5279
-1.04	0.1492	-0.48	0.3156	0.08	0.5319
-1.03	0.1515	-0.47	0.3192	0.09	0.5359
-1.02	0.1539	-0.46	0.3228	0.10	0.5398
-1.01	0.1562	-0.45	0.3264	0.11	0.5438
-1.00	0.1587	-0.44	0.3300	0.12	0.5478
-0.99	0.1611	-0.43	0.3336	0.13	0.5517
-0.98	0.1635	-0.42	0.3372	0.14	0.5557
-0.97	0.1660	-0.41	0.3409	0.15	0.5596
-0.96	0.1685	-0.40	0.3446	0.16	0.5636
-0.95	0.1711	-0.39	0.3483	0.17	0.5675
-0.94	0.1736	-0.38	0.3520	0.18	0.5714
-0.93	0.1762	-0.37	0.3557	0.19	0.5753
-0.92	0.1788	-0.36	0.3594	0.20	0.5793
-0.91	0.1814	-0.35	0.3632	0.21	0.5832
-0.90	0.1841	-0.34	0.3669	0.22	0.5871
-0.89	0.1867	-0.33	0.3707	0.23	0.5910
-0.88	0.1894	-0.32	0.3745	0.24	0.5948
-0.87	0.1922	-0.31	0.3783	0.25	0.5987
-0.86	0.1949	-0.30	0.3821	0.26	0.6026
-0.85	0.1977	-0.29	0.3859	0.27	0.6064
-0.84	0.2005	-0.28	0.3897	0.28	0.6103
-0.83	0.2033	-0.27	0.3936	0.29	0.6141

z	A(z)	z	A(z)	z	A(z)
0.30	0.6179	0.86	0.8051	1.42	0.9222
0.31	0.6217	0.87	0.8078	1.43	0.9236
0.32	0.6255	0.88	0.8106	1.44	0.9251
0.33	0.6293	0.89	0.8133	1.45	0.9265
0.34	0.6331	0.90	0.8159	1.46	0.9279
0.35	0.6368	0.91	0.8186	1.47	0.9292
0.36	0.6406	0.92	0.8212	1.48	0.9306
0.37	0.6443	0.93	0.8238	1.49	0.9319
0.38	0.6480	0.94	0.8264	1.50	0.9332
0.39	0.6517	0.95	0.8289	1.51	0.9345
0.40	0.6554	0.96	0.8315	1.52	0.9357
0.41	0.6591	0.97	0.8340	1.53	0.9370
0.42	0.6628	0.98	0.8365	1.54	0.9382
0.43	0.6664	0.99	0.8389	1.55	0.9394
0.44	0.6700	1.00	0.8413	1.56	0.9406
0.45	0.6736	1.01	0.8438	1.57	0.9418
0.46	0.6772	1.02	0.8461	1.58	0.9429
0.47	0.6808	1.03	0.8485	1.59	0.9441
0.48	0.6844	1.04	0.8508	1.60	0.9452
0.49	0.6879	1.05	0.8531	1.61	0.9463
0.50	0.6915	1.06	0.8554	1.62	0.9474
0.51	0.6950	1.07	0.8577	1.63	0.9484
0.52	0.6985	1.08	0.8599	1.64	0.9495
0.53	0.7019	1.09	0.8621	1.65	0.9505
0.54	0.7054	1.10	0.8643	1.66	0.9515
0.55	0.7088	1.11	0.8665	1.67	0.9525
0.56	0.7123	1.12	0.8686	1.68	0.9535
0.57	0.7157	1.13	0.8708	1.69	0.9545
0.58	0.7190	1.14	0.8729	1.70	0.9554
0.59	0.7224	1.15	0.8749	1.71	0.9564
0.60	0.7257	1.16	0.8770	1.72	0.9573
0.61	0.7291	1.17	0.8790	1.73	0.9582
0.62	0.7324	1.18	0.8810	1.74	0.9591
0.63	0.7357	1.19	0.8830	1.75	0.9599
0.64	0.7389	1.20	0.8849	1.76	0.9608
0.65	0.7422	1.21	0.8869	1.77	0.9616
0.66	0.7454	1.22	0.8888	1.78	0.9625
0.67	0.7486	1.23	0.8907	1.79	0.9633
0.68	0.7517	1.24	0.8925	1.80	0.9641
0.69	0.7549	1.25	0.8944	1.81	0.9649
0.70	0.7580	1.26	0.8962	1.82	0.9656
0.71	0.7611	1.27	0.8980	1.83	0.9664
0.72	0.7642	1.28	0.8997	1.84	0.9671
0.73	0.7673	1.29	0.9015	1.85	0.9678
0.74	0.7704	1.30	0.9032	1.86	0.9686
0.75	0.7734	1.31	0.9049	1.87	0.9693
0.76	0.7764	1.32	0.9066	1.88	0.9699
0.77	0.7794	1.33	0.9082	1.89	0.9706
0.78	0.7823	1.34	0.9099	1.90	0.9713
0.79	0.7852	1.35	0.9115	1.91	0.9719
0.80	0.7881	1.36	0.9131	1.92	0.9726
0.81	0.7910	1.37	0.9147	1.93	0.9732
0.82	0.7939	1.38	0.9162	1.94	0.9738
0.83	0.7967	1.39	0.9177	1.95	0.9744
0.84	0.7995	1.40	0.9192	1.96	0.9750
0.85	0.8023	1.41	0.9207	1.97	0.9756

z	A(z)	z	A(z)	z	A(z)
1.98	0.9761	2.33	0.9901	2.67	0.9962
1.99	0.9767	2.34	0.9904	2.68	0.9963
2.00	0.9772	2.35	0.9906	2.69	0.9964
2.01	0.9778	2.36	0.9909	2.70	0.9965
2.02	0.9783	2.37	0.9911	2.71	0.9966
2.03	0.9788	2.38	0.9913	2.72	0.9967
2.04	0.9793	2.39	0.9916	2.73	0.9968
2.05	0.9798	2.40	0.9918	2.74	0.9969
2.06	0.9803	2.41	0.9920	2.75	0.9970
2.07	0.9808	2.42	0.9922	2.76	0.9971
2.08	0.9812	2.43	0.9925	2.77	0.9972
2.09	0.9817	2.44	0.9927	2.78	0.9973
2.10	0.9821	2.45	0.9929	2.79	0.9974
2.11	0.9826	2.46	0.9931	2.80	0.9974
2.12	0.9830	2.47	0.9932	2.81	0.9975
2.13	0.9834	2.48	0.9934	2.82	0.9976
2.14	0.9838	2.49	0.9936	2.83	0.9977
2.15	0.9842	2.50	0.9938	2.84	0.9977
2.16	0.9846	2.51	0.9940	2.85	0.9978
2.17	0.9850	2.52	0.9941	2.86	0.9979
2.18	0.9854	2.53	0.9943	2.87	0.9979
2.19	0.9857	2.54	0.9945	2.88	0.9980
2.20	0.9861	2.55	0.9946	2.89	0.9981
2.21	0.9864	2.56	0.9948	2.90	0.9981
2.22	0.9868	2.57	0.9949	2.91	0.9982
2.23	0.9871	2.58	0.9951	2.92	0.9982
2.24	0.9875	2.59	0.9952	2.93	0.9983
2.25	0.9878	2.60	0.9953	2.94	0.9984
2.26	0.9881	2.61	0.9955	2.95	0.9984
2.27	0.9884	2.62	0.9956	2.96	0.9985
2.28	0.9887	2.63	0.9957	2.97	0.9985
2.29	0.9890	2.64	0.9959	2.98	0.9986
2.30	0.9893	2.65	0.9960	2.99	0.9986
2.31	0.9896	2.66	0.9961	3.00	0.9987
2.32	0.9898				

Answers to the Odd Exercises

CHAPTER 1

Exercise Set 1-1

1.

x	-2	-1	0	1	2	3
y	-1	1	3	5	7	9
point	(-2, -1)	(-1, 1)	(0, 3)	(1, 5)	(2, 7)	(3, 9)

3.

x	-2	-1	0	1	2	3
y	3	2	1	0	-1	-2
point	(-2, 3)	(-1, 2)	(0, 1)	(1, 0)	(2, -1)	(3, -2)

5.

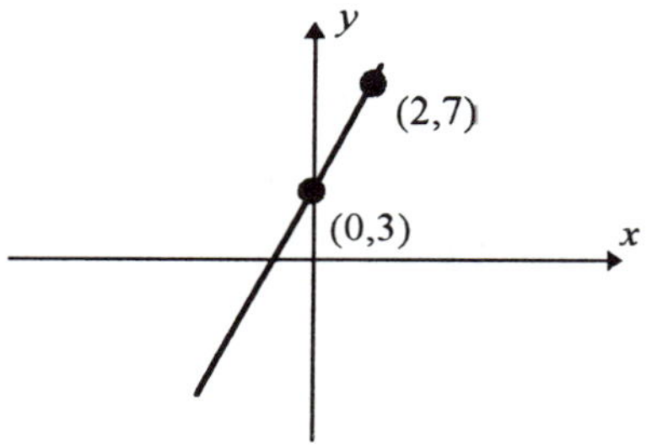

7.

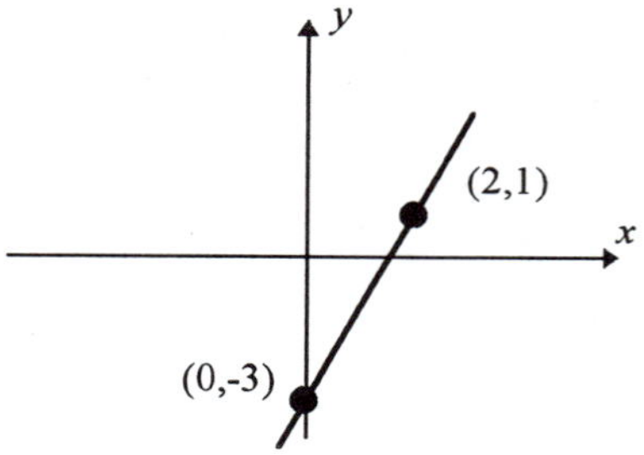

9.

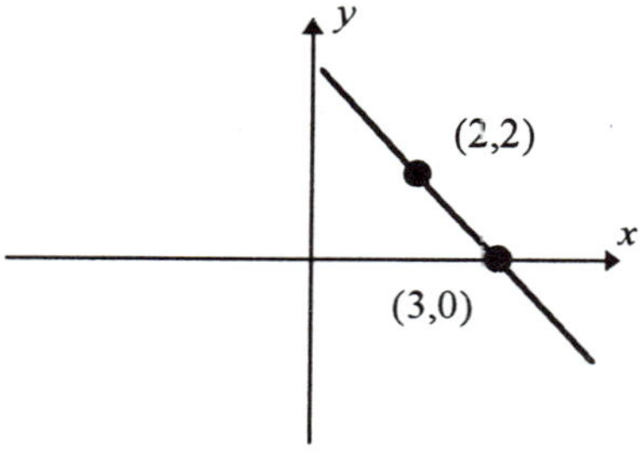

11.

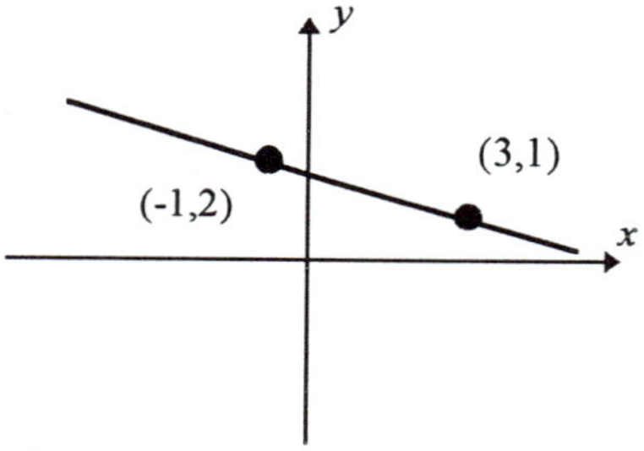

13.

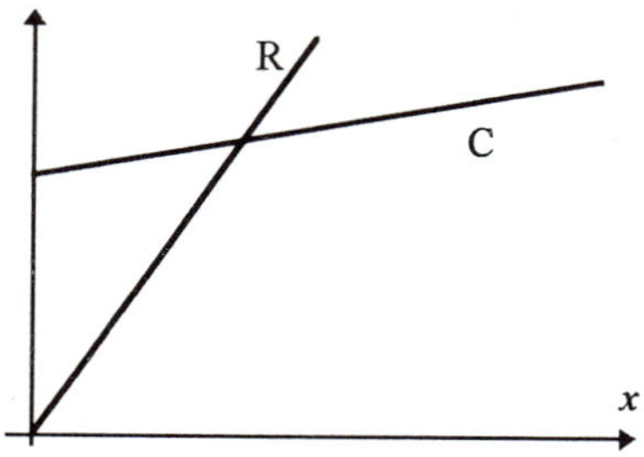

15.

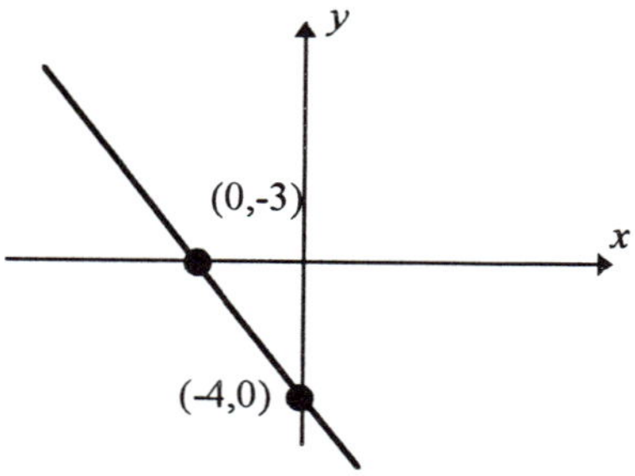

17.

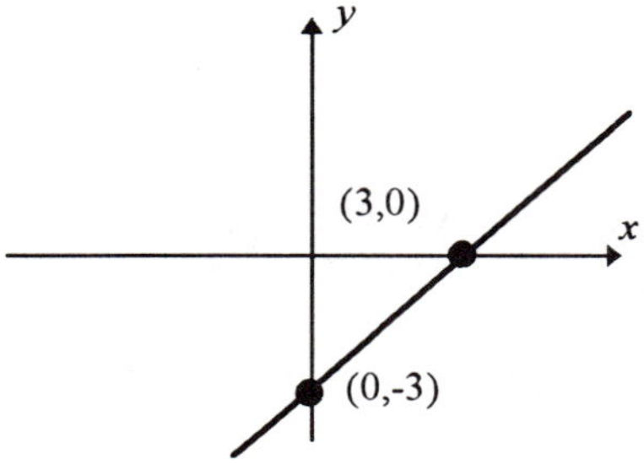

19. (a) $82,800, (b) 30%

21.

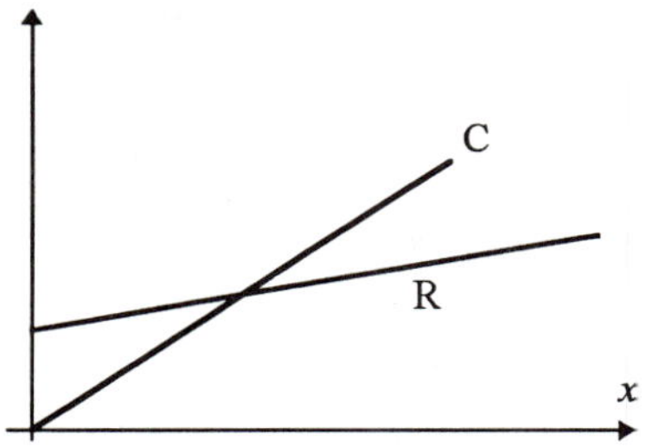

23.

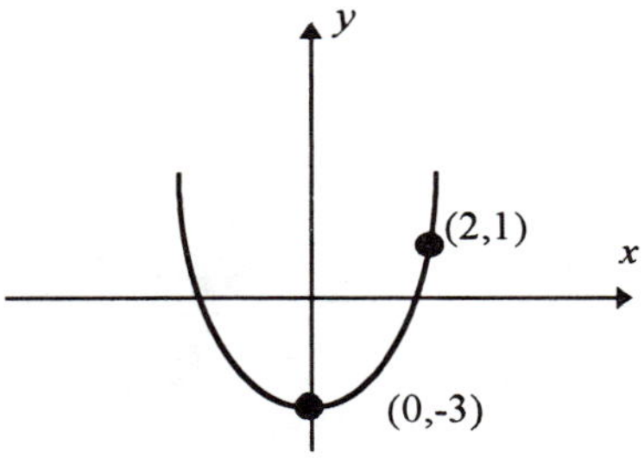

25.

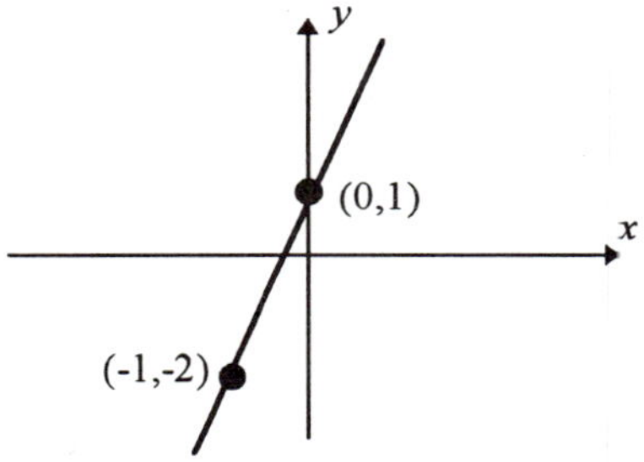

27.

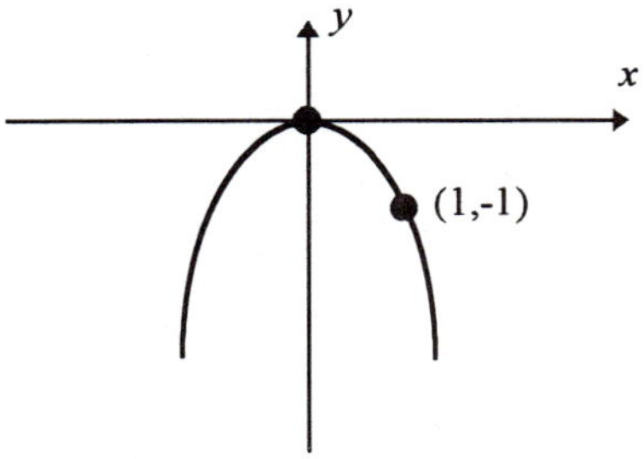

29.

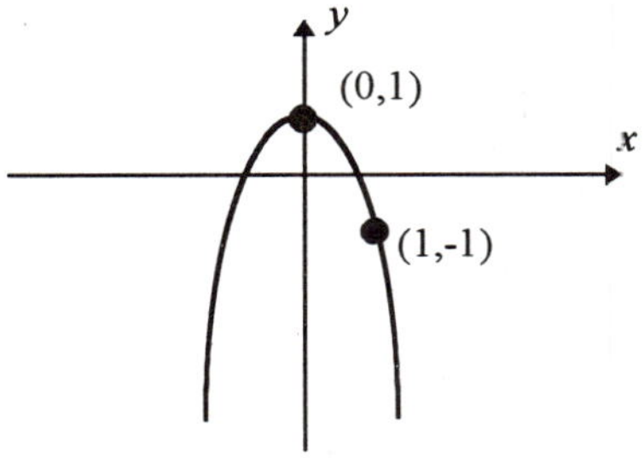

31.

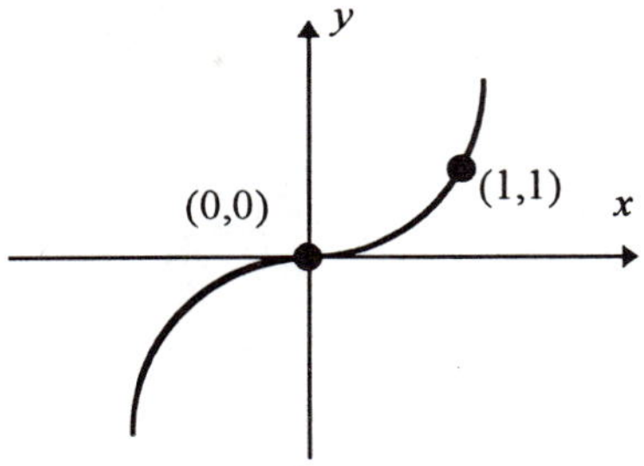

33.

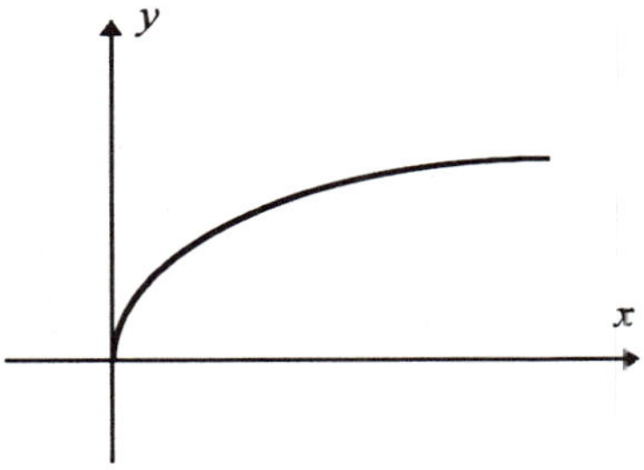

35.

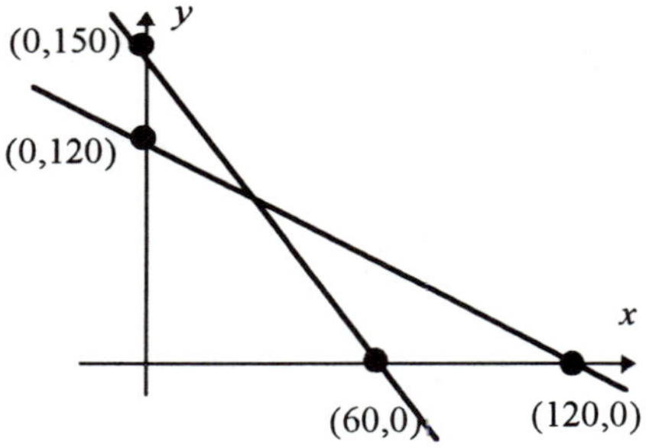

37.

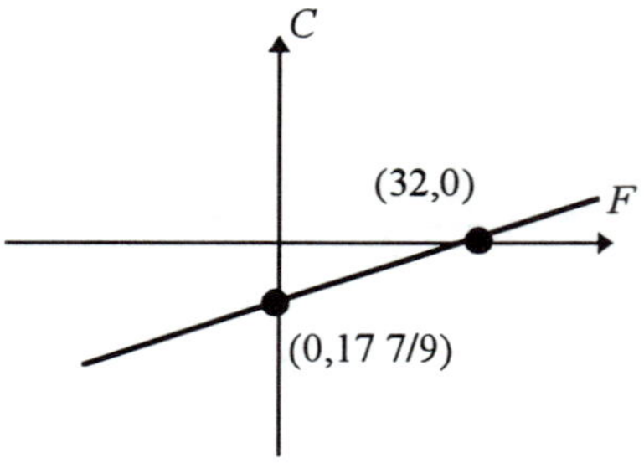

39. 37 degrees

41. $50,000

Referenced Exercise Set 1-1

1. (a) 72.4% (b) 1978

3. (a) 2 hours, (b) 3 hours, (c) 1 hr, 20 min

5. (a) T = 0.03x (b) T = 0.05x

Exercise Set 1-2

1.

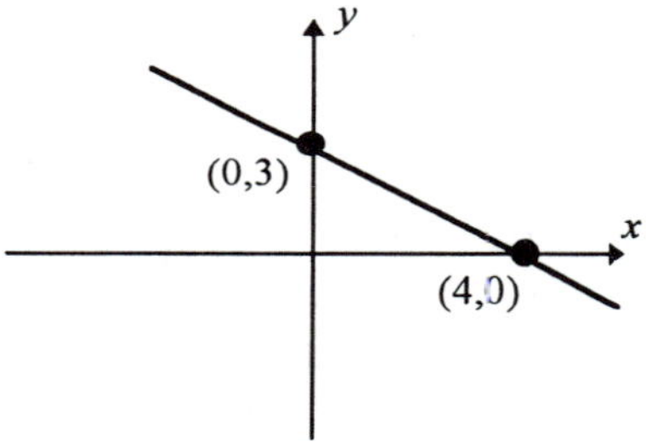

5.

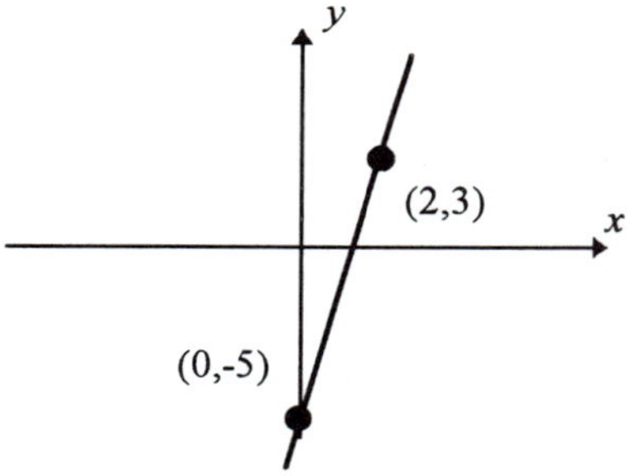

3.

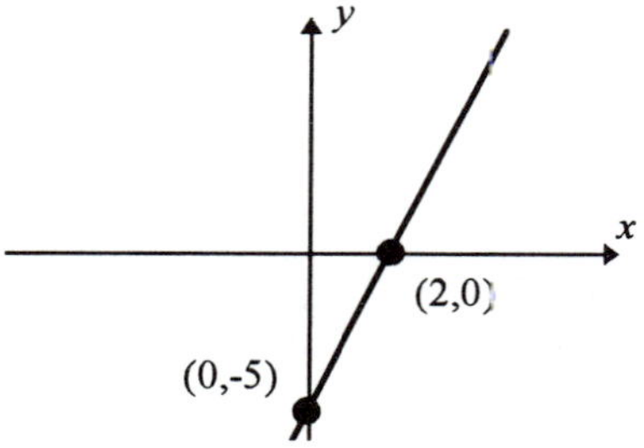

7.

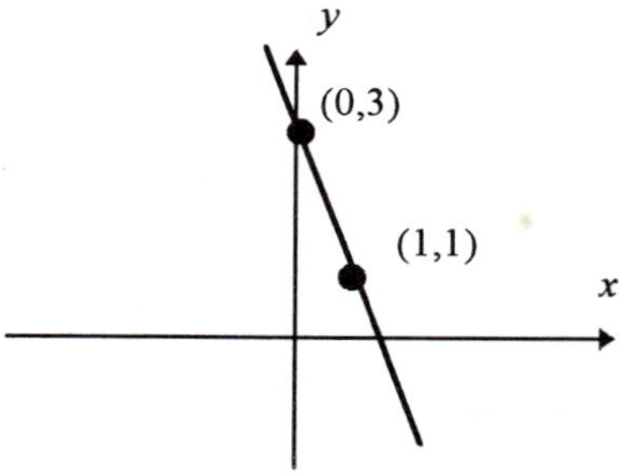

9. -3

11. 2

13. -6

15. 2

17. They are not parallel.

19. (a) undefined, (b) m =0

21. $y = 2x$

23. $y = -6x + 8$

25. $y = -7x + 13$

27. $y = 2x + 7$

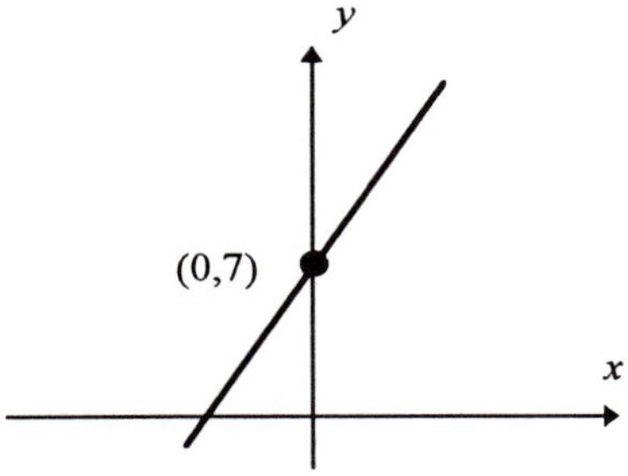

29. $y = -2x - 7$

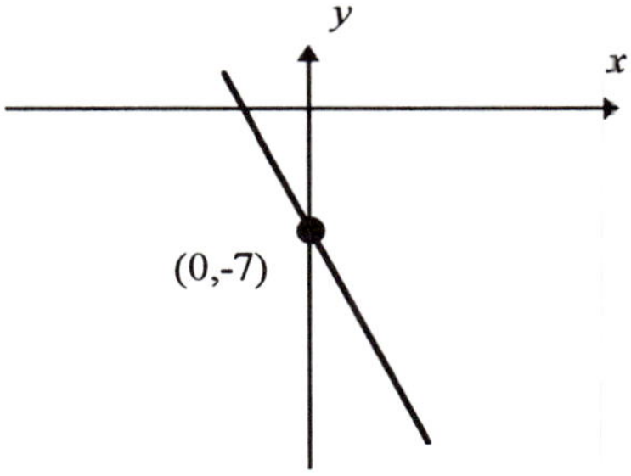

31. $y + 1 = -2(x - 6)$

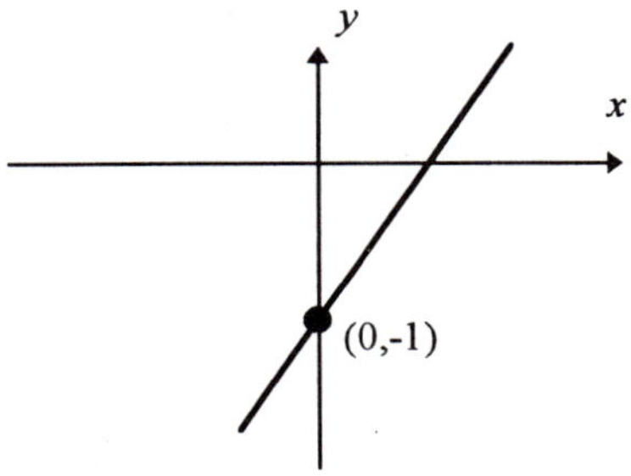

33. $y = 2x - 1$

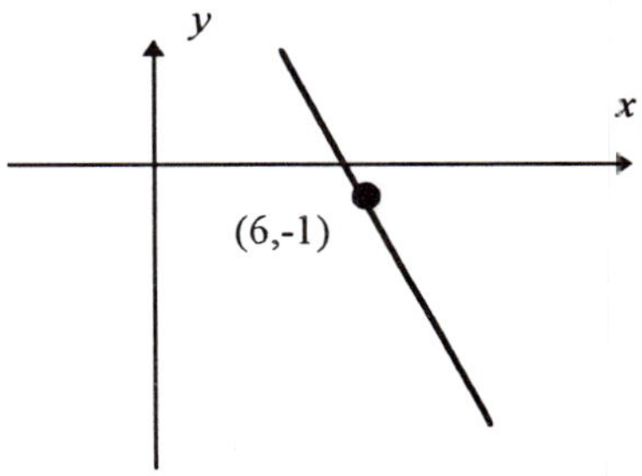

35. $y = x + 2$

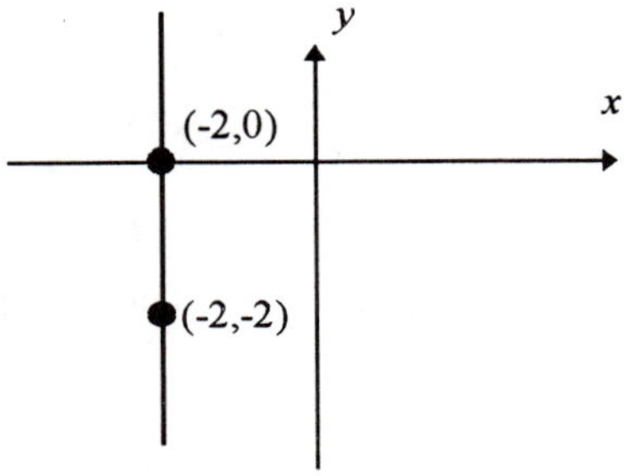

37. $x = -2$

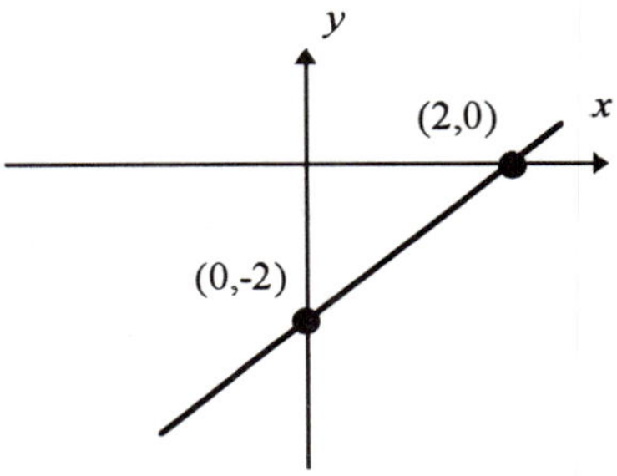

39. $y = -3$

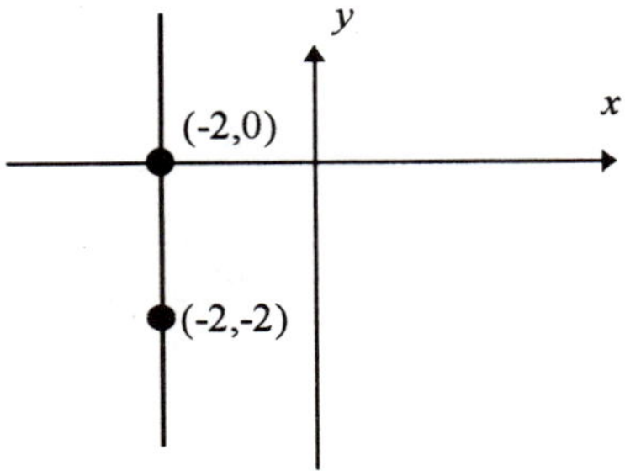

41. $x = -2$

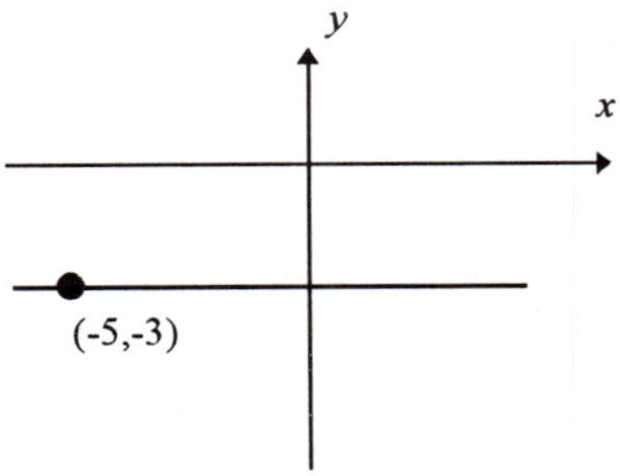

43. (a) $y = -\frac{3}{2}x + 4$, (b) $-\frac{3}{2}$, (c) 4

45. $y = -4x + 1$

47. (a)

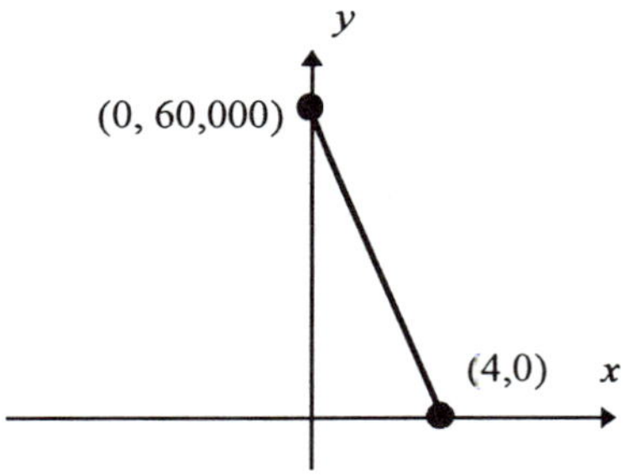

(b) 15,000, (c) Initial cost

49. (a) $y = 48643.3x + 17558072$, (b) 18,044,505
51. The population decreased from 1980 to 1990
53. No, for example, $xy = 1$

Referenced Exercise Set 1-2

1. (a) $y = 0.5x + 7.5$

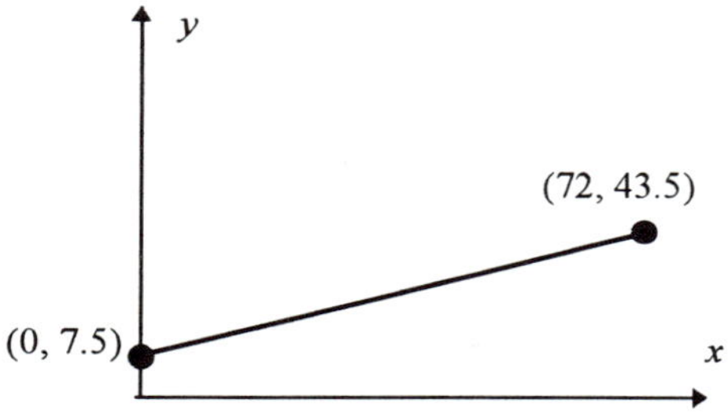

(c) 43.5

3. (a) $y = 0.2x + 15$, (b) $y = 0.1x + 10$, (c) $y = 0.1x + 3$
5. (a) During the first few minutes of running the percentage of energy supplied to the body by carbohydrates increases, (b) After a few minutes of running the percentage of energy supplied to the body by carbohydrates decreases.

Cumulative Exercise Set 1-2

1.

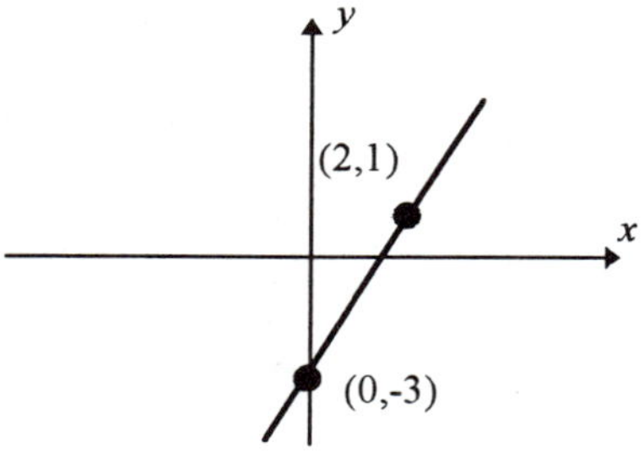

3.

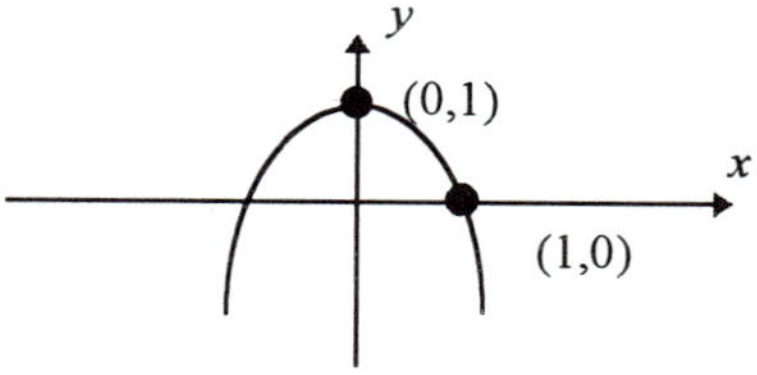

5. $y = -4x$

7. $2x - 47y = -20$

9. $27,000

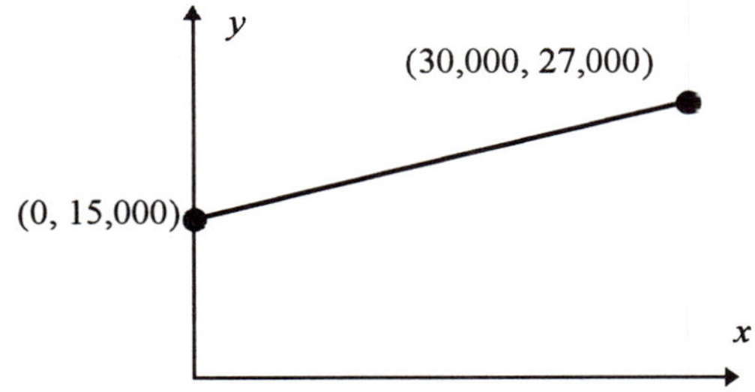

Exercise Set 1-3

1. (3, 2)

3. (2,0)

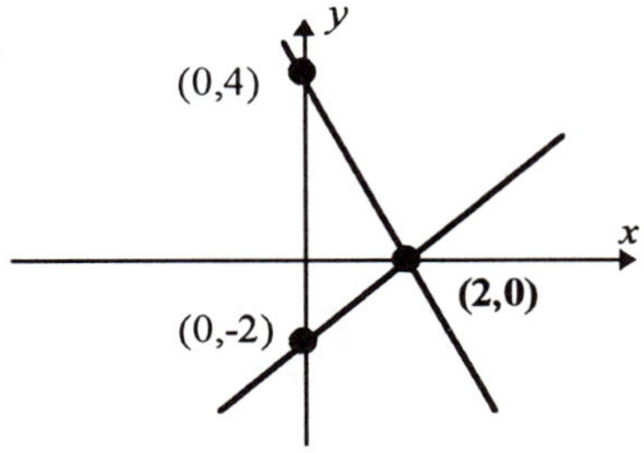

5. (2, 1) **7.** (1, -1) **9.** (2, -1)
11. (3, 2) **13.** (10, 2) **15.** (50, 60) **17.** (20, 40) **19.** (5, 10)
21. (50, 50) **23.** (1, -3) **25.** Inconsistent
27. Dependent **29.** 20,000 boxes of mints **31.** (2/3, -5/3) **33.** (4, 2)
35. (20, 100) **37.** (2, 0) **39.** (100, 20)
41. (0, 0) **43.** No solution **45.** (10, 4)
47. (2, 4) **49.** (1500, 3500) **51.** April, 1983

Referenced Exercise Set 1-3

1. (a) 12.2 million cars, (b) 8.9 million cars, (c) 3.3 million cars
3. (a) 9%, (b) 9%

Cumulative Exercise Set 1-3

1. -1

3.

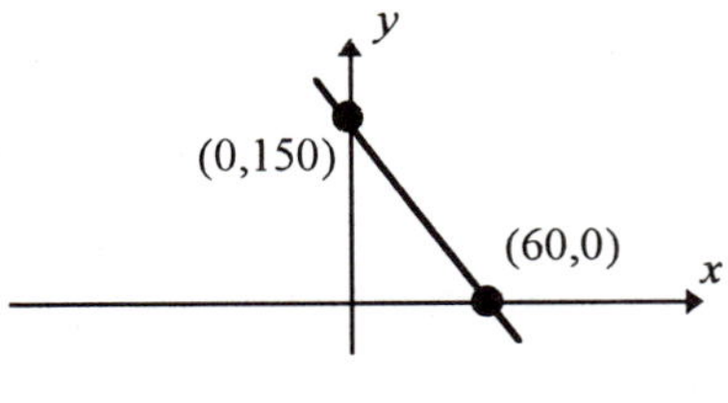

5. $y = -3x + 1$

7. y = 3x - 7 **9.** (-2, 5) **11.** Inconsistent

13. (a) Positive
(b) Positive
(c) Negative

15. (a) D = .24(x - 1) + .23
(b) S = .11(x - 1) + .1
(c) A caller saves $.11 each minute after the first minute

Exercise Set 1-4

1. y = 0.3x + 2 **3.** y = -0.6x + 4 **5.** y = 2x + 2.75

7. y = -3.5x + 22.8 **9.** y = 0.316x + 0.163 **11.** y = 2.036x + 6.214
13. y = -0.46x - 0.57 **15.** y = (12/5)x + 92/3 **17.** 71.5 inches
19. Not linear **21.** Not linear

Referenced Exercise Set 1-4

1. (a)

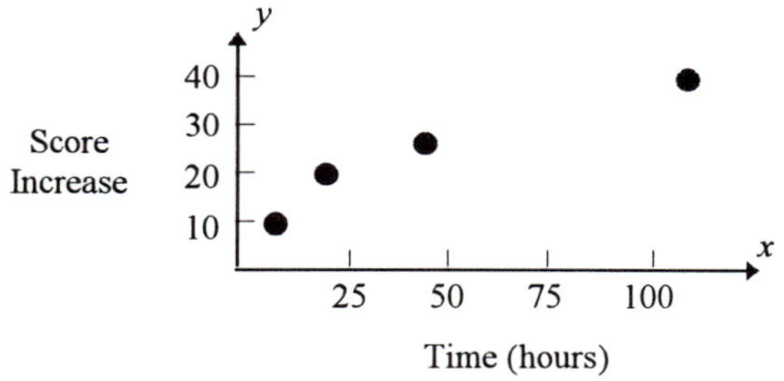

(b) No

(c) $4b + 17m = 100$
$170b + 13899m = 6090$

(d) $y = \frac{1292}{4711}c + \frac{59958}{4711}$ (e) $\frac{318{,}358}{4711} \approx 67.6$ (f) $\frac{411{,}162}{1292} \approx 318.2$

3. The points are linear: $y = -0.00355x + 58.56$
5. Least squares, (b) $y = 1.25x + 20.5$, (c) 45.5, (d) 10
7. (a) $y = 0.8204x - 1608.0351$, (b) 33.59% (or about one-third)

Cumulative Exercise Set 1-4

1. -7 **3.** $y = 3x$ **5.** -4/3
7. (-4, -8) **9.** $y = 1.04x + 2.94$ **11.** $C = 2.5x + 3$

Exercise Set 1-5

1. $x + y = 10$ **3.** $6000x + 8000y = 70{,}000$ **5.** $50x + 100y = 750$
7. $3x + 5y = 40$ **9.** $2000x + 3000y = 25{,}000$
11. (a) $150x + 100y = 12{,}000$, $x + y = 100$, (b) (40,60)
13. (a) $10x + 6y = 4000$, $x + y = 500$, (b) (250,250)
15. (a) $2x + y = 80$, $3x + 2y = 140$, (b) (20,40)
17. (a) $2x + 3y = 40$, $100x + 200y = 2500$, (b) (5,10)
19. (a) $0.5x + 0.4y = 45$, $0.3x + 0.6y = 45$, (b) (50,50)
21. (a) $x + y = 100$, $200x + 350y = 26{,}000$, (b) (60,40)
23. (a) $x + 2y = 400$, $25x + 30y = 8000$, (b) (200,100)
25. $x + y \geq 100$ **27.** $x + y \leq 200$ **29.** $300x + 400y \leq 30{,}000$
31. $5x + 10y \leq 1000$ **33.** $2x + 5y \geq 600$
35. $x + y \geq 100$
$150x + 100y \leq 12{,}000$
37. $2x + y \geq 80$
$3x + 2y \leq 140$
39. $x + y \leq 6$
$50x + 100y \geq 400$

Cumulative Exercise Set 1-5

1. (0, -8), (6. 0) **3.** $y = 3x - 1$ **5.** (20, 50)
7. $2x + 3y \leq 35$ **9.** $40x + 35y \geq 575$

Review Exercises

1.

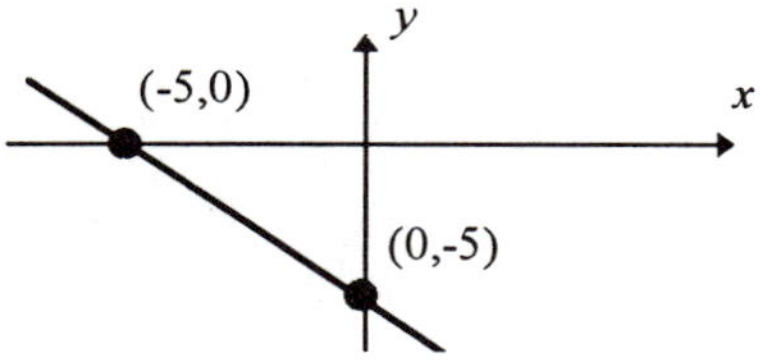

3.

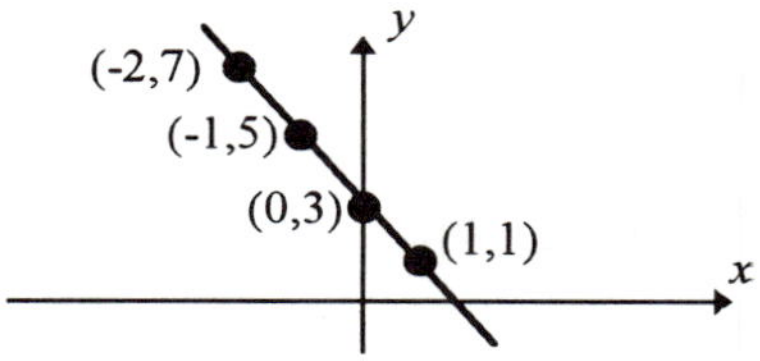

5.

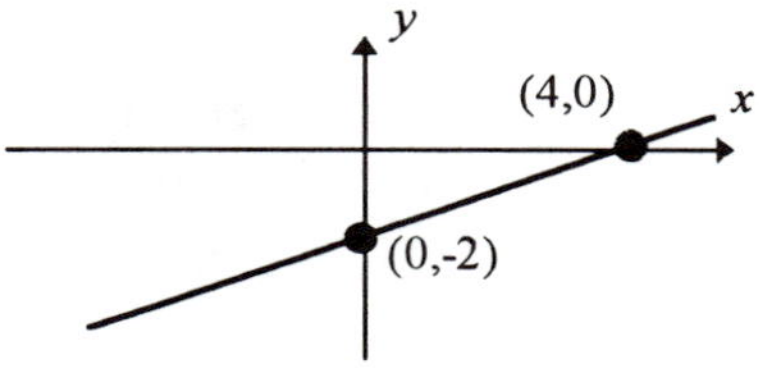

7. $y = 3x - 3$ **9.** $C = .22x + 25$ **11.** (-5/3, 4/3)

13. For example, (2,0), (0, -2) **15.** $y = 0.3x + 4.6$

CHAPTER 2

Exercise Set 2-1

1. Maximize $P = 40x + 30y$
Subject to $150x + 100y \le 12{,}000$
$x + y \le 100$
$x \ge 0, y \ge 0$

3. Minimize $F = 4x + 2y$
Subject to $45x + 9y \ge 45$
$9x + 6y \ge 15$
$x \ge 0, y \ge 0$

5. Minimize $C = 132x + 80y$
Subject to $3x + 6y \le 36$
$3x + 2y \ge 24$
$x \ge 0, y \ge 0$

7. Maximize $P = 500x + 800y$
Subject to $100x + 200y \le 2{,}000$
$3x + 4y \le 48$
$x \ge 0, y \ge 0$

9. Maximize $P = .5x + .7y$
Subject to $.3x + .4y \le 120$
$.2x + .1y \le 60$
$x \ge 0, y \ge 0$

11. Maximize $P = 2000x + 6000y$
Subject to $4x + 5y \ge 40$
$200x + 500y \le 3000$
$x \ge 0, y \ge 0$

13. Minimize $C = 500x + 600y$
Subject to $x + y \ge 10$
$40x + 60y \le 480$
$x \ge 0, y \ge 0$

15. Minimize $C = 20x + 30y$
Subject to $3x + 2y \ge 600$
$10x + 20y \le 4000$
$x \ge 0, y \ge 0$

17. Minimize $C = 100x + 120y$
Subject to $20x + 50y \ge 9000$
$x + 2y \ge 400$
$x \ge 0, y \ge 0$

19. Minimize $C = 100x + 120y$
Subject to $20x + 50y \ge 9000$
$x + 2y \ge 360$
$x \ge 0, y \ge 0$

21. Maximize $P = 4x + 5y$
Subject to $.5x + .3y \le 960$
$.4x + .6y \le 1200$
$x \ge 0, y \ge 0$

23. Maximize $P = 1800x + 2100y$
Subject to $800x + 1000y \le 16000$
$x + y \le 180$
$x \ge 0, y \ge 0$

25. Maximize $R = 1000x + 1200y$
Subject to $8000x + 10000y \le 9{,}000{,}000$
$x + y \le 1000$

27. Maximize $R = 1000x + 1200y$
Subject to $8000x + 10000y \le 12{,}000{,}000$
$x + y \le 1000$

$x \ge 0, y \ge 0$

29. Minimize $C = 20{,}000x + 15{,}000y$
Subject to $150x + 100y \ge 1000$
$50x + 50y \ge 450$
$x \ge 0, y \ge 0$

33. Maximize $F = 600x + 900y$
Subject to $x + y \le 12$
$8x + 16y \le 160$
$6x + 3y \le 66, x \ge 0, y \ge 0$

37. Table b

$x \ge 0, y \ge 0$

31. Minimize $C = 9000x + 15000y$
Subject to $x + y \le 12$
$2x + 3y \le 30$
$100x + 300y \ge 1800, x \ge 0, y \ge 0$

35. Maximize $P = 140x + 90y$
Subject to $x + y \le 100$
$60x + 30y \le 3600$
$10x + 15y \le 1200, x \ge 0, y \ge 0$

39. Table a

Referenced Exercise Set 2-1

1. Minimize $C = .07x + .03y$
Subject to $9.2x + 4.6y \ge 165{,}000{,}00$
$7.6x + 3.1y \le 230{,}000{,}000$
$x \ge 0, y \ge 0$

3. Minimize $P = .34x + .22y$
Subject to $x + y \le 2000$
$x - 3y \ge 0$
$x \ge 0, y \ge 0$

Case StudyExercises

1. Maximize $N = x + y$
Subject to $200x + 600y \le 35{,}000$
$42{,}000x + 6000y \ge 672{,}000$
$x \ge 0, y \ge 0, x \ge 20$

3. Maximize $N = x + y$
Subject to $1400x + 500y \le 35{,}000$
$40{,}000x + 550y \ge 672{,}000$
$x \ge 20, y \ge 0$

Exercise Set 2-2

1. (0, 3), (4, 2), (10, 0), (0, 0)

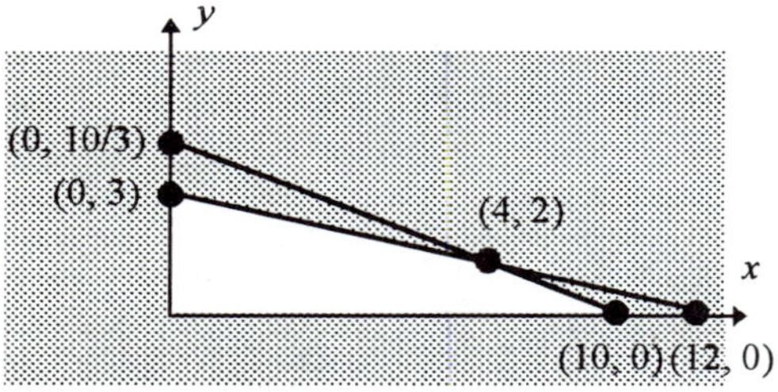

3. (2, 8/3), (6, 0), (18, 0)

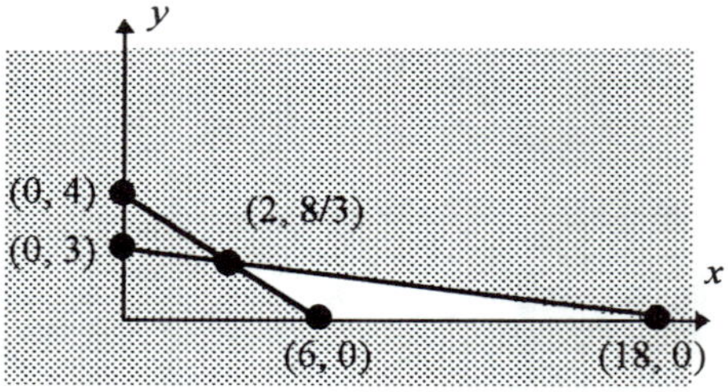

5. (24/5, 9/5), (6, 0), (12, 0)

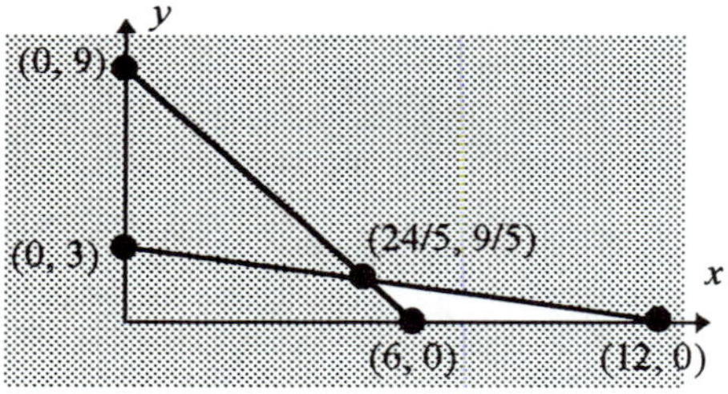

7. (10, 0)
9. (2, 8/3)
11. No solution
13. (40, 60)
15. (5/7, 10/7)
17. (6, 3)
19. (8, 6)
21. (0, 300)
23. (5, 4)
25. (0, 8)
27. (200, 0)
29. (200, 100)
31. (0, 180)
33. (120, 120)
35. (100, 80)
37. (5, 10) in millions
39. (0, 1000)
41. (2,7)
43. (0, 6)
45. (0, 10)
47. (30, 60)

Referenced Exercise Set 2-2

1. (0, 90)

3. (200, 0)

Cumulative Exercise Set 2-2

1. $250x + 125y \le 2000$
$x + y \le 100$

3. Minimize $C = 900x + 1000y$
Subject to $50x + 75y \ge 450$
$x + y \le 8$ $x \ge 0, y \ge 0$

5. Maximize $R = 1000x + 1200y$
Subject to $x + y \le 2000$
$10000x + 12000y \le 21{,}000{,}000$
$x \ge 0, y \ge 0$

7. (6, 2)

9. (1500, 500)

Case Study Exercises

1. $x = 0, y = 44$

3. $x = 32, y = 0$

Review Exercises

1. $280x + 330y \le 4800$
$x + y \le 100$

3. Minimize $C = 12000x + 15000y$
Subject to $150x + 200y \ge 2500$
$x + y \le 15$
$x \ge 0, y \ge 0$

5. Minimize $C = 40x + 60y$
Subject to $3x + 2y \ge 800$
$60x + 80y \le 24000$
$x \ge 0, y \ge 0$

7. (0, 11), (0, 4), (3, 2), and (0, 11) maximizes the objective function

9. (0, 12.5)

11. (800/3, 0)

CHAPTER 3

Exercise Set 3-1

1. Not well-defined
3. Not well-defined
5. {-6, -5, -4, -3, -2, -1, 0, 1}
7. {1, 4, 9, 16, 25, 36, 49)
9. $A \not\subset X$
11. $A \subset X$
13. {2, 3, 4, 5, 9, 10, 11, 12}
15. {50, 51, 52, ... }
17. {0, 1, 2, 3, 4, 6, 8}
19. Z = { ... , -2, -1, 0, 1, 2, ... }
21. {0, 2}
23. N = {1, 2, 3, ... }
25. $A \subset X$
27. $A \subset X$
29. {2, 6, 10, 14, ... }
31. $\{x \in Z : x \ge 1\}$
33. Z
35. $\{\sqrt{4}, \sqrt{16}, \sqrt{64}, \ldots\}$
37. {1, 2, 3, 4, 5, 6}
39. {12, 24, 36, 48, ... }
41. (a) {a, e, i}, (b) {a, e, i , o, u, m, t, h, c, s}, (c) {m, t, h, c, s}
43. (a) {2, 4, 6, 8, 10}, (b) {1, 2, 3, ..., 10, 12, 14, ..., 100}, (c) {1, 3, 5, 7, 9}
45. (a) true, (b) false, (c) true

Referenced Exercise Set 3-1

1. {Denver, Phoenix, Columbus}

3. Each of the major cities has cost of living more than \$100 or the cost of miscellaneous services is more than \$100.

5. (a) {Brazil, Canada, China}, (b) {Canada, U.S., U.S.S.R.}

7. ∅ contains no elements, {∅ } contains one element.

9. (a) {Boston, Los Angeles, Miami, New York, San Francisco, Washington}
(b) {Chicago, Los Angeles, Miami, New York, San Francisco, Washington}
(c) Then A = B, so each city with a highest rate inside the CBDs also has a highest rate outside.
(d) no.

Exercise Set 3-2

1. 21 **3.** 8 **5.** 3 **7.** 29
9. 9 **11.** 44 **13.** 5000
15. 580

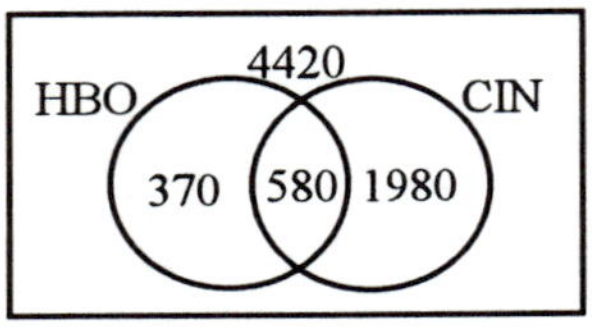

17. (a) 5, (b) 136, (c) 82, (d) 85

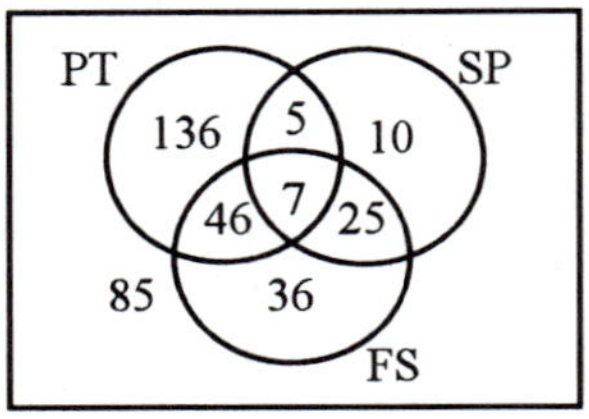

19. (a) 125, (b) 75, (c) 150, (d) 1650

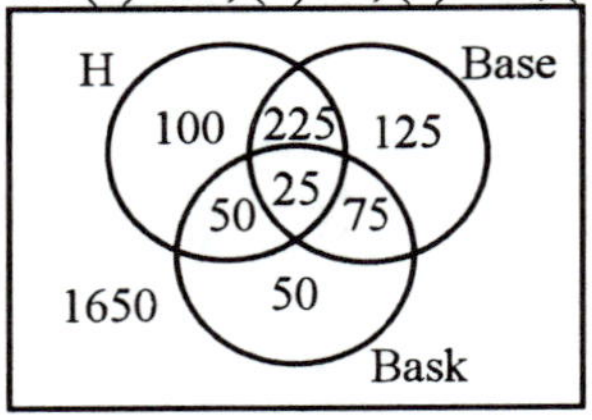

21. 155

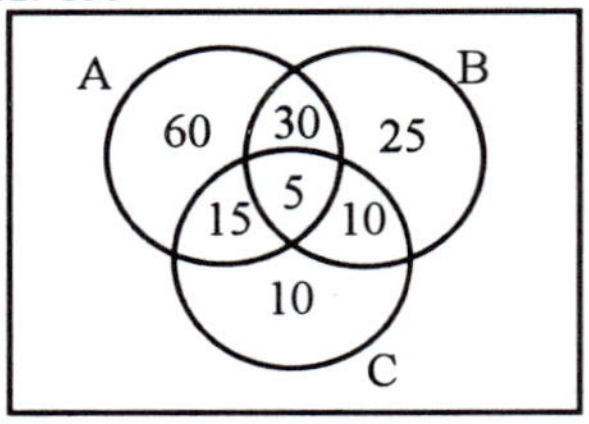

23. n(A) = 36, n(B) = 6
25. 44

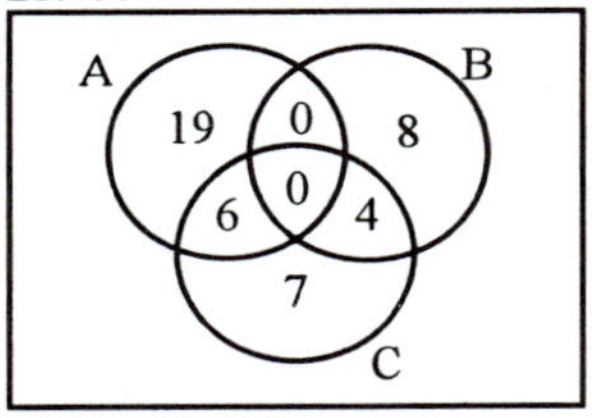

27.

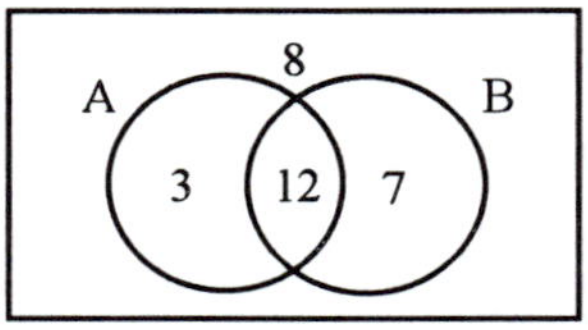

29. 61 **31.** yes, if A = B **33.** (a) 30, (b) 1 **35.** 102
37. $n(A) + n(B) + n(C) - n(A \cap B) - n(A \cap C) - n(B \cap C) + n(A \cap B \cap C)$
39. 8 **41.** $n(\varnothing) = 0$, $n(\{\varnothing\}) = 1$

Referenced Exercise Set 3-2

1. 1429 **3.** 719 **5.** 30%
7. (a) black, (b) magenta, (c) cyan, (d) black, (e) white **9.** 64 regions

Cumulative Exercise Set 3-2

1. S = {0, 1, 2, 3, 4, 5, 6, 7, 8,} **3.** (a) $A \cup B = \{1, 2, 3, 4, 5, 6\}$, $A \cap B' = \{1\}$,
(b) $A \cup B = \{a, b, c, d, g$", $A \cap B' = \{a\}$
5. 32 **7.** 16 **9.** 7

Exercise Set 3-3

1. 12

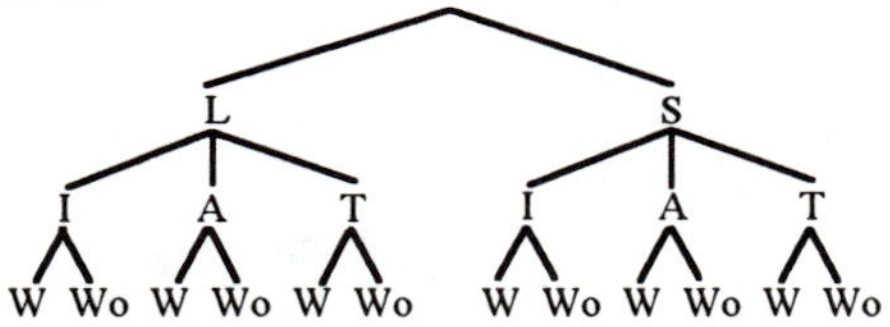

3. 18

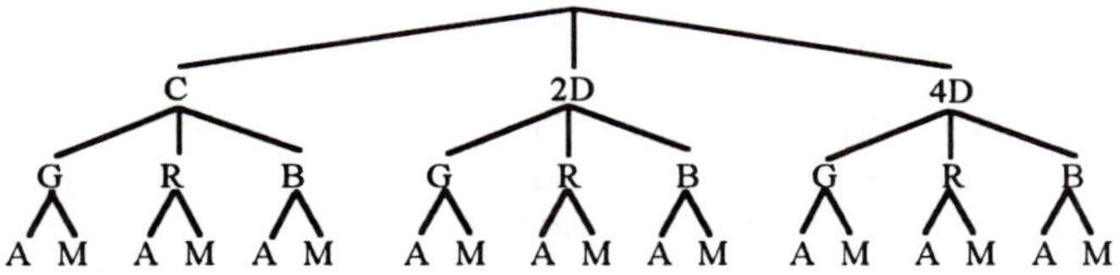

5. 12

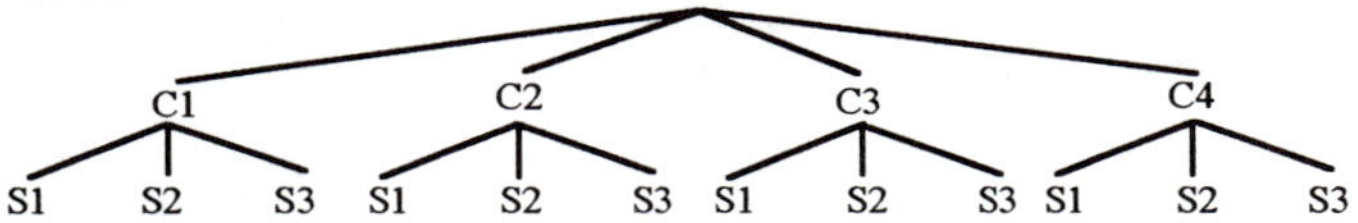

7. 32
9. 125
11. 15,600
13. 676,,000
15. 1000
17. 900
19.

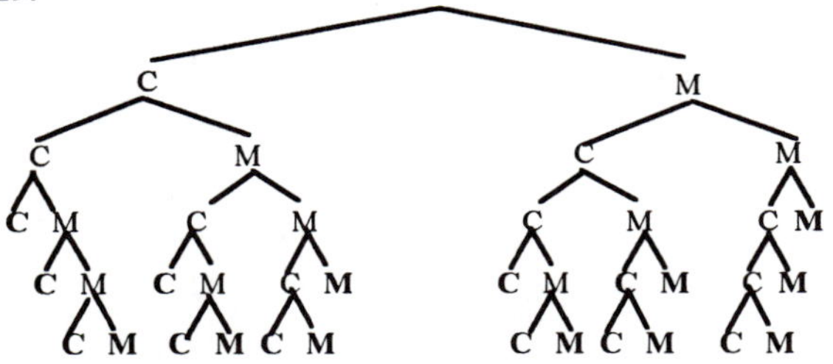

21. 36
23. 1,679,616
25. 9
27. 59,049
29. 36^6
31. 899,991
33. 24
35. 12 (PTEE, TPEE, PEET, TEEP, EEPT, EETP, EPTE, ETPE, ETEP, EPET, PETE, TEPE)
37. 67,600
39.

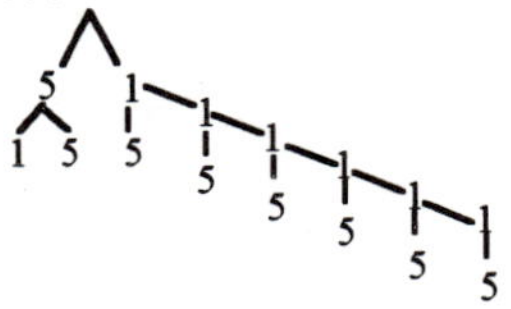

41. 5^{20}
43. 72
45. 1,189,760
47. 8400
49. 1976

Referenced Exercise Set 3-3

1. 3,628,800 (10·9·8·7·6·5·4·3·2·1) **3**. 64

Cumulative Exercise Set 3-3

1. S = {2, 4, 6, 8}
3. (a) {Jan, Feb, Mar, Apr, May, Jun, Jul, Aug, Sep, Oct, Nov, Dec}
(b) {Jun, Jul, Aug, Sep} , (c) {Jan, Feb, Mar, May, Jul, Aug, Oct, Dec}
(d) The set of months not having 30 days, (e) {Jun, Sep}
(f) {Apr, Jun, Jul, Aug, Sep, Nov}
5. $B \subset A$ **7.** 8 **9.** (a) 0, (b) 28, (c) 15
11. (a) 17,576, (b) yes, (c) 15,600 **13.** 72

Exercise Set 3-4

1. (a) 20 (b) 10 **3.** (a) 30 (b) 45 **5.** (a) 1 (b) 1
7. (a)1 (b) 120 **9.** 90 **11.** 13,800 (=25x24x23)
13. 12 **15.** 720 **17.** 48
19. 15,504 **21.**(a) 21 (b) 42 **23.** 24
25. 144 **27.** 64 **29.** (a) 792 (b) 120
31. 1950 **33.** 845,000 **35.** 1000
37. 10,000 **39.** 220 **41.** (a) 256 (b) 24
43. (a) 21 (b) 15 (c) 30 **45.** 72
47. (a) because 5 + 9 = 14 (b) because n + (n - r) = n

Referenced Exercise Set 3-4

1. 8! = 40,320

Cumulative Exercise Set 3-4

1. (a) yes (b) no **3.** 9
5. (H,1), (H,2), (H,3), (H,4), (H,5), (H,6) , (T,1), (T,2), (T,3), (T,4), (T,5), (T,6)
7. (a) 16 (b) 216 **9.** 10^6 = 1,000,000 **11.** (a) 1320 (b) 220
13. (a) 1 (b) 252 (c) 90

Case Study Exercises

1. 1,352,000 **3.** 16,208,000 **5.** 12,000,000
7. No

Review Exercises

1. (a) Not well-defined, (b) well-defined **3.** (a) A′ = {f, g, h,...,z}, (b) V′ ∩ A′ = {f, g, h, j, k, l, m, n, p, q, r, s, t, v, w, x, y, z}, (c) (V ∪ A)′ = V′ ∩ A′ **5.** 10 **7.** The set of irrational numbers **9.** (a) 3, (b) 17, (c) 13, (d) 22 **11.** 324 **13.** (a) 20,736, (b) 11,880
15. 40,320 **17.** 635,376
19. (a) 6840, (b) 680 **21.** 15,600

CHAPTER 4

Exercise Set 4-1

1. {1, 2, 3, 4 } **3.** {l, e, t, r} **5.** {0, 1, 2, 3, 4}
7. {2, 3, 4, 5, 6, 7, 8} **9.** {0, 1, 2, 3}
11. {HHH, HHT, HTH, HTT, THH, THT, TTH, TTT}
13. {GGG, GGR, GGB, GRB, GBB, BBR} **15.** {GGG, GGR, GRG, RGG, GGB, GBG, BGG, GRB, GBR, BRG, BGR, RBG, RGB, GBB, BGB, BBG, BBR, BRB, RBB}
17. {2, 3, 4, 5, 6} **19.** {7, 8, 9, 10, 11, 12} **21.** {(1,1), (1,2), (2,1), (6,6)}
23. {(3,6), (6,3), (4,5), (5,4), (4,6), (6,4), (5,5), (5,6), (6,5), (6,6)}

25. (a) {(1,1), (1,2), (2,1), (2,2), (3,1), (1,3), (6,6)}, (b) {(1,1), (1,2), (2,1)}, (c) {(6,6)}
27. (a) {(1,1), (2,2), (3,3), (4,4), (5,5), (6,6), (3,6), (6,3), (4,5), (5,4), (6,4), (4,6), (6,5), (5,6)}
(b) {(5,5), (6,6), (c) {(1,1), (2,2), (3,3), (4,4)}
29. (a) {(4,1), (4,2), (4,3), (4,4), (4,5), (4,6)}
(b) ∅, (c) {(1,1), (6,6), (4,1), (4,2), (4,3), (4,4), (4,5), (4,6)}
31. {2, 3, 4, 5, 6, 7, 8, 9, 10} **33.** {(1,6), (2,5), (3,4), (4,3)}
35. E = {(A,B), (A,C), (A,R), (A,S), (A,T), (B,C), (B,R), (B,S), (B,T), (C,R), (C,S), (C,T)}
F = {(R,S), (R,T), (S,T), (C,R), (C,S), (C,T)}
G = {(C,A), (C,B), (C,R), (C,S), (C,T), (S,A), (S,B), (S,R), (S,T)}
37. { (R,T) }
39. {(C,A), (C,B), (C,R), (C,S), (C,T), (S,B), (S,R), (S,R), (S,T), (R,T)}
41. { (R,T) } **43.** {C,R), (C,S), (C,T)} **45.** {JUL, AUG}
47. ∅ **49.** {JAN, OCT,NOV, DEC}**51**. yes
53. no **55**. yes, because $E \cap E' = \varnothing$ **57.** yes

Exercise Set 4-2

1. 1/3 **3**. 1/2 **5**. 1/2
7. 2/3 **9**. 1/3 **11**. 2/3
13. 5/6 **15**. 3/4 **17**. 1/2
19. 3/4 **21**.

R	B
1/6	5/6

23.

R	G	W	B
1/6	1/6	1/3	1/3

25. 1/18 **27**. 13/36 **29**. 5/12
31.1/4 **33**. 23/36 **35**. 1/2
37. 1/13 **39**. 3/13 **41**. 4/11
43. 2/7 **45**. 2 to 7, 4 to 3 **47**. 13 to 37, 31 to 19
49. 1/8 **51**.1/2 **53**. 1/2
55. 1/2 **57**. 1/8 **59**. 7/8
61. 1/4 **63**.1/2 **65**. 1/8
67. 7/8 **69**. 1/16 **71**. (a) 7 to 13, (b) 7 to 13

Cumulative Exercise Set 4-2

1. {2, 3, 4, 5, 6, 7, 8, 9, 10, 11, 12} **3.** {0, 1, 2, 3, 4}
5. {1, 2, 3, 4, 5, 6, 8, 10} **7**. (a) {Jan, Feb, Mar, Apr, May, Jun}, (b) {Mar, Apr},
(c) {Feb, Apr, May, Jun, Jul, Aug, Sep, Nov, Dec}, (d) {Feb, Apr}
9. 1/4 **11**. 33/36 = 11/12

Exercise Set 4-3

1. .115 **3**. .359 **5**. .418
7. 3/4 **9**. 3/52 **11**. not mutually exclusive
13. yes **15**. no, the sum of the probabilities is not equal to one
17. no **19**. .907 **21**. .702
23. 334/365 **25**. .17 **27**. .39
29. .57**31**. .05 **33**. .75 **35**. .25
37. .7 **39**. (a) 9/41, (b) 7/41, (c) 16/41

41.

a	b	c
6/9	2/9	1/9

43.

a	b	c	d
1/7	1/7	2/7	3/7

45. 10/19
47. not mutually exclusive
49. not mutually exclusive

Cumulative Exercise Set 4-3

1. {(1, red), (2, red), (3, red), (4, red), (5, red), (6, red), (1, blue), (2, blue), (3, blue), (4, blue), (5, blue), (6, blue)}

3. S = { 0, 1, 2, 3}, (a) {1, 2, 3}, (b) {1, 3}, (c) {0, 2}

5. (a) .8 (= .15 + .25 + .10 + .05 + .25), (b) .65 (= .15 + .25 + .25), (c) .6 (= 1 - .4 = 1 - Pr(F)), (d) .2 (= 1 - .8)

7. $Pr(E \cup F) = Pr(E) + Pr(F) - Pr(E \cap F) = Pr(2, 4, 6, 8, 10, 12) + Pr(2, 3, 4, 5, 6, 7) - Pr(2, 4, 6)$ $= 18/36 + 21/36 - 9/36 = 30/36$

9. (a) .35 (b) .85 (c) .65 (d) .15

Exercise Set 4-4

1. 8/13
3. 4/13
5. 4/13
7. 11/18
9. 1/6
11. .5
13. .45
15. .25
17. 90%
19. 10%
21.

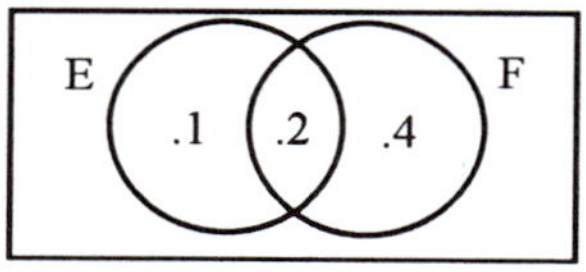

23. .1
25. .1
27. .2
29. .6
31. (a) 15/500 = .03, (b) 435/500 = .87, (c) 290/500 = .58
33. 690/800 = .8625
35. 710/800 = .8875
37. 50/318
39. 166/318
41.

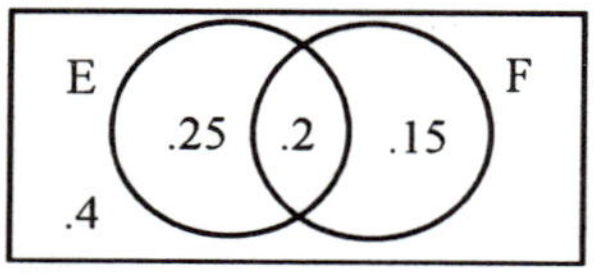

43. .6
45.

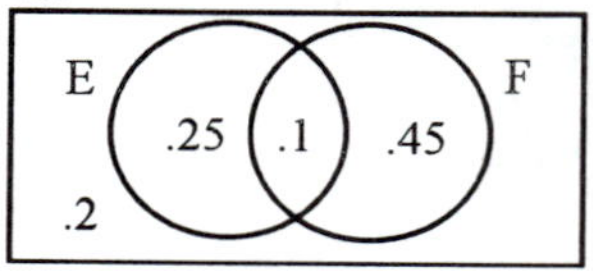

47. .1

Referenced Exercise Set 4-4

1. .68 and .44

Exercise Set 4-5

1. 10/36 = 5/18
3. 2/3
5. 8/13
7. 4/13
9. 2/3
11. 1
13. .75
15. .33
17. .11
19. 4/11
21. 2/7
23
25.
27. 1
29.
31.
33. (a) 7 to 13, (b) 7 to 13
35. .63
37. 15/100 = .15
39. .3

Referenced Exercise Set 4-5

1. .68 and .44

Cumulative Exercise Set 4-5

1.
3.
5.
7.
9. (a) .29, (b) .68, .32, (c) 1
11. (a) .63, (b) .94

Exercise Set 4-6

1.
3.
5.
7.
9. .9366...
11. 2/3
13. 64/67≈ .96
15. 19/83 ≈ .23
17. Not independent, 7/18
19. Independent, 1/2
21. .005
23. 1/3; 2/3
25. 79/180 ≈ .44
27. (a) 1/7, (b) 12/35, (c) 3/7
29. (a) 8%, (b) 36%
31. 18/43 ≈ .42
33. 12/57 ≈ .21
35.
37. (a) and (c) Use Example 1 to disprove; (b) use $Pr(E \cap F) + Pr(E' \cap F) = Pr(F)$
39. Use the addition principle, the rule for complements and two factorization.

Referenced Exercise Set 4-6

1. (a) 3/13, (b) 4/26 = 2/13, (c) 17/26

Cumulative Exercise Set 4-6

1. (a) {0, 2, 4, 5}, (b) { 4 }, (c) {0, 1, 2, 3}
3. (a) .3, (b) .55
5. .143
7. (a) 2/13, (b) 3/4
9. No, Pr(E|F) = 12/14
11. 3/13

CHAPTER 5

Exercise Set 5-1

1. .0577
3. .4121
5. .2060
7. .7583
9. .4211
11. .1095
13. .5165
15. .8709
17. .8560
19. .496
21..4186
23. .0405
25. .2148
27. .0211
29. .0475
31. .00000174
33. .000297
35. .002194
37. .7692
39. .0674
41. .4196

Referenced Exercise Set 5-1

1. 3/91

Cumulative Exercise Set 5-1

1. (a) {0, 1, 2}, (b) {less than 2, at least 2}, (c) {0, 1, 2, 3}
3. 3/8
5. 1/13
7. 35/36
9. 264/560 = 33/70
11. (a) 15/45 = 1/3, (b) 48/90 = 8/15, (c) 6/45 = 2/15
13. (a) 1 · 19/20 · 18/20 · ... · 11/20 ≈0.9345, (b) 0.3698

Exercise Set 5-2

1. .8953
3. .1111
5. .3913
7. .04
9. .06
11. .1509
13. .0816
15. .04, .15, .325
17. .0777
19. .6311
21. .3093
23. .1739
25. .5217
27. 1/3
29. 4/11

31. .2857 **33**..25 **35**. .625
37. .6770 **33** .3614

Referenced Exercise Set 5-2

1. 55/92 ≈ .60

Cumulative Exercise Set 5-2

1. 1/4 **3**. 14/99 **5**. .17
7. .24 **9**. .4

Exercise Set 5-3

1. 15/64 **3**. 22/64 = 11/32 **5**. 28/256 = 7/64
7. .1445 **9**. .0146 **11**. .3560
13. .7373 **15**. .9011 **17**. .3292
19. .2561 **21**. .7901 **23**. .0453
25. 2508 **27**. 9536 **29**. (a).2936, (b) .7969
31. (a) .1172, (b) 1719 **33**. (a) 1956, (b) .2217 **35**. (a).0042, (b) .9996
37. (a) .2013, (b) .3222 **39**. (a) .1323, (b) .9692 **41**. .8732

Referenced Exercise Set 5-3

1. .0017

Cumulative Exercise Set 5-3

1. (a) 5/9, (b) 1/18 **3**. 3/4 **5**. 35/75 = 7/25
7. (a) .00025, (b) .999981 **9**. (a) .1877, .5256 **11**. (a) 4/10 = 2/5, (b) 13/40

Exercise Set 5-4

1. 25 **3**. 2 **5**. 2.4
7. 2/13, no **9**. 4/9, no **11**. 1/3, no
13. Change 4 to 2 **15**. Site 1 **17**. Property A
19. Property B **21**. 3.5 **23**. .2
25. 1.25 **27**. 3 **29**. -.499
31. -.0389 **33**. 10 **35**. 1.0
37. .32

Referenced Exercise Set 5-4

1. E(American) = -1/19
E(european) = -1/74

Cumulative Exercise Set 5-4

1. (a) 15/36, (b) 5/36 **3**. .589 **5**. 17/40
7. 50 **9**. (a) 45/1024, (b) 386/1024 **11**. (a) .3, (b) 60
13. (a) 1/12, (b) 1/144

Review Exercises

1. 0.07 **3**. 2/11 **5**. 4/19
7. 0.166 **9**. .0.436 **11**. 2.9

CHAPTER 6

Exercise Set 6-1

1. .2856 **3.** .4283 **5.** 3/8

7. 1/2

9.

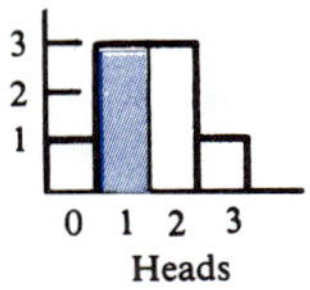

11.

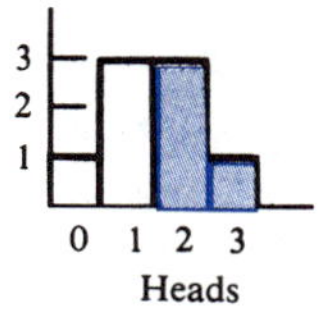

13.

x	f
0	1
1	2
2	1

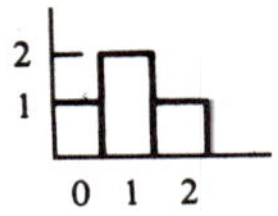

15.

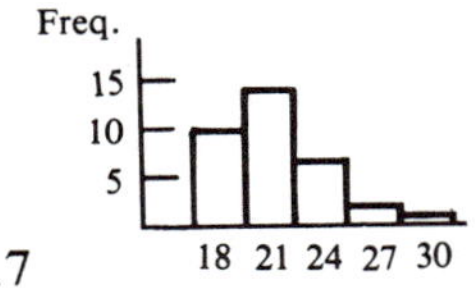

17. 15/16 **19.** 4/9

21. .992 **23.** .79

25. .29 **27.**

29..7

31. .7 **33.** .5

35. .3

Referenced Exercise Set 6-1

1.

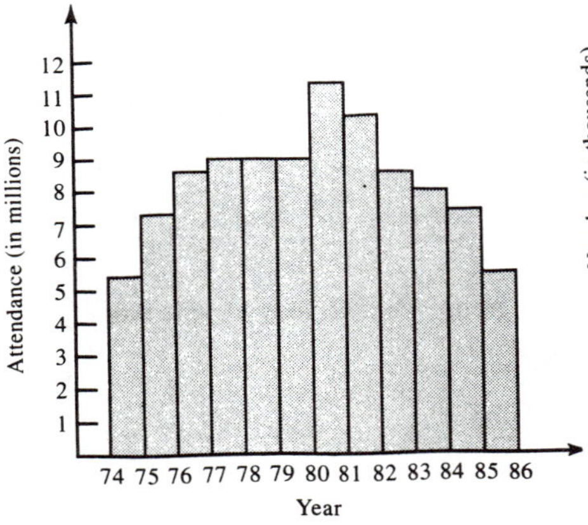

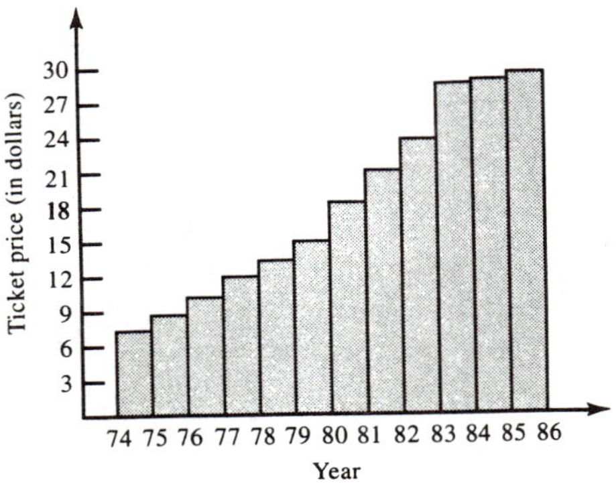

3. Each histogram is close to a normal curve.

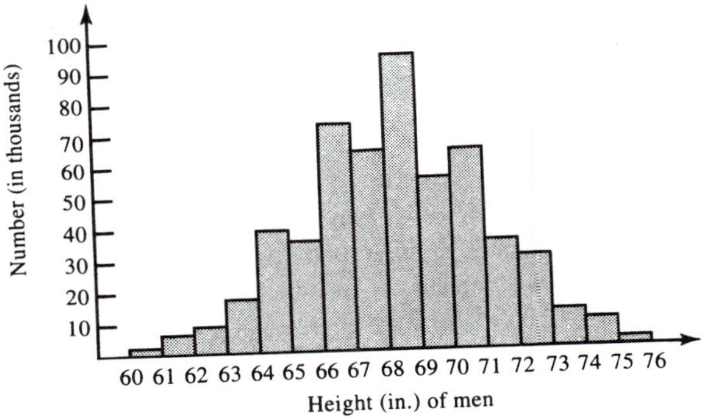

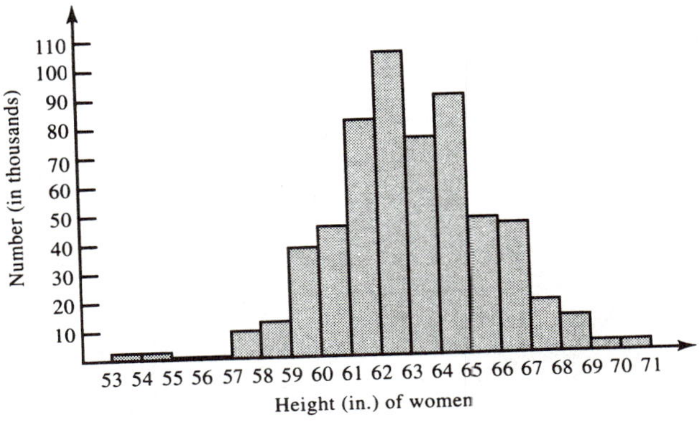

5. Fear of 13.

7. no

9. No, the sum of the probabilities is not 1.

Exercise Set 6-2

1. Mean = 161/10, Median = 8
3. Mean = 0.32/6 = 0.053, Median = 0.055
5. Mean = 23/14, Median = 1.5
7. Mean = 47/7 ≈ 6.7, Median = 9
9. E(X) = 31/8
11. E(X) = 3.3
13. Mean = 37/22 ≈ 1.68
15. E = 13/8 = 1.625
17. Mean = -0.5, median = 0
19. Mean = 6.695, median = 7.3
21. Mean = 0, median 0
23. E(X)= -0.1
25. E(X)= -2
27. E(X)= 1.5
29. Mean = 230.3 lb., median = 225 lb
31. Mean = 43/13 ≈ 3.3
33. (a) Each entry is 0, (b) each entry is 2, (c) no
35. (a) 4.3/11 ≈ 0.39, (b) reward
37. Mean = 4.6, median = 4
39. 1.6

Referenced Exercise Set 6-2

1. (a) 12.09, (b) 8, (c) people argued for the median, the bank argued for the mean
   (d) The payment via the median is (20)(7/8)$9500) =$166,250, the payment via the mean is (20)(7/12)($9500) = $110,833.33; the difference is $55,416.67.
3. The expected value is -2 so the expected death month is very close to the birthday.
5. Mean = 62.57 in., median = 62 in.
7. 5.908, or about 6 games.
9. (a) 1, (b) 1.2, (c) yes

Cumulative Exercise Set 6-2

1.

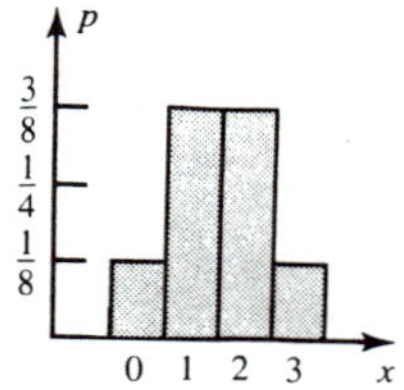

3.

X	2	3	4	5	6	7	8	9	10	11	12
p	$\frac{1}{36}$	$\frac{2}{36}$	$\frac{3}{36}$	$\frac{4}{36}$	$\frac{5}{36}$	$\frac{6}{36}$	$\frac{5}{36}$	$\frac{4}{36}$	$\frac{3}{36}$	$\frac{2}{36}$	$\frac{1}{36}$

$Pr(6 \le X \le 9) = \frac{5}{9}$

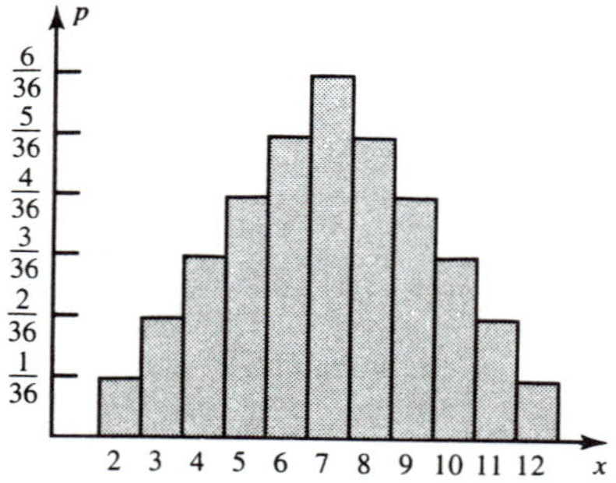

5. 3.2, .2

7. 0.65

Case Study Exercises

1. The mean family size in the Lenin study is approximately 7.3. The mean family size in the second study is approximately 6.7.
3. The mean number of cattle owned in the Lenin study is approximately 7.7. The mean number of cattle owned in the second study is approximately 4.4.
5. (a) 12.4, (b) 5.1, (c) 8.3, (d) 5.8

Exercise Set 6-3

1. $V(X) = 5.05, \sigma = 2.25$
3. $V(X) = 11.89, \sigma = 3.45$
5. $V(X) = 107.4, \sigma = 10.36$
7. $V(X) = 11.37, \sigma = 3.37$
9. $V(X) = 7.28, \sigma = 2.7$
11. $V(X) = 2.36, \sigma = 1.54$
13. $V(X) = 0.61, \sigma = 0.78$
15. \$13,800, $\sigma = 3763$
17. $Pr(40 \le X \le 60) = 0.75$
19. $V(X) = 0.108, \sigma = 0.329$
21. $V(X) = 0.836, \sigma = 0.914$
23. $V(X) = 2883.796, \sigma = 53.701$
25. $V(X) = 3.29, \sigma = 1.814$
27. $V(X) = 41, \sigma = 6.403$
29. $\sigma = 1.26$
31. $\sigma = 0.992$
33. $\sigma = 0.87$
35. (a) $\mu = 1.87$, (b) $\sigma = 1.26$
37. (a) 2, (b) 2, (c) 2, (d) 2
39. (a) 2, (b) 1
41. 8/9
45. $\sqrt{n}/2$

Referenced Exercise Set 6-3

1. 3.4
3..

	EF	TS	C	AC
mean	204.1	200.3	198.6	200.0
σ	11.6	8.5	6.4	10.4

This method of comparison is not conclusive, a more exhaustive analysis must be performed.

5. (a) 3.87%, (b) 65.84%, (c) the standard line
7. The authors are not convinced
9. 2

Cumulative Exercise Set 6-3

1.

x	0	1	2	3	4
f	1	4	6	4	1

5. 201/24 = 8.375
7. $2076/34 \approx 61.06$
9. 6, 6, 6, 6, 7
11. 2

Exercise Set 6-4

1. About 0.475
3. About 0.34
5. About 0.16
7. About 0.025
9. 0.0668
11. 0.6826
13. 44.35%
15. 11.51%
17. 24.17%
19. 59.87%
21. About 0.135
23. About 0.815
25. About 0.84
27. 0.6480
29. 0.6541
31. 27.61%
33. 13.57%

Referenced Exercise Set 6-4

1. 38.21%
3. 0.3812
5. 0.99% (less than 1%)
7. (a) 9.81%, (b) 84.13%
9. (a) 0.8413, (b) 0.7088
11. 0.6826

Exercise Set 6-5

1. $\mu = 72, \sigma = 6$
3. $\mu = 40.5, \sigma = 4.5$
5. .1841
7. .4041
9. .1210
11. .6757
13. .9370
15. .1635
17. .1841
19. .7286
21. .9945
23. .9328
25. .7088
27. .6886
29. .8485
31. .2643
33. .9576
35. .9279
37. .5516
39. Virtually zero
41. .1788
43. .2420
45. .3830

Referenced Exercise Set 6-5

1. (a) .0028, (b) 166 **3.** (a) .9987, (b) .0013

Cumulative Exercise Set 6-5

1. (a)

X	0	1	2	3
p	1/8	3/8	3/8	1/8

3. 0 **5.** (a) 12/5 **7.** 0.6247

9. (a) .3, (b) 5 **11.** 0.1841

Case Study Exercises

1.

	H	M	D
$\bar{x}$	.44	.73	.80
σ	.33	.59	.35

not strong enough

3.

	H	M	D
$\bar{x}$	7.04	11.52	4
σ	2.43	3.44	1.73

not strong enough

5.

	H	M	D
$\bar{x}$	3.46	7.68	8.5
σ	1.74	2.70	2.47

Madison

7.

	H	M	D
$\bar{x}$	7.67	5.96	6.83
σ	2.13	2.16	2.07

not strong enough

9.

	H	M	D
$\bar{x}$	.58	1.11	.77
σ	.43	.83	.45

Hamilton

Review Exercises

1.

3.

5.

x	3	2	1	0
p	.0129	.1376	.4359	.4135

7. .9871 **9.** Mean = 2.4, median = 2 **11.** - 1/4

13. (a) $15,000, (b) $17,916.67 **15.** Avg cost = $148.90, σ = 37.31

17. (a) .6924, (b) .8531 **19.** .0668 **21.** .1294

23. σ = 2 **25.** 1200 **27.** .9525